T0344501

DARK ENERGY
Observational and Theoretical Approaches

'Dark energy' is the name given to the unknown cause of the Universe's accelerating expansion, one of the most significant discoveries in recent cosmology. Understanding this enigmatic ingredient of the Universe and its gravitational effects is a very active, and growing, field of research.

In this volume, twelve world-leading authorities on the subject present the basic theoretical models that could explain dark energy, and the observational and experimental techniques employed to measure it. Covering the topic from its origin, through recent developments, to its future perspectives, this book provides a complete and comprehensive introduction to dark energy for a range of readers. It is ideal for physics graduate students who have just entered the field and for researchers seeking an authoritative reference on the topic.

PILAR RUIZ-LAPUENTE is Associate Professor in Astrophysics at the University of Barcelona. She is also a member of one of the two collaborations that discovered the acceleration of the Universe, for which she shared the Gruber Prize for Cosmology.

DARK ENERGY

Observational and Theoretical Approaches

Edited by

PILAR RUIZ-LAPUENTE

University of Barcelona

Shaftesbury Road, Cambridge CB2 8EA, United Kingdom

One Liberty Plaza, 20th Floor, New York, NY 10006, USA

477 Williamstown Road, Port Melbourne, VIC 3207, Australia

314–321, 3rd Floor, Plot 3, Splendor Forum, Jasola District Centre, New Delhi – 110025, India

103 Penang Road, #05–06/07, Visioncrest Commercial, Singapore 238467

Cambridge University Press is part of Cambridge University Press & Assessment,
a department of the University of Cambridge.

We share the University's mission to contribute to society through the pursuit of
education, learning and research at the highest international levels of excellence.

www.cambridge.org
Information on this title: www.cambridge.org/9780521518888

© Cambridge University Press & Assessment 2010

This publication is in copyright. Subject to statutory exception and to the provisions
of relevant collective licensing agreements, no reproduction of any part may take
place without the written permission of Cambridge University Press & Assessment.

First published 2010
First paperback edition 2013

A catalogue record for this publication is available from the British Library

Library of Congress Cataloging-in-Publication data
Dark energy : observational and theoretical approaches / edited by Pilar Ruiz-Lapuente.
p. cm.
ISBN 978-0-521-51888-8 (Hardback)
1. Dark energy (Astronomy)   2. Dark matter (Astronomy)
I. Ruiz-Lapuente, P.   II. Title.
QB791.3.D366 2010
523.01–dc22

2009043580

ISBN   978-0-521-51888-8   Hardback
ISBN   978-1-107-64702-2   Paperback

Cambridge University Press & Assessment has no responsibility for the persistence
or accuracy of URLs for external or third-party internet websites referred to in this
publication and does not guarantee that any content on such websites is, or will
remain, accurate or appropriate.

Contents

Contributors

Bruce Bassett
University of Cape Town, Rondebosch, Cape Town 7700, and South African Astronomical Observatory, Observatory, Cape Town, South Africa

Ruth Durrer
Department of Theoretical Physics, University of Geneva, 24, Quai E. Ansermet, 1211 Geneva 4, Switzerland

Alan Heavens
SUPA Institute for Astronomy, ROE Blackford Hill, Edinburgh, EH9 3HJ, UK

Renée Hlozek
Department of Astrophysics, University of Oxford, Keble Road, Oxford, OX1 3RH, UK

Alex Kim
Physical Sciences Division, Lawrence Berkeley National Laboratory, Berkeley, CA 94720, USA

Robert P. Kirshner
Harvard–Smithsonian Center for Astrophysics, Cambridge, MA 02138, USA

Roy Maartens
Institute of Cosmology and Gravitation, University of Portsmouth, Portsmouth, PO1 2EG, UK

Thanu Padmanabhan
Inter-University Centre for Astronomy and Astrophysics, Post Bag 4, Ganeshkhind, Pune 411007, India

David Polarski
Laboratory of Theoretical Physics and Astroparticles, CNRS, University of Montpellier II, 34095 Montpellier, France

Pilar Ruiz-Lapuente
Department of Astronomy, University of Barcelona, Martí i Franqués 1, 08028 Barcelona, Spain

Jean-Philippe Uzan
Institute of Astrophysics of Paris, UMR-7095 CNRS, University Pierre and Marie Curie, 98 bis bd Arago, 75014 Paris, France

W. Michael Wood-Vasey
Department of Physics and Astronomy, University of Pittsburgh, Pittsburgh, PA 15260, USA

Preface

We commonly denote as dark energy the physics yet to be determined that causes the present acceleration of the universe. Dark energy is a very wide subject residing in unkowns of very different kinds, since the origin of the cosmic acceleration can be of very diverse nature. It can be linked to the presence of a new component of the universe in the form of a field with similarities to the field that causes inflation. It can be a departure of gravity from general relativity or it can be related to vacuum energy.

The importance of the subject, which involves the understanding of gravity and the present components of the universe, will make it a very active field of research in the next decades.

Many experiments are planned to try to identify the nature of dark energy. They involve the use of ground-based facilities or space missions designed specifically to study this phenomenon. There is a flurry of results coming every year, which put new constraints on what dark energy can be.

In the present volume, we have not tried to be exhaustive with such an open field, but we give diverse points of view, both theoretical and observational.

The theoretical perspectives try to present the various possibilities under discussion on what dark energy can be. The observational chapters present the field now and in the future with the advent of multiple experiments and new facilities.

In his chapter "Dark energy, gravitation and the Copernican principle", Jean-Philippe Uzan introduces the ingredients of our cosmological description of the universe and how we include dark energy. Proposals for dark energy are scrutinized. The models of dark energy are classified into: those in which acceleration is driven by the gravitational effect of new fields; models in which fields do not dominate the matter content but are coupled to photons and affect the observations; models in which a finite number of new fields are introduced and couple to the standard field models, such as in the case of scalar–tensor theories; and models in which drastic modifications of general relativity are required, such as the models introducing extra dimensions. The author discusses the constraints in modifying general relativity found in various regimes as well as the options to modify the Einstein–Hilbert action. He reconsiders our assumption of the place that we occupy in the universe. Usually in cosmology one works within what is called the Copernican principle, which merely states that we do not live in a special place (the center) in the universe. This principle is in general very minimal in its requirements when compared with the frequently used cosmological principle, which supposes that the universe is spatially isotropic and homogeneous. Within the Copernican principle, the cosmological principle is recovered if we assume that isotropy around the observer holds. Uzan asks which possibilities are open if we go beyond the Copernican principle; for instance, the possibility that we may be living close to the center of a large underdense region of the universe. Non-homogeneous models of the universe are considered. Many possibilities are open if one relaxes the Copernican principle. But their cosmological extensions are testable by the data and future surveys can further constrain them.

In their chapter, Ruth Durrer and Roy Maartens examine why most models proposed for dark energy are not good candidates. As candidates within general relativity, they consider quintessence models. They find that some aspects of quintessence models are not satisfactory: they require a strong fine tuning of the parameters of the Lagrangian to secure recent dominance of the field. More generally, the quintessence potential, like the inflaton potential, remains arbitrary until fundamental physics selects the potential. The authors look at other models involving scalar fields. Those with a non-standard kinetic term on the Lagrangian can be ruled out because they break causality.

Out of the frame of general relativity other possibilities can be considered, as argued by Durrer and Maartens, but present equal problems. One of those alternatives is $f(R)$ or scalar–tensor theories of gravity where the Lagrangian is described with an arbitrary function of the Ricci scalar. $f(R)$ models are the only acceptable low-energy generalization of the Einstein–Hilbert action of general relativity. In these theories gravity is mediated by an extra spin-0 degree of freedom. The requirement of late acceleration leads to a very light mass for the scalar. However, on solar system scales, the light scalar induces strong deviations from the weak field limit of general relativity in contradiction with the observations.

Another class of dark energy models that requires going beyond general relativity is the so-called braneworld dark energy models. Durrer and Maartens examine these candidates and pay attention to the Dvali, Gabadadze and Porrati (DGP) braneworld model, which modifies general relativity at low energies. This model produces "self-acceleration" of the late-time universe due to a weakening of gravity at low energies. Despite the good features of the DGP model to provide a simplified explanation for the acceleration, the predictions of the model are in tension with supernova observations and with WMAP5 data on large scales. From their overview, Durrer and Maartens conclude that none of the contenders for the ΛCDM model appears better than ΛCDM. However, the need to explore the realm of possibilities makes it worth testing models outside general relativity and other alternatives.

The chapter by David Polarski entitled "Some views on dark energy" reviews alternatives to dark energy and finds some of the problems mentioned by Durrer and Maartens. Polarski devotes significant attention to the scalar–tensor dark energy models. The author discusses limits placed on these models by the solar system experiments. He discusses the $f(R)$ models that could in principle provide late acceleration of the universe without the need to invoke a cosmological constant. He points out that many $f(R)$ models cannot produce a standard matter-dominated stage before the late-time accelerated expansion. However, there is still room for viable $f(R)$ candidates. Polarski addresses the question of how to disentangle, from observations, dark energy candidates that are within general relativity from those that are outside. He outlines how the dynamics of the matter perturbation can give a clear test.

In his chapter entitled "Emergent gravity and dark energy" Thanu Padmanabhan gives a presentation of the cosmological constant problem, followed by a criticism of most current theoretical approaches to the dark energy problem. According to this author, the ultimate explanation for a non-zero cosmological constant should come from quantum gravity, but even in the absence of such a microscopic description of the structure of spacetime, new insights would be provided by an intermediate description, playing a role similar to that of thermodynamics, which, while being only phenomenological, depends on the existence of microscopic degrees of freedom and reveals them in a way that escapes the continuum mechanics description. The cosmological constant would be a low-energy relic of quantum gravitational physics. It introduces a fundamental length L_Λ, which would determine the lowest possible energy density due to quantum spacetime fluctuations. The relevant degrees of freedom, in calculating that, should scale as the surface area and not as the volume of the corresponding region of spacetime. Einstein's equations can be obtained from an action principle that uses only the surface term of the Einstein–Hilbert action and the virtual displacements of horizons. The usual interpretation of the surface term as the entropy of horizons thus links spacetime dynamics and horizon thermodynamics. In such a framework, the bulk value of the cosmological constant decouples from the dynamical degrees of freedom and what is observable through gravitational effects should just be the fluctuations in the vacuum energy. The scale relevant to the structure of the gravitational vacuum should be the size L_Λ of the cosmological horizon. The size of the fluctuations in the energy density then gives the right magnitude for the cosmological constant: it is small because it is a quantum relic.

As observational accounts on the state of dark energy, we have several chapters devoted to dark energy from supernovae, which have been at the origin of the discovery of dark energy and are providing most of the data for determining the equation of state. There is, unavoidably, some overlap among them, but each chapter reflects different perspectives.

In his chapter, Robert P. Kirshner presents the history of how SNe Ia became such good cosmological distance indicators. He recalls the very first attempts to use them when the collected data were sparse and did not allow one to establish that the universe had a low Ω_m and moreover that there were indications that the cosmological constant was non-zero. As founder of the Harvard supernova team that has been the core of the High-Z Team, ESSENCE and now PARITEL, his recollections on the discovery of acceleration of the universe are most valuable. In this book, we have not had the opportunity to have a parallel presentation by Saul Perlmutter describing the successes and difficulties in getting the high-z supernova method to work, but his talk at the Gruber Prize ceremony in 2007 gives a summary of the long path towards the discovery of the acceleration of the universe. Recently, Gerson Goldhaber published his memoirs on the subject in the Dark Matter 2008 conference. He focused on year 1997 when many of us were attending the Santa Barbara Workshop on Supernovae. Being linked first to Harvard and afterwards to the Supernova Cosmology Project, I have a sense of how things developed. I agree that the year 1997 was crucial and at the Supernova Cosmology Project meeting in Berkeley that I attended, it seemed clear that Gerson's results were already pointing to the direction that ended in the discovery of non-zero Λ. The paper took a long time to get written but the seeds were sown early. I conclude that the teams led by both Robert P. Kirshner and colleagues and Saul Perlmutter and colleagues were necessary to the discovery.

The chapter by Pilar Ruiz-Lapuente gives an introduction to the method of supernovae to measure the expansion rate of the universe and the equation of state of dark energy. The basic relation that allows one to calibrate supernovae as distance indicators is presented. The author reviews how different samples of SNe Ia have allowed us to improve the determination of the index of the equation of state. The sources of error in the method are discussed. The progress in controlling those errors is shown by looking at their size at the time of discovery of the acceleration of the universe and their expected magnitude in planned missions to gather thousands of supernovae from space. Prospects for testing the isotropy of the universe with SNe Ia and how one will be able to assess FRW cosmologies are pointed out. The chapter mentions some dark energy models that have been tested with supernovae and the results that have been gathered. The complementarity of the supernova test with other probes is emphasized.

The chapter by Michael Wood-Vasey entitled "The future of supernova cosmology" gives an update on the diversity of questions that can be addressed with the new facilities opened to do large surveys of the sky, gathering hundreds of thousands of supernovae. Tests of the Copernican principle and precision tests on the equation of state of dark energy can be achieved by the Large Synoptic Survey Telescope and the Panoramic Survey Telescope and Rapid Response System, which will scan all the sky enabling a very large number of data to be collected.

In his chapter, "The space advantage for measuring dark energy with Type Ia supernovae", Alex Kim examines why a space mission is needed to be able to measure dark energy to the precision required to identify its nature. Space basically provides more potential for discovery. The author reviews the physics behind the supernova data points. He describes what goes into the so-called magnitude of a supernova. Ultimately, the advantages of going to space are presented. Rest-frame multichannel optical data of supernovae out to $z\sim1.8$ that extend redwards to $0.63\,\mu$m (and preferentially even further) can provide the required information on SNe Ia, foreground dust and ultimately dark energy models. The relative yields of surveys from the ground and surveys from space are evaluated. It is shown that for a ground-based survey to be competitive with a space-based one, the extent in angle covered and collecting areas must be 10^5 times larger than for a 2-m space telescope. It is seen how surveys such as those planned for the Large Synoptic Survey Telescope cannot be competitive at high z with a space mission based on a 2-m telescope.

Bassett and Hlozek review the method of baryon acoustic oscillations to measure dark energy and the ongoing projects that have begun to measure the Hubble parameter and the angular distances at various z with this method. Their chapter starts with a historical overview of methods that have used angular distances as cosmic distance indicators. Then it reviews the physics of baryon acoustic oscillations and their cosmological uses. As Bassett and Hlozek explain, before recombination and decoupling the universe consisted of a hot plasma of photons and baryons which were tightly coupled via Thompson scattering. The competing forces of radiation pressure and gravity set up oscillations in the photon fluid. The tightly coupled baryon–photon plasma underwent perturbations that propagated outwards as an acoustic wave. At recombination the cosmos becomes neutral and the pressure on the baryons is removed. The baryon wave stalls and quickly slows down while the photons propagate freely away forming what we now observe as the cosmic microwave background. The characteristic scale of the spherical shell formed when the baryon wave stalled is imprinted on the distribution of the baryons as an excess density transmitted through gravity to dark matter, which preferentially clumps on this shell. This leaves the imprint of the original baryon acoustic oscillation and expands as the universe does. Once the origin of the method has been reviewed, the authors take a close look at the observational challenges encountered by this approach. Bruce Bassett and Renée Hlozek discuss issues of non-linearity within the method. Various observational possibilities are discussed about where the measurement of the baryon acoustic oscillations could be made. Finally, forecasts for the use of this approach to measure dark energy are given.

As another observational approach to determine the nature of dark energy, Alan Heavens presents the method of weak lensing. Lensing, as noted by the author, is very appealing theoretically as the physics involved is simple and robust and a direct connection can be made between weak lensing observables and the statistical properties of the matter distribution. Those statistical properties depend on cosmological parameters in a known way, thus weak lensing can be employed as a cosmological tool. Heavens introduces the fundamental concepts in gravitational lensing such as shear and magnification. He points out that when distance information for the sources is available, it is possible to do a 3D reconstruction of the mass distribution. The procedure is to divide the survey into slices at different distances, and perform a study of the shear pattern on each slice.

Another very interesting test is the shear ratio test, an approach that has a simpler dependence on cosmological parameters and presents the advantage of probing cosmology with significant independence from the growth rate of fluctuations.

As models that could be explored by these gravitational lensing approaches, we have the contenders to ΛCDM such as braneworld models and other modified gravity approaches. In a section entitled "Testing gravity models", Heavens introduces Bayesian evidence as a useful way to select between alternatives for dark energy. The author shows the expected Bayesian evidence for models and experiments planned. Heavens finds that a weak lensing experiment could distinguish between general relativity and modified gravity before *Planck*. The expected evidence will depend on the number of galaxies surveyed.

The chapters above give an idea of the most active methods for determining the nature of dark energy. The major activity in the coming years will likely arise from the three selected methods above. Plenty of observational projects will bring results that hopefully will clarify the nature of dark energy.

I could not finish this preface without acknowledging Cambridge University Press and editor Vince Higgs for the support given to the project. I would also like to thank all the authors for providing their insights on this timely subject.

Pilar Ruiz-Lapuente

Part I
Theory

1

Dark energy, gravitation and the Copernican principle

JEAN-PHILIPPE UZAN

1.1 Cosmological models and their hypotheses

1.1.1 Introduction

The progress of physical cosmology during the past ten years has led to a "standard" cosmological model in agreement with all available data. Its parameters are measured with increasing precision but it requires the introduction of a dark sector, including both dark matter and dark energy, attracting the attention of both observers and theoreticians.

Among all the observational conclusions, the existence of a recent acceleration phase of the cosmic expansion has become more and more robust. The quest for the understanding of its physical origin is however just starting (Peebles and Ratra, 2003; Peter and Uzan, 2005; Copeland *et al.*, 2006; Uzan, 2007). Models and speculations are flourishing and we may wonder to what extent the observations of our local universe may reveal the physical nature of the dark energy. In particular, there exist limitations to this quest intrinsic to cosmology, related to the fact that most observations are located on our past light-cone (Ellis, 1975), and to finite volume effects (Bernardeau and Uzan, 2004) that can make many physically acceptable possibilities indistinguishable in practice.

This text discusses the relations between the cosmic acceleration and the theory of gravitation and more generally with the hypotheses underlying the construction of our cosmological model, such as the validity of general relativity on astrophysical scales and the Copernican principle. We hope to illustrate that cosmological data now have the potential to test these hypotheses, which go beyond the measurements of the parameters.

Dark Energy: Observational and Theoretical Approaches, ed. Pilar Ruiz-Lapuente. Published by Cambridge University Press. © Cambridge University Press 2010.

1.1.2 Cosmology, physics and astronomy

Cosmology sits at the crossroads between theoretical physics and astronomy.

Theoretical physics, based on physical laws, tries to describe the fundamental components of nature and their interactions. These laws can be probed locally by experiments. These laws need to be extrapolated to construct cosmological models. Hence, any new idea or discovery concerning these laws can naturally call for an extension of our cosmological model (e.g. introducing massive neutrinos in cosmology is now mandatory).

Astronomy confronts us with phenomena that we have to understand and explain consistently. This often requires the introduction of hypotheses beyond those of the physical theories (Section 1.1.3) in order to "save the phenomena" (Duhem, 1908), as is actually the case with the dark sector of our cosmological model. Needless to say, even if a cosmological model is in agreement with all observations, whatever their accuracy, it does not prove that it is the "correct" model of the universe, in the sense that it is the correct cosmological extrapolation and solution of the local physical laws.

Dark energy confronts us with a compatibility problem since, in order to "save the phenomena" of the observations, we have to include new ingredients (cosmological constant, matter fields or interactions) beyond those of our established physical theories. However, the required value for the simplest dark energy model, i.e. the cosmological constant, is more than 60 orders of magnitude smaller than what is expected from theoretical grounds (Section 1.1.6). This tension between what is required by astronomy and what is expected from physics reminds us of the twenty-centuries long debate between Aristotelians and Ptolemaeans (Duhem, 1913), that was resolved not only by the Copernican model but more importantly by a better understanding of the physics, since Newton's gravity was compatible only with one of these three models that, at the time, could not be distinguished observationally.

1.1.3 Hypotheses of our cosmological model

The construction of any cosmological model relies on four main hypotheses:

(H1) a theory of gravity,

(H2) a description of the matter contained in the universe and its non-gravitational interactions,

(H3) symmetry hypotheses, and

(H4) a hypothesis on the global structure, i.e. the topology, of the universe.

These hypotheses are not on the same footing, since H1 and H2 refer to the physical theories. These two hypotheses are, however, not sufficient to solve the field equations and we must make an assumption on the symmetries (H3) of the solutions describing our universe on large scales, while H4 is an assumption on some global properties of these cosmological solutions, with the same local geometry.

Our reference cosmological model is the ΛCDM model. It assumes that gravity is described by general relativity (H1), that the universe contains the fields of the standard model of particle physics plus some dark matter and a cosmological constant, the last two having no physical explanation at the moment. Note that in the cosmological context this involves an extra assumption, since what will be required by the Einstein equations is the effective stress–energy tensor averaged on large scales. It thus implicitly refers to a, usually not explicit, averaging procedure (Ellis and Buchert, 2005). It also deeply involves the Copernican principle as a symmetry hypothesis (H3), without which the Einstein equations usually cannot be solved, and usually assumes that the spatial sections are simply connected (H4). H2 and H3 imply that the description of standard matter reduces to a mixture of a pressureless fluid and a radiation perfect fluid.

1.1.4 Copernican principle

The *cosmological principle* supposes that the universe is spatially isotropic and homogeneous. In particular, this implies that there exists a privileged class of observers, called fundamental observers, who all see an isotropic universe around them. It implies the existence of a cosmic time and states that all the properties of the universe are the same everywhere at the same cosmic time. It is supposed to hold for the smoothed-out structure of the universe on large scales. Indeed, this principle has to be applied in a statistical sense since there exist structures in the universe.

We can distinguish it from the *Copernican principle* which merely states that we do not live in a special place (the center) in the universe. As long as isotropy around the observer holds, this principle actually leads to the same conclusion as the cosmological principle.

The cosmological principle makes definite predictions about all unobservable regions beyond the observable universe. It completely determines the entire structure of the universe, even for regions that cannot be observed. From this point of view, this hypothesis, which cannot be tested, is very strong. On the other hand, it leads to a complete model of the universe. The Copernican principle has more modest consequences and leads to the same conclusions but only for the observable universe where isotropy has been verified. It does not make any prediction on the structure of the universe for unobserved regions (in particular, space could be homogeneous and non-isotropic on scales larger than the observable universe). We refer to Bondi (1960), North (1965) and Ellis (1975) for further discussions on the definition of these two principles.

We emphasize that, as will be discussed in the next section, our reference cosmological model includes a primordial phase of inflation in order to explain the origin of the large-scale structures of the universe. Inflation gives a theoretical prejudice in favor of the Copernican principle since it predicts that all classical (i.e. non-quantum) inhomogeneities (curvature, shear, etc.) have been washed-out during this phase. If it is sufficiently long, we expect the principle to hold on scales much larger than those of the observable universe, hence backing-up the cosmological principle, since unobservable regions today arise from the same causal process that affected the conditions in our local universe. While the standard predictions of inflation are in agreement with all astronomical data, we should not forget that it is only a theoretical argument to which we shall come back if we find observable evidence against isotropy (Pereira *et al.*, 2007; Pitrou *et al.*, 2008), curvature (Uzan *et al.*, 2003), and homogeneity (e.g. a spatial topology of the universe).

These principles lead to a Robertson–Walker (RW) geometry with metric

$$ds^2 = -dt^2 + a^2(t)\gamma_{ij}\, dx^i\, dx^j, \tag{1.1}$$

where t is the cosmic time and γ_{ij} is the spatial metric on the constant time hypersurfaces, which are homogeneous and isotropic, and thus of constant curvature. It follows that the metric is reduced to a single function of time, the scale factor, a. This implies that there is a one-to-one mapping between the cosmic time and the redshift z:

$$1 + z = \frac{a_0}{a(t)}, \tag{1.2}$$

if the expansion is monotonous.

1.1.5 ΛCDM reference model

The dynamics of the scale factor can be determined from the Einstein equations which reduce for the metric (1.1) to the Friedmann equations:

$$H^2 = \frac{8\pi G}{3}\rho - \frac{K}{a^2} + \frac{\Lambda}{3}, \tag{1.3}$$

$$\frac{\ddot{a}}{a} = -\frac{4\pi G}{3}(\rho + 3P) + \frac{\Lambda}{3}. \tag{1.4}$$

$H \equiv \dot{a}/a$ is the Hubble function and $K = 0, \pm 1$ is the curvature of the spatial sections. G and Λ are the Newton and cosmological constants. ρ and P are respectively the energy density and pressure of the cosmic fluids and are related by

$$\dot{\rho} + 3H(\rho + P) = 0.$$

Defining the dimensionless density parameters as

$$\Omega = \frac{8\pi G\rho}{3H^2}, \quad \Omega_\Lambda = \frac{\Lambda}{3H^2}, \quad \Omega_K = -\frac{K}{H^2 a^2}, \tag{1.5}$$

respectively for the matter, the cosmological constant and the curvature, the first Friedmann equation can be rewritten as

$$E^2(z) \equiv \left(\frac{H}{H_0}\right)^2$$

$$= \Omega_{\text{rad}0}(1+z)^4 + \Omega_{\text{mat}0}(1+z)^3 + \Omega_{K0}(1+z)^2 + \Omega_{\Lambda 0}, \tag{1.6}$$

with $\Omega_{K0} = 1 - \Omega_{\text{rad}0} - \Omega_{\text{mat}0} - \Omega_{\Lambda 0}$. All background observables, such as the luminosity distance, the angular distance, etc., are functions of $E(z)$ and are thus not independent.

Besides this background description, the ΛCDM also accounts for an understanding of the large-scale structure of our universe (galaxy distribution, cosmic microwave background anisotropy) by using the theory of cosmological perturbations at linear order. In particular, in the sub-Hubble regime, the growth rate of the density perturbations is also a function of $E(z)$.

One must, however, extend this minimal description by a primordial phase in order to solve the standard cosmological problems (flatness, horizon, etc.). In our reference model, we assume that this phase is described by an inflationary period during which the expansion of the universe is almost exponentially accelerated. In such a case, the initial conditions for the gravitational dynamics that will lead to the large-scale structure are also determined so that our model is completely predictive. We refer to Chapter 8 of Peter and Uzan (2005) for a detailed description of these issues that are part of our cosmological model but not directly related to our actual discussion.

In this framework, the dark energy is well defined and reduces to a single number equivalent to a fluid with equation of state $w = P/\rho = -1$. This model is compatible with all astronomical data, which roughly indicates that

$$\Omega_{\Lambda 0} \simeq 0.73, \quad \Omega_{\text{mat}0} \simeq 0.27, \quad \Omega_{K0} \simeq 0.$$

1.1.6 The cosmological constant problem

This model is theoretically well-defined, observationally acceptable, phenomenologically simple and economical. From the perspective of general relativity the value of Λ is completely free and there is no argument allowing us to fix it, or equivalently, the length scale $\ell_\Lambda = |\Lambda_0|^{-1/2}$, where Λ_0 is the astronomically deduced value of the cosmological constant. Cosmology roughly imposes that

$$|\Lambda_0| \leq H_0^2 \iff \ell_\Lambda \leq H_0^{-1} \sim 10^{26}\,\mathrm{m} \sim 10^{41}\,\mathrm{GeV}^{-1}.$$

In itself this value is no problem, as long as we only consider classical physics. Notice, however, that it is disproportionately large compared to the natural scale fixed by the Planck length:

$$\ell_\Lambda > 10^{60}\ell_\mathrm{P} \iff \frac{\Lambda_0}{M_\mathrm{Pl}^2} < 10^{-120} \iff \rho_{\Lambda_0} < 10^{-120} M_\mathrm{Pl}^4 \sim 10^{-47}\,\mathrm{GeV}^4,$$

(1.7)

when expressed in terms of energy density.

The main problem arises from the interpretation of the cosmological constant. The local Lorentz invariance of the vacuum implies that its energy–momentum tensor must take the form (Zel'dovich, 1988) $\langle T_{\mu\nu}^{\mathrm{vac}} \rangle = -\langle \rho \rangle g_{\mu\nu}$, which is equivalent to that of a cosmological constant. From the quantum point of view, the vacuum energy receives a contribution of the order of

$$\langle \rho \rangle_\mathrm{vac}^\mathrm{EW} \sim (200\,\mathrm{GeV})^4, \qquad \langle \rho \rangle_\mathrm{vac}^\mathrm{Pl} \sim (10^{18}\,\mathrm{GeV})^4, \tag{1.8}$$

arising from the zero point energy, respectively fixing the cutoff frequency of the theory to the electroweak scale or to the Planck scale. This contribution implies a disagreement of respectively 60 to 120 orders of magnitude with astronomical observations!

This is the cosmological constant problem (Weinberg, 1989). It amounts to understanding why

$$|\rho_{\Lambda_0}| = |\rho_\Lambda + \langle \rho \rangle_\mathrm{vac}| < 10^{-47}\,\mathrm{GeV}^4 \tag{1.9}$$

or equivalently,

$$|\Lambda_0| = |\Lambda + 8\pi G \langle \rho \rangle_\mathrm{vac}| < 10^{-120} M_\mathrm{Pl}^2, \tag{1.10}$$

i.e. why ρ_{Λ_0} is so small today, but non-zero.

Today, there is no known solution to this problem and two approaches have been designed. One the one hand, one sticks to this model and extends the cosmological model in order to explain why we observe such a small value of the cosmological constant (Garriga and Vilenkin, 2004; Carr and Ellis, 2008). We shall come back to this approach later. On the other hand, one hopes that there should exist a physical mechanism to exactly cancel the cosmological constant and looks for another mechanism to explain the observed acceleration of the universe.

1.1.7 The equation of state of dark energy

The equation of state of the dark energy is obtained from the expansion history, assuming the standard Friedmann equation. It is thus given by the general expression (Martin *et al.*, 2006)

$$3\Omega_{de}w_{de} = -1 + \Omega_K + 2q, \tag{1.11}$$

q being the deceleration parameter,

$$q \equiv -\frac{a\ddot{a}}{\dot{a}^2} = -1 + \frac{1}{2}(1+z)\frac{d\ln H^2}{dz}. \tag{1.12}$$

This expression (1.11) does not assume the validity of general relativity or any theory of gravity, but gives the relation between the dynamics of the expansion history and the property of the matter that would lead to this acceleration if general relativity described gravity. Thus, the equation of state, as defined in Eq. (1.11), reduces to the ratio of the pressure, P_{de}, to the energy density, ρ_{de}, of an effective dark energy fluid under this assumption only, that is if

$$H^2 = \frac{8\pi G}{3}(\rho + \rho_{de}) - \frac{K}{a^2}, \tag{1.13}$$

$$\frac{\ddot{a}}{a} = -\frac{4\pi G}{3}(\rho + \rho_{de} + 3P + 3P_{de}). \tag{1.14}$$

All the background information about dark energy is thus encapsulated in the single function $w_{de}(z)$. Most observational constraints on the dark energy equation of state refer to this definition.

1.2 Modifying the minimal ΛCDM

The *Copernican principle* implies that the spacetime metric reduces to a single function, the scale factor $a(t)$, which can be Taylor expanded as $a(t) = a_0 + H_0(t - t_0) - \frac{1}{2}q_0H_0^2(t - t_0)^2 + \cdots$. It follows that the conclusion that the cosmic expansion is accelerating ($q_0 < 0$) does not involve any hypothesis about the theory of gravity (other than that the spacetime geometry can be described by a metric) or the matter content, as long as this principle holds.

The assumption that the Copernican principle holds, and the fact that it is so central in drawing our conclusion on the acceleration of the expansion, splits our investigation into two avenues. Either we assume that the Copernican principle holds and we have to modify the laws of fundamental physics or we abandon the

Copernican principle, hoping to explain dark energy without any new physics but at the expense of living in a particular place in the universe. While the first solution is more orthodox from a cosmological point of view, the second is indeed more conservative from a physical point of view. It will be addressed in Section 1.2.4. We are thus faced with a choice between "simple" cosmological solutions with new physics and more involved cosmological solutions of standard physics.

This section focuses on the first approach. If general relativity holds then Eq. (1.4) tells us that the dynamics has to be dominated by a dark energy fluid with $w_{de} < -\frac{1}{3}$ for the expansion to be accelerated. The simplest solution is indeed the cosmological constant Λ for which $w_{de} = -1$ and which is the only model not introducing new degrees of freedom.

1.2.1 General classification of physical models

1.2.1.1 General relativity

Einstein's theory of gravity relies on two independent hypotheses.

First, the theory rests on the Einstein equivalence principle, which includes the universality of free-fall, the local position and local Lorentz invariances in its weak form (as other metric theories) and is conjectured to satisfy it in its strong form. We refer to Will (1981) for a detailed explanation of these principles and their implications. The weak equivalence principle can be mathematically implemented by assuming that all matter fields are minimally coupled to a single metric tensor $g_{\mu\nu}$. This metric defines the length and times measured by laboratory clocks and rods so that it can be called the *physical metric*. This implies that the action for any matter field, ψ say, can be written as $S_{matter}[\psi, g_{\mu\nu}]$. This so-called *metric coupling* ensures in particular the validity of the universality of free-fall.

The action for the gravitational sector is given by the Einstein–Hilbert action

$$S_{gravity} = \frac{c^3}{16\pi G} \int d^4x \sqrt{-g_*}R_*, \tag{1.15}$$

where $g_{\mu\nu}^*$ is a massless spin-2 field called the *Einstein metric*. The second hypothesis states that both metrics coincide

$$g_{\mu\nu} = g_{\mu\nu}^*.$$

The underlying physics of our reference cosmological model (i.e. hypotheses H1 and H2) is thus described by the action

$$S_{gravity} = \frac{c^3}{16\pi G} \int d^4x \sqrt{-g}(R - 2\Lambda) + \sum_{\text{standard model}+\textbf{CDM}} S_{matter}[\psi_i, g_{\mu\nu}], \tag{1.16}$$

which includes all known matter fields plus two unknown components (in bold face).

1.2.1.2 Local experimental constraints

The assumption of a metric coupling is well tested in the solar system. First, it implies that all non-gravitational constants are spacetime independent, and have been tested to a very high accuracy in many physical systems and for various fundamental constants (Uzan, 2003, 2004; Uzan and Leclercq, 2008), e.g. at the 10^{-7} level for the fine structure constant on time scales ranging to 2–4 Gyrs. Second, the isotropy has been tested from the constraint on the possible quadrupolar shift of nuclear energy levels (Prestage *et al.*, 1985; Chupp *et al.*, 1989; Lamoreaux *et al.*, 1986) proving that matter couples to a unique metric tensor at the 10^{-27} level. Third, the universality of free-fall of test bodies in an external gravitational field at the 10^{-13} level has been tested in the laboratory (Baessler *et al.*, 1999; Adelberger *et al.*, 2001). The Lunar Laser ranging experiment (Williams *et al.*, 2004), which compares the relative acceleration of the Earth and Moon in the gravitational field of the Sun, also probes the strong equivalence principle at the 10^{-4} level. Fourth, the Einstein effect (or gravitational redshift) states that two identical clocks located at two different positions in a static Newton potential U and compared by means of electromagnetic signals will exhibit a difference in clock rates of $1 + [U_1 - U_2]/c^2$, where U is the gravitational potential. This effect has been measured at the 2×10^{-4} level (Vessot and Levine, 1978).

The parameterized post-Newtonian formalism (PPN) is a general formalism that introduces 10 phenomenological parameters to describe any possible deviation from general relativity at the first post-Newtonian order (Will, 1981). The formalism assumes that gravity is described by a metric and that it does not involve any characteristic scale. In its simplest form, it reduces to the two Eddington parameters entering the metric of the Schwartzschild metric in isotropic coordinates:

$$g_{00} = -1 + \frac{2Gm}{rc^2} - 2\beta^{\mathrm{PPN}}\left(\frac{2Gm}{rc^2}\right)^2, \qquad g_{ij} = \left(1 + 2\gamma^{\mathrm{PPN}}\frac{2Gm}{rc^2}\right)\delta_{ij}.$$

Indeed, general relativity predicts $\beta^{\mathrm{PPN}} = \gamma^{\mathrm{PPN}} = 1$. These two phenomenological parameters are constrained by: (1) the shift of the Mercury perihelion (Shapiro *et al.*, 1990), which implies that $|2\gamma^{\mathrm{PPN}} - \beta^{\mathrm{PPN}} - 1| < 3 \times 10^{-3}$; (2) the Lunar Laser ranging experiment (Williams *et al.*, 2004) which implies $|4\beta^{\mathrm{PPN}} - \gamma^{\mathrm{PPN}} - 3| = (4.4 \pm 4.5) \times 10^{-4}$ and (3) the deflection of electromagnetic signals, which are all controlled by γ^{PPN}. For instance, very long baseline interferometry (Shapiro *et al.*, 2004) implies that $|\gamma^{\mathrm{PPN}} - 1| = 4 \times 10^{-4}$, while measurement of the time delay variation to the Cassini spacecraft (Bertotti *et al.*, 2003) sets $\gamma^{PPN} - 1 = (2.1 \pm 2.3) \times 10^{-5}$.

The PPN formalism does not allow us to test finite range effects that could be caused e.g. by a massive degree of freedom. In that case one expects a Yukawa-type deviation from the Newton potential,

$$V = \frac{Gm}{r} \left(1 + \alpha e^{-r/\lambda}\right),$$

that can be probed by "fifth force" experimental searches. λ characterizes the range of the Yukawa deviation while its strength α may also include a composition dependence (Uzan, 2003). The constraints on (λ, α) are summarized in Hoyle *et al.* (2004), which typically shows that $\alpha < 10^{-2}$ on scales ranging from the millimeter to the solar system size.

In general relativity, the graviton is massless. One can, however, give it a mass, but this is very constrained. In particular, around a Minkowski background, the mass term must have the very specific form of the Pauli–Fierz type in order to avoid ghosts (see below for a more precise definition) being excited. This mass term is, however, inconsistent with solar system constraints because there exists a discontinuity (van Dam and Veltman, 1970; Zakharov, 1970) between the case of a strictly massless graviton and a very light one. In particular, such a term can be ruled out from the Mercury perihelion shift.

General relativity is also tested with pulsars (Damour and Esposito-Farèse, 1998; Esposito-Farèse, 2005) and in the strong field regime (Psaltis, 2008). For more details we refer to Will (1981), Damour and Lilley (2008) and Turyshev (2008). Needless to say, any extension of general relativity has to pass these constraints. However, deviations from general relativity can be larger in the past, as we shall see, which makes cosmology an interesting physical system to extend these constraints.

1.2.1.3 Universality classes

There are many possibilities to extend this minimal physical framework. Let us start by defining universality classes (Uzan, 2007) by restricting our discussion to field theories. This helps in identifying the new degrees of freedom and their couplings.

The first two classes assume that gravitation is well described by general relativity and introduce new degrees of freedom beyond those of the standard model of particle physics. This means that one adds a new term $S_{de}[\psi; g_{\mu\nu}]$ in the action (1.16) while keeping the Einstein–Hilbert action and the coupling of all the fields (standard matter and dark matter) unchanged. They are:

1. *Class A* consists of models in which the acceleration is driven by the gravitational effect of the new fields. They thus must have an equation of state smaller than $-\frac{1}{3}$. They are

not coupled to the standard matter fields or to dark matter so that one is adding a new sector

$$S_{\text{de}}[\phi; g_{\mu\nu}]$$

to the action (1.16), where ϕ stands for the dark energy field (not necessarily a scalar field). Standard examples include *quintessence* models (Wetterich, 1988; Ratra and Peebles, 1988) which invoke a canonical scalar field slow-rolling today, *solid dark matter* models (Battye *et al.*, 1999) induced by frustrated topological defects networks, *tachyon* models (Sen, 1999), *Chaplygin gas* (Kamenshchik *et al.*, 2001) and *K-essence* (Armendariz-Picon *et al.*, 2000; Chiba *et al.*, 2000) models invoking scalar fields with a non-canonical kinetic term.

2. *Class B* introduces new fields which do not dominate the matter content so that they do not change the expansion rate of the universe. They are thus not required to have an equation of state smaller than $-\frac{1}{3}$. These fields are however coupled to photons and thus affect the observations. An example (Csaki *et al.*, 2002; Deffayet *et al.*, 2002) is provided by *photon–axion oscillations*, which aims at explaining the dimming of supernovae not by an accelerated expansion but by the fact that some of the photons have oscillated into invisible axions. In that particular case, the electromagnetic sector is modified according to

$$S_{\text{em}}[A_\mu; g_{\mu\nu}] \rightarrow S_{\text{em}}[A_\mu, a_\mu; g_{\mu\nu}].$$

A specific signature of these models would be a violation of the distance duality relation (see Section 1.3.3.1).

Then come models with a modification of general relativity. Once such a possibility is considered, many new models arise (Will, 1981). They are:

3. *Class C* includes models in which a finite number of new fields are introduced. These fields couple to the standard model fields and some of them dominate the matter content (at least at late time). This is the case in particular for scalar–tensor theories in which a scalar field couples universally and leads to the class of extended quintessence models, chameleon models or $f(R)$ models depending on the choice of the coupling function and potential (see Section 1.2.3). For these models, one has a new sector

$$S_\varphi[\varphi; g_{\mu\nu}]$$

and the couplings of the matter fields will be modified according to

$$S_{\text{matter}}[\psi_i; g_{\mu\nu}] \rightarrow S_{\text{matter}}[\psi_i; A_i^2(\varphi)g_{\mu\nu}].$$

If the coupling is not universal, a signature may be the variation of fundamental constants and a violation of the universality of free-fall. This class also offers the possibility to have $w_{\text{de}} < -1$ with a well-defined field theory and includes models in which a scalar field couples differently to the standard matter field and dark matter.

4. *Class D* includes more drastic modifications of general relativity with e.g. the possibility to have more types of gravitons (massive or not and most probably an infinite

number of them). This is the case for models involving extra dimensions such as multi-brane models (Gregory *et al.*, 2000), multigravity (Kogan *et al.*, 2000), brane-induced gravity (Dvali *et al.*, 2000) or simulated gravity (Carter *et al.*, 2001). In these cases, the new fields modified the gravitational interaction on the large scale but do not necessarily dominate the matter content of the universe. Some of these models may also offer the possibility to mimic an equation of state $w_{de} < -1$.

These various modifications, summarized in Fig. 1.1, can be combined to get more exotic models.

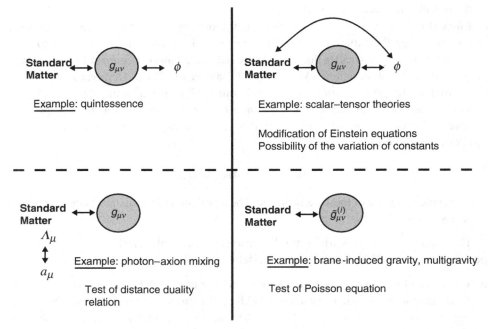

Fig. 1.1. Summary of the different classes of physical dark energy models. As discussed in the text, various tests can be designed to distinguish between them. The classes differ according to the nature of the new degrees of freedom and their couplings. Left column accounts for models where gravitation is described by general relativity while right column models describe a modification of general relativity. In the upper classes, the new fields dominate the matter content of the universe at low redshift. Upper-left models (class A) consist of models in which a new kind of gravitating matter is introduced. In the upper-right models (class C), a light field induces a long-range force so that gravity is not described by a massless spin-2 graviton only. In this class, Einstein equations are modified and there may be a variation of the fundamental constants. The lower-right models (class D) correspond to models in which there may exist an infinite number of new degrees of freedom, such as in some classes of braneworld scenarios. These models predict a modification of the Poisson equation on large scales. In the last class (lower-left, class B), the distance duality relation may be violated. (From Uzan, 2007.)

1.2.1.4 "Modified gravity" vs. new matter

The different models in the literature are often categorized as "modified gravity" or "new matter". This distinction may, however, be subtle.

First, we shall define *gravity* as the long-range force that cannot be screened. We are used to describing this interaction by general relativity so that it is associated with a massless spin-2 graviton. In our view, gravity cannot be modified but only its description, i.e. general relativity. As an example, scalar–tensor theories (see Section 1.2.3) extend general relativity by a spin-0 interaction which can be long range according to the mass of the scalar field. In this case, the interaction is even universal so that it does not imply any violation of the weak Einstein equivalence principle.

Note also that whatever the model, it requires the introduction of new fields beyond those of the standard model. The crucial difference is that in models with "new matter" (e.g. class A), the amount of dark energy is imposed by initial conditions and its gravitational effect induces the acceleration of the universe. In a "modified gravity" model (e.g. classes C and D) the standard matter and cold dark matter generate an effective dark energy component. The acceleration may thus be a consequence of the fact that the gravitational interaction is weaker than expected on large scales. But, it may be that the energy density of the new field also dominates the dynamics but is still determined by the energy density of the standard field.

1.2.2 Modifying general relativity

1.2.2.1 In which regime?

Before investigating gravity beyond general relativity, let us try to sketch the regimes in which these modifications may (or will) appear. We can distinguish the following regimes.

- *Weak–strong field regimes* can be characterized by the amplitude of the gravitational potential. For a spherical static spacetime, $\Phi = GM/rc^2$. It is of order of $\Phi_\odot \sim 2 \times 10^{-6}$ at the surface of the Sun and equal to $\frac{1}{2}$ for a black hole.
- *Small–large distances*. Such modifications can be induced by a massive degree of freedom that will induce a Yukawa-like coupling. While constrained on the scale of the solar system, we have no constraints on scales larger than $10h^{-1}$ Mpc.
- *Low–high acceleration regimes* are of importance in discussion of galaxy rotation curves and (galactic) dark matter, as suggested by the MOND phenomenology (Milgrom, 1983). In particular, the kind of modification of the gravitation theory that could account for the dark matter cannot occur at a characteristic distance because of the Tully–Fischer law.
- *Low–high curvature regimes* will distinguish the possible extensions of the Einstein–Hilbert action. For instance a quadratic term of the form αR^2 becomes significant

compared to R when $GM/r^3c^2 \gg \alpha^{-1}$ even if Φ remains small. In the solar system, $R_\odot \sim 4 \times 10^{-28}\,\text{cm}^{-2}$.

In cosmology, we can suggest various possible regimes in which to modify general relativity. The dark matter problem can be accounted for by a modification of Newton gravity below the typical acceleration $a_0 \sim 10^{-8}\,\text{cm\,s}^{-2}$. It follows that the regime for which a dark matter component is required can be characterized by

$$\Phi R < a_0^2 \sim 3 \times 10^{-31} R_\odot. \tag{1.17}$$

Concerning the homogeneous universe, one can sort out from the Friedmann equations that

$$R_{\text{FL}}(z) = 3H_0^2[\Omega_{m0}(1+z)^3 + 4\Omega_{\Lambda 0}], \tag{1.18}$$

from which we deduce that $R_{\text{FL}} \sim 10^{-5} R_\odot$ at the time of nucleosynthesis, $R_{\text{FL}} \sim 10^{-20} R_\odot$ at the time of decoupling and $R_{\text{FL}} \sim 10^{-28} R_\odot$ at $z=1$. The curvature scale associated with a cosmological constant is $R_\Lambda = \frac{1}{6}\Lambda$ and the cosmological constant (or dark energy) problem corresponds to a low curvature regime,

$$R < R_\Lambda \sim 1.2 \times 10^{-30} R_\odot. \tag{1.19}$$

The fact that the limits (1.17) and (1.19) intersect illustrates the coincidence problem, that is $a_0 \sim cH_0$ and $\Omega_{m0} \sim \Omega_{\Lambda 0}$. Note that both arise on curvature scales much smaller than those probed in the solar system.

Let us now turn to the cosmological perturbations. The gravitational potential at the time of the decoupling ($z \sim 10^3$) is of the order of $\Phi \sim 10^{-5}$. During the matter era, the Poisson equation imposes that $\Delta\Phi \propto \delta\rho_m a^2$, which is almost constant. It follows that we never expect a potential larger than $\Phi \sim 10^{-5}$ on cosmological scales. We are thus always in a weak field regime. The characteristic distance scale is fixed by the Hubble radius c/H_0. The curvature perturbation associated with the large-scale structures is, in the linear theory, of the order

$$\delta R = \frac{6}{a^2}\Delta\Phi \sim 3H_0^2 \Omega_{m0}(1+z)^3 \delta_m(z).$$

Since at redshift zero, $\langle \delta_m^2 \rangle = \sigma_8 \sim 1$ in a ball of radius of 8 Mpc, we conclude that $\langle \delta R^2 \rangle^{1/2} \sim 3H_0^2 \Omega_{m0}\sigma_8$ while $R_{\text{FL}} = 3H_0^2 \Omega_{m0}$ if $\Lambda = 0$. This means that the curvature perturbation becomes of the order of the background curvature at a redshift $z \sim 0$, even if we are still in the weak field limit. This implies that the effect of the large-scale structures on the background dynamics may be non-negligible. This effect has been argued to be at the origin of the acceleration of the universe (Ellis and Buchert, 2005; Ellis, 2008) but no convincing formalism to describe this backreaction has been constructed yet. Note that in this picture the onset of the acceleration phase will be determined by the amplitude of the initial power spectrum.

In conclusion, to address the dark energy or dark matter problem by a modification of general relativity, we are interested in modifications on large scales (typically Hubble scales), low acceleration (below a_0) or small curvature (typically R_Λ).

1.2.2.2 General constraints

In modifying general relativity, we shall demand that the new theory:

- *does not contain ghosts*, i.e. degrees of freedom with negative kinetic energy. The problem with such a ghost is that the theory would be unstable. In particular, the vacuum can decay into an arbitrary amount of positive energy (standard) gravitons whose energy would be balanced by negative energy ghosts.
- *has a Hamiltonian bounded from below*. Otherwise, the theory would be unstable, even if one cannot explicitly identify a ghost degree of freedom.
- *the new degrees of freedom are not tachyon*, i.e. do not have a negative mass.
- *is compatible with local tests* of deviation from general relativity, in particular in the solar system described in Section 1.2.1.2.

Then, starting from the action (1.16), we see that we can either modify the Einstein–Hilbert action while leaving the coupling of all matter fields to the metric unchanged or modify the coupling(s) in the matter action. The possibilities are numerous (Will, 1981; Esposito-Farèse and Bruneton, 2007; Uzan, 2007) and we cannot start an extensive review of the models here. We shall thus consider some examples that will illustrate the constraints cited above, but with no goal of exhaustivity.

1.2.2.3 Modifying the Einstein–Hilbert action

Let us start with the example of higher-order gravity models based on the quadratic action (here we follow the very clear analysis of Esposito-Farèse and Bruneton (2007) for our discussion):

$$S_{\text{gravity}} = \frac{c^3}{16\pi G} \int d^4x \sqrt{-g} \left[R + \alpha C_{\mu\nu\rho\sigma}^2 + \beta R^2 + \gamma \text{GB} \right], \tag{1.20}$$

where $C_{\mu\nu\rho\sigma}$ is the Weyl tensor and $\text{GB} \equiv R_{\mu\nu\rho\sigma}^2 - 4R_{\mu\nu}^2 + R^2$ is the Gauss–Bonnet term. α, β and γ are three constants with dimension of an inverse mass square. Since GB does not contribute to the local field equations of motion, we will not consider it further. The action (1.20) gives a renormalizable theory of quantum gravity at all orders provided α and β are non-vanishing (Stelle, 1978). However, such theories contain ghosts. This can be seen from the graviton propagator, which takes the form $1/(p^2 + \alpha p^4)$. It can indeed be decomposed in irreducible fractions as

$$\frac{1}{p^2 + \alpha p^4} = \frac{1}{p^2} - \frac{1}{p^2 + \frac{1}{\alpha}}.$$

The first term is nothing but the standard propagator of the usual massless graviton. The second term corresponds to an extra-massive degree of freedom with mass α^{-1} and its negative sign indicates that it carries negative energy: it is a ghost. Moreover, if α is negative, this ghost is also a tachyon! The only viable such modification arises from βR^2, which introduces a massive spin-0 degree of freedom.

These considerations can be extended to more general theories involving an arbitrary function of the metric invariants, $f(R, R_{\mu\nu}, R_{\mu\nu\rho})$, which also generically (Hindawi *et al.*, 1996; Tomboulis, 1996) contain a massive spin-2 ghost. They are thus not stable theories with the exception of $f(R)$ theories, discussed in Section 1.2.3.3.

A possibility may be to consider models designed such that their second-order expansion never shows any negative energy kinetic term. As recalled in Esposito-Farèse and Bruneton (2007) and Woodard (2006), these models still exhibit instabilities, the origin of which can be related to a theorem by Ostrogradsky (1850) showing that their Hamiltonian is generically not bounded from below.

We summarize this theorem following the presentation by Woodard (2006). Consider a Lagrangian depending on a variable q and its first two time derivatives $\mathcal{L}(q, \dot{q}, \ddot{q})$ and assume that it is not degenerate, i.e. that $\ddot{q}$ cannot be eliminated by an integration by parts. Then the definition $p_2 \equiv \partial \mathcal{L}/\partial \ddot{q}$ can be inverted to get $\ddot{q}$ as a function q, $\dot{q}$ and p_2, $\ddot{q}[q, \dot{q}, p_2]$, and the initial data must be specified by two pairs of conjugate momenta defined by $(q_1, p_1) \equiv (q, \partial \mathcal{L}/\partial \dot{q} - d(\partial \mathcal{L}/\partial \ddot{q})/dt)$ and $(q_2, p_2) \equiv (\dot{q}, \partial \mathcal{L}/\partial \ddot{q})$. The Hamiltonian defined as $\mathcal{H} = p_1 \dot{q}_1 + p_2 \dot{q}_2 - \mathcal{L}$ can be shown to be the generator of time translations and the Hamilton equations that derive from $\mathcal{H}$ are indeed equivalent to the Euler–Lagrange equations derived from $\mathcal{L}$. In terms of q_i and p_i, the Hamiltonian takes the form

$$\mathcal{H} = p_1 q_2 + p_2 \ddot{q}[q_1, q_2, p_2] - \mathcal{L}(q_1, q_2, \ddot{q}[q_1, q_2, p_2]).$$

This expression is however linear in p_1 so that the Hamiltonian is not bounded from below and the theory is necessarily unstable. Let us note that this constraint can be avoided by non-local theories, that is if the Lagrangian depends on an infinite number of derivatives, as e.g. string theory, even though its expansion may look pathological.

1.2.2.4 Modifying the matter action

Many other possibilities, known as bi-metric theories of gravity, arise if one assumes that $g_{\mu\nu} \neq g^*_{\mu\nu}$. One can postulate that the physical metric is a combination of various fields, e.g.

$$g_{\mu\nu}[g^*_{\mu\nu}, \varphi, A_\mu, B_{\mu\nu}, \ldots] = A^2(\varphi) \left[g^*_{\mu\nu} + \alpha_1 A_\mu A_\nu + \alpha_2 g^*_{\mu\nu} g_*^{\alpha\beta} A_\alpha A_\beta + \cdots \right].$$

As long as these new fields enter quadratically, their field equation is generically of the form $(\nabla_\mu \nabla^\mu) A = AT$ where T is the matter source. It follows that matter cannot generate them if their background value vanishes. On the other hand, if their background value does not vanish then these fields define a preferred frame and Lorentz invariance is violated.

Such modifications have however drawn some attention, especially in the attempts to construct a field theory reproducing the MOND phenomenology (Milgrom, 1983). In particular, in order to increase light deflection in scalar–tensor theories of gravity, a *disformal coupling* (Bekenstein, 1993), $g_{\mu\nu} = A^2(\varphi) g^*_{\mu\nu} + B(\varphi) \partial_\mu \varphi \partial_\nu \varphi$, was introduced. It was generalized to *stratified theory* (Sanders, 1997) by replacing the gradient of the scalar field by a dynamical unit vector field $(g^*_{\mu\nu} A^\mu A^\nu = -1)$, $g_{\mu\nu} = A^2(\varphi) g^*_{\mu\nu} + B(\varphi) A_\mu A_\nu$. This is the basis of the TeVeS theory proposed by Bekenstein (2004). The mathematical consistency and the stability of these field theories were investigated in depth in the excellent analysis of Esposito-Farèse and Bruneton (2007). It was shown that no present theory passes all available experimental constraints while being stable and admitting a well-posed Cauchy problem.

Esposito-Farèse and Bruneton (2007) also notice that while couplings of the form $g_{\mu\nu}[g^*_{\mu\nu}, R^*_{\mu\nu}, R^*_{\mu\nu\alpha\beta}, \ldots]$ seem to lead to well-defined theories in vacuum (in particular) when linearizing around a Minkowsky background, they are unstable inside matter, because the Ostrogradsky theorem strikes back.

The case in which only a scalar partner, $g_{\mu\nu} = A^2(\varphi) g^*_{\mu\nu}$, is introduced leads to consistent field theories and is the safest way to modify the matter coupling. We shall discuss these scalar–tensor theories of gravity in Section 1.2.3.

1.2.2.5 Higher-dimensional theories

Higher-dimensional models of gravity, including string theory (see e.g. Damour and Lilley (2008)), predict non-metric coupling as discussed in the previous section. Many scalar fields, known as *moduli*, appear in the dimensional reduction to four dimensions.

As a simple example, let us consider a five-dimensional spacetime and assume that gravity is described by the Einstein–Hilbert action

$$S = \frac{1}{12\pi^2 G_5} \int \bar{R}\sqrt{|\bar{g}|}\, \mathrm{d}^5 x, \tag{1.21}$$

where we denote by a bar quantities in five dimensions to distinguish them from the analogous quantities with no bar in four dimensions. The aim is to determine the independent elements of the metric $\bar{g}_{AB}$, of which these are 15 in five dimensions. We decompose the metric into a symmetric tensor part $g_{\mu\nu}$, with 10 independent components, a vector part, A_α, with 4 components and finally a scalar field, ϕ, to complete the counting of the number of degrees of freedom ($15 = 10 + 4 + 1$). The metric is thus decomposed as

$$\bar{g}_{AB} = \begin{pmatrix} g_{\mu\nu} + \dfrac{1}{M^2}\phi^2 A_\mu A_\nu & \dfrac{1}{M}\phi^2 A_\mu \\[2ex] \dfrac{1}{M}\phi^2 A_\nu & \phi^2 \end{pmatrix}, \tag{1.22}$$

where the different components depend a priori both on the usual spacetime coordinates x^α and on the coordinate in the extra-dimension y. The constant M has dimensions of mass, so that A_α also has dimensions of mass, whereas the scalar field ϕ is here dimensionless. Finally, while capital latin indices vary in the entire five-dimensional spacetime, $A, B = 0, \ldots, 4$, greek indices span the four-dimensional spacetime, namely $\mu, \nu = 0, \ldots, 3$. Compactifying on a circle and assuming that none of the variables depends on the transverse direction y (*cylinder condition*), the action (1.21) reduces to the four-dimensional action

$$S = \frac{1}{16\pi G} \int \mathrm{d}^4 x \sqrt{-g}\, \phi \left(R - \frac{\phi^2}{4M^2} F_{\alpha\beta} F^{\alpha\beta} \right), \tag{1.23}$$

where $F_{\alpha\beta} \equiv \partial_\alpha A_\beta - \partial_\beta A_\alpha$ and where we have set

$$G = \frac{3\pi \bar{G}_5}{4V_{(5)}},$$

and factored out the finite volume of the fifth dimension, $V_{(5)} = \int \mathrm{d}y$. The scalar field couples explicitly to the kinetic term of the vector field. It can be checked that this coupling cannot be eliminated by a redefinition of the metric, whatever the function $A(\phi)$: this is the well-known conformal invariance of electromagnetism in four dimensions. Such a term induces a variation of the fine structure constant as well as a violation of the universality of free-fall (Uzan, 2003). Such dependencies of the masses and couplings are generic for higher-dimensional theories and in particular string theory.

The cylinder condition is justified as long as we consider the fifth dimension to be topologically compact with the topology of a circle. In this case, all the fields that are defined in this space, i.e. the four-dimensional metric $g_{\mu\nu}$, the vector A_{α} and the dilaton ϕ, and any additional matter fields that the theory should describe, are periodic functions of the extra-dimension and can therefore be expanded into Fourier modes. The radius R of this dimension then turns out to be naturally $R \sim M^{-1}$. For large enough M, the radius is too small to have observable consequences: to be sensitive to the fifth dimension, the energies involved must be comparable to M. Decomposing all the fields in Fourier modes, e.g.

$$\phi\left(x_{\mu}, y\right) = \sum_{n=-\infty}^{+\infty} \phi_n\left(x_{\mu}\right) e^{inMy}, \qquad \text{with} \qquad \phi_{-n} = \phi_n^{\star} \qquad (1.24)$$

(ϕ real), we conclude that the four-dimensional theory will also contain a infinite tower of modes of increasing mass.

While these tree-level predictions of string theory are in contradiction with experimental constraints, many mechanisms can reconcile it with experiment. In particular, it has been claimed that quantum loop corrections to the tree-level action may modify the coupling in such a way that it has a minimum (Damour and Polyakov, 1994). The scalar field can thus be attracted toward this minimum during the cosmological evolution so that the theory is attracted toward general relativity. Another possibility is to invoke an environmental dependence, as can be implemented in scalar–tensor theories by the chameleon mechanism (Khoury and Weltman, 2004) which invokes a potential with a minimum not coinciding with that of the coupling function.

In higher dimensions, the Einstein–Hilbert action can also be modified by adding the Gauss–Bonnet term, GB, since it does not enter the field equations only in four dimensions. The D-dimensional Einstein–Hilbert action can then be modified to include a term of the form αGB. In particular, this is the case in the low-energy limit of heterotic string theory (Gross and Sloan, 1987). In various configurations, in particular with branes, it has been argued that the Gauss–Bonnet invariant can also couple to a scalar field (Amendola *et al.*, 2006), i.e. $\alpha(\varphi)$GB. As long as the modification is linear in GB, it is ghost-free.

In the context of braneworld, it was shown that some models with infinite volume extra-dimension can produce a modification of general relativity leading to an acceleration of the expansion. In the DGP model (Dvali *et al.*, 2000), one considers beside the five-dimensional Einstein–Hilbert a four-dimensional term induced on the brane

$$S = \frac{M_5^2}{2} \int \bar{R}_5 \sqrt{|\bar{g}_5|}\, \mathrm{d}^5 x + \frac{M_4^2}{2} \int R_4 \sqrt{|g_4|}\, \mathrm{d}^4 x. \tag{1.25}$$

There is a competition between these two terms and the five-dimensional term dominates on scales larger than $r_c = M_4^2/2M_5^3$. The existence or absence of ghost in this class of models is still under debate. Some of these models (Deffayet, 2005) have also been claimed to describe massive gravitons without being plagued by the van Dam–Veltman–Zakharov discontinuity (see Section 1.2.1.2).

As a conclusion, higher-dimensional models offer a rich variety of possibilities among which some may be relevant to describe a modification of general relativity on large scales.

1.2.3 Example: scalar–tensor theories

As discussed in Section 1.2.2.4, the case in which only a scalar partner to the graviton is introduced leads to consistent field theories and is the safest way to modify the matter coupling.

1.2.3.1 Formulation

In scalar–tensor theories, gravity is mediated not only by a massless spin-2 graviton but also by a spin-0 scalar field that couples universally to matter fields (this ensures the universality of free-fall). In the Jordan frame, the action of the theory takes the form

$$S = \int \frac{\mathrm{d}^4 x}{16\pi G_*} \sqrt{-g}\,[F(\varphi)R - g^{\mu\nu}Z(\varphi)\varphi_{,\mu}\varphi_{,\nu} - 2U(\varphi)]$$
$$+ S_{\mathrm{matter}}[\psi; g_{\mu\nu}], \tag{1.26}$$

where G_* is the bare gravitational constant. This action involves three arbitrary functions (F, Z, and U) but only two are physical since there is still the possibility to redefine the scalar field. F needs to be positive to ensure that the graviton carries positive energy. S_{matter} is the action of the matter fields that are coupled minimally to the metric $g_{\mu\nu}$. In the Jordan frame, the matter is universally coupled to the metric so that the length and time as measured by laboratory apparatus are defined in this frame.

It is useful to define an Einstein frame action through a conformal transformation of the metric

$$g_{\mu\nu}^* = F(\varphi)g_{\mu\nu}. \tag{1.27}$$

In the following all quantities labelled by a star (*) will refer to the Einstein frame. Defining the field φ_* and the two functions $A(\varphi_*)$ and $V(\varphi_*)$ (see e.g. Esposito-Farèse and Polarski, 2001) by

$$\left(\frac{d\varphi_*}{d\varphi}\right)^2 = \frac{3}{4}\left(\frac{d\ln F(\varphi)}{d\varphi}\right)^2 + \frac{1}{2F(\varphi)}, \tag{1.28}$$

$$A(\varphi_*) = F^{-1/2}(\varphi), \tag{1.29}$$

$$2V(\varphi_*) = U(\varphi)F^{-2}(\varphi), \tag{1.30}$$

the action (1.26) reads as

$$S = \frac{1}{16\pi G_*}\int d^4x\sqrt{-g_*}\left[R_* - 2g_*^{\mu\nu}\,\partial_\mu\varphi_*\,\partial_\nu\varphi_* - 4V(\varphi_*)\right]$$
$$+ S_{\text{matter}}[A^2(\varphi_*)\,g_{\mu\nu}^*;\,\psi]. \tag{1.31}$$

The kinetic terms have been diagonalized so that the spin-2 and spin-0 degrees of freedom of the theory are perturbations of $g_{\mu\nu}^*$ and φ_* respectively.

In this frame, the field equations take the form

$$G_{\mu\nu}^* = 8\pi G_* T_{\mu\nu}^*$$
$$+2\partial_\mu\varphi_*\partial_\nu\varphi_* - g_{\mu\nu}^*\,(\partial_\alpha\varphi_*)^2 - 2g_{\mu\nu}^*V, \tag{1.32}$$

$$(\nabla_\mu\nabla^\mu)_*\varphi_* = V_{,\varphi_*} - 4\pi G_*\alpha(\varphi_*)T_{\mu\nu}^*g_*^{\mu\nu}, \tag{1.33}$$

$$\nabla_\mu T_*^{\mu\nu} = \alpha(\varphi_*)T_{\sigma\rho}^*g_*^{\sigma\rho}\partial^\nu\varphi_*, \tag{1.34}$$

where we have defined the Einstein frame stress-energy tensor

$$T_*^{\mu\nu} \equiv \frac{2}{\sqrt{-g_*}}\frac{\delta S_{\text{matter}}}{\delta g_{\mu\nu}^*},$$

related to the Jordan frame stress-energy tensor by $T_{\mu\nu}^* = A^2 T_{\mu\nu}$. The function

$$\alpha(\varphi_*) \equiv \frac{d\ln A}{d\varphi_*} \tag{1.35}$$

characterizes the coupling of the scalar field to matter (we recover general relativity with a minimally coupled scalar field when it vanishes). For completeness, we also introduce

$$\beta(\varphi_*) \equiv \frac{d\alpha}{d\varphi_*}. \tag{1.36}$$

Note that in the Einstein frame the Einstein Eqs. (1.32) are the same as those obtained in general relativity with a minimally coupled scalar field.

The action (1.26) defines an effective gravitational constant $G_{\rm eff} = G_*/F = G_* A^2$. This constant does not correspond to the gravitational constant effectively measured in a Cavendish experiment. The Newton constant measured in this experiment is

$$G_{\rm cav} = G_* A_0^2 (1 + \alpha_0^2), \tag{1.37}$$

where the first term, $G_* A_0^2$ corresponds to the exchange of a graviton while the second term $G_* A_0^2 \alpha_0^2$ is related to the long-range scalar force.

1.2.3.2 Cosmological signatures

The post-Newtonian parameters can be expressed in terms of the values of α and β today as

$$\gamma^{\rm PPN} - 1 = -\frac{2\alpha_0^2}{1 + \alpha_0^2}, \qquad \beta^{\rm PPN} - 1 = \frac{1}{2}\frac{\beta_0 \alpha_0^2}{(1 + \alpha_0^2)^2}. \tag{1.38}$$

The solar system constraints discussed in Section 1.2.1.2 imply α_0 to be very small, typically $\alpha_0^2 < 10^{-5}$ while β_0 can still be large. Binary pulsar observations (Esposito-Farèse, 2005) impose that $\beta_0 > -4.5$.

The previous constraints can be satisfied even if the scalar–tensor theory was far from general relativity in the past. The reason is that these theories can be attracted toward general relativity (Damour and Nordtvedt, 1993) if their coupling function or potential has a minimum. This can be illustrated in the case of a massless ($V = 0$) dilaton with quadratic coupling ($a \equiv \ln A = \frac{1}{2}\beta\varphi_*^2$). The Klein–Gordon equation (1.33) can be rewritten in terms of the number of e-folds in the Einstein frame as

$$\frac{2}{3 - \varphi_*'^2}\varphi_*'' + (1 - w)\varphi_*' = -\alpha(\varphi_*)(1 - 3w). \tag{1.39}$$

As emphasized by Damour and Nordtvedt (1993), this is the equation of motion of a point particle with a velocity-dependent inertial mass $m(\varphi_*) = 2/(3 - \varphi_*'^2)$ evolving in a potential $\alpha(\varphi_*)(1 - 3w)$ and subject to a damping force $-(1 - w)\varphi_*'$. During the cosmological evolution the field is driven toward the minimum of the coupling function. If $\beta > 0$, it drives φ_* toward 0, that is $\alpha \to 0$, so that the scalar–tensor theory becomes closer and closer to general relativity. When $\beta < 0$, the theory is driven away from general relativity and is likely to be incompatible with local tests unless φ_* was initially arbitrarily close to 0.

During the radiation era, $w = \frac{1}{3}$ and the coupling is not efficient so that φ_* freezes to a constant value. Then, during the matter era, the coupling acts as a

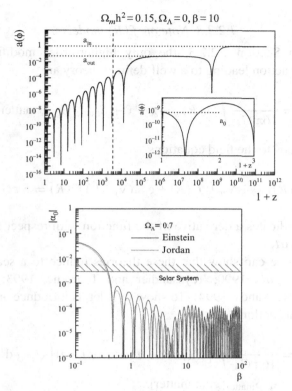

Fig. 1.2. *Top*: Evolution of the dilaton as a function of redshift. In the radiation era the dilaton freezes to a constant value and is then driven toward the minimum of the coupling function during the matter era. *Bottom*: constraints on scalar–tensor theories of gravity with a massless dilaton with quadratic coupling in the (α_0, β) plane. At large β the primordial nucleosynthesis sets more stringent constraints than the solar system. (From Coc *et al.*, 2006.)

potential with a minimum in zero, hence driving φ_* towards zero and the theory towards general relativity (see Fig. 1.2).

This offers a rich phenomenology for cosmology and in particular for the dark energy question. It has been shown that quintessence models can be extended to a scalar–tensor theory of gravity (Uzan, 1999; Bartolo and Pietroni, 2000) and that it offers the possibility to have an equation of state smaller than -1 with a well-defined theory (Martin *et al.*, 2006). The constraints on the deviations from general relativity can also be sharpened by the use of cosmological observations such as cosmic microwave background anisotropies (Riazuelo and Uzan, 2002), weak gravitational lensing (Schimd *et al.*, 2005), and big-bang nucleosynthesis (Coc *et al.*, 2006). Figure 1.2 summarizes the constraints that can be obtained from primordial nucleosynthesis.

1.2.3.3 Note on f(R) models

As discussed in Section 1.2.2.3, the only higher-order modifications of the Einstein–Hilbert action leading to a well-defined theory are

$$S = \frac{1}{16\pi G_*} \int f(R)\sqrt{-g}\, \mathrm{d}^4 x + S_{\text{matter}}[g_{\mu\nu}; \text{matter}]. \qquad (1.40)$$

Such a theory leads to the field equations

$$f'(R)R_{\mu\nu} - \frac{1}{2}f(R)g_{\mu\nu} - \nabla_\mu\partial_\nu f'(R) + g_{\mu\nu}(\nabla_\mu\nabla^\mu)f'(R) = 8\pi G_* T_{\mu\nu}, \qquad (1.41)$$

where a prime indicates a derivative of the function with respect to its argument, i.e. $f'(R) \equiv \mathrm{d}f/\mathrm{d}R$.

Interestingly, one can show that these theories reduce to a scalar–tensor theory (Gottlöber *et al.*, 1990; Teyssandier and Tourrenc, 1993; Mangano and Sokolowski, 1994; Wands, 1994). To show this, let us introduce an auxiliary field φ and consider the action

$$S = \frac{1}{16\pi G_*} \int \left[f'(\varphi)R + f(\varphi) - \varphi f'(\varphi)\right]\sqrt{-g}\,\mathrm{d}^4 x$$
$$+ S_{\text{matter}}[g_{\mu\nu}; \text{matter}]. \qquad (1.42)$$

The variation of this action with respect to the scalar field indeed implies, if $f''(\varphi) \neq 0$ (the case $f'' = 0$ is equivalent to general relativity with a cosmological constant), that

$$R - \varphi = 0. \qquad (1.43)$$

This constraint permits Eq. (1.41) to be rewritten in the form

$$f'(\varphi)G_{\mu\nu} - \nabla_\mu\partial_\nu f'(\varphi) + g_{\mu\nu}(\nabla_\mu\nabla^\mu)f'(\varphi) + \frac{1}{2}[\varphi f'(\varphi) - f(\varphi)]g_{\mu\nu} = 8\pi G_* T_{\mu\nu}, \qquad (1.44)$$

which then reduces to Eq. (1.32) after the field redefinitions necessary to shift to the Jordan frame. Note that, even if the action (1.42) does not possess a kinetic term for the scalar field, the theory is well defined since the true spin-0 degree of freedom clearly appears in the Einstein frame, and with a positive energy.

The change of variable (1.28) implies that we can choose $\varphi_* = \frac{\sqrt{3}}{2}\ln f'(\varphi)$ so that the theory in the Einstein frame is defined by

$$A^2 \propto e^{-\frac{4\varphi_*}{\sqrt{3}}}, \qquad V = \frac{1}{4}\left\{\varphi(\varphi_*)e^{\frac{2\varphi_*}{\sqrt{3}}} - f[\varphi(\varphi_*)]\right\}e^{-\frac{4\varphi_*}{\sqrt{3}}}. \qquad (1.45)$$

Note that α_0 cannot be made arbitrarily small since the form of the coupling function A arises from the function f. In order to make these models compatible with solar system constraints, the potential should be such that the scalar field is massive enough, while still being bounded from below.

This example highlights the importance of looking for the true degrees of freedom of the theory. A field redefinition can be a useful tool to show that two theories are actually equivalent. This result was generalized (Wands, 1994) to theories involving $f[R, (\nabla_\mu \nabla^\mu)R, \ldots, (\nabla_\mu \nabla^\mu)^n R]$ which were shown to be equivalent to $(n + 1)$ scalar–tensor theories.

This equivalence between $f(R)$ and scalar–tensor theories assumes that the Ricci scalar is a function of the metric and its first derivatives. There is a difference when one considers $f(R)$ theories in the Palatini formalism (Flanagan, 2004), in which the metric and the connections are assumed to be independent fields, since while still being equivalent to scalar–tensor theories, the scalar field does not propagate because it has no kinetic term in the Einstein frame. It thus reduces to a Lagrange parameter whose field equation sets a constraint.

1.2.3.4 Extensions

The previous set-up can easily be extended to include n scalar fields (Damour and Esposito-Farèse, 1992) in which case the kinetic term will contain an $n \times n$ symmetric matrix, $g_*^{\mu\nu} \gamma_{ab}(\varphi_c) \, \partial_\mu \varphi^a \, \partial_\nu \varphi^b$.

Another class of models arises when one considers more general kinetic terms of the form $f(s, \varphi)$ where $s = g_*^{\mu\nu} \partial_\mu \varphi \, \partial_\nu \varphi$. When the coupling function reduces to $A = 1$, these models are known as K-essence (Armendariz-Picon *et al.*, 2000; Chiba *et al.*, 2000). We refer to Esposito-Farèse and Bruneton (2007) and Bruneton (2007) for a discussion of the conditions to be imposed on f in order for such a theory to be well defined.

1.2.3.5 Reconstructing theories

This section has illustrated the difficulty of consistently modifying general relativity. Let us emphasize that most of the models we discussed contain several free functions and general relativity in some continuous limit. It is clear that most of them cannot be excluded observationally.

It is important to remember that we hope these theories go beyond a pure description of the data. In particular, it is obvious that the function $E(z)$ defined in Eq. (1.6) for a ΛCDM model can be reproduced by many different models. In particular, one can always design a scalar field model inducing an energy density $\rho_{\mathrm{de}}(z)$, obtained from the observed function $H^2(z)$ by subtracting the contributions of the matter we know (i.e. pressureless matter and radiation). Its potential is given by (Uzan, 2007)

$$V(a) = \frac{H(1-X)}{16\pi G}\left(6H + 2aH' - \frac{aHX'}{1-X}\right),$$

$$Q(a) = \int \frac{d\ln a}{\sqrt{8\pi G}}\left[aX' - 2(1-X)a\frac{H'}{H}\right], \qquad (1.46)$$

with $X(a) \equiv 8\pi G\rho_{de}(a)/3H^2(a)$ in order to reproduce $\{H(a), \rho_{de}(a)\}$.

The background dynamics provides only one observable function, namely $H(z)$, so that it can be reproduced by many theories having at least one free function. To go further, we must add independent information, which can be provided e.g. by the growth rate of the large-scale structure. An illustrative game was presented in Uzan (2007) in which it was shown that while the background dynamics of the DGP model (Dvali *et al.*, 2000) can be reproduced by a quintessence model, both models did not share the same growth rate and can be distinguished, in principle, at this level. However, both the background and sub-Hubble perturbation dynamics of the DGP model can be reproduced by a well-defined scalar–tensor theory, which has two arbitrary functions. The only way to distinguish the two models is then to add local information, since the scalar–tensor theory that reproduces the cosmological dynamics of the DGP model would induce a time variation of the gravitational constant above acceptable experimental limits.

This shows the limit of the model-dependent approach in which a reconstructed theory could simply be seen as a description of a set of data if its number of free functions is larger than the observable relations provided by the data. The reconstruction method can, however, lead to interesting conclusions and to the construction of counter-examples. For instance, it was shown (Esposito-Farèse and Polarski, 2001) that a scalar–tensor theory with $V = 0$ cannot reproduce the background dynamics of the ΛCDM.

This should encourage us to consider the simplest possible extension, namely with the minimum number of new degrees of freedom and arbitrary functions. In that sense the ΛCDM model is very economical since it reproduces all observations at the expense of a single new constant.

1.2.4 Beyond the Copernican principle

As explained above, the conclusion that the cosmic expansion is accelerating is deeply related to the Copernican principle. Without such a uniformity principle, the reconstruction of the geometry of our spacetime becomes much more involved.

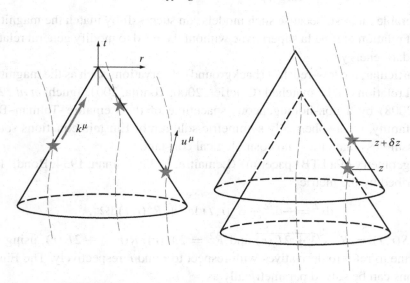

Fig. 1.3. *Left*: Most low-redshift data are localized on our past light-cone. In a non-homogeneous spacetime there is no direct relation between the redshift that is observed and the cosmic time needed to reconstruct the expansion history. *Right*: The time drift of the redshift allows information about two infinitely close past light-cones to be extracted. δz depends on the proper motions of the observer and the sources as well as the spacetime geometry.

Indeed, most low redshift observations provide the measurements of some physical quantities (luminosity, size, shape, etc.) as a function of the position on the celestial sphere and the redshift. In any spacetime, the redshift is defined as

$$1 + z = \frac{\left(u^\mu k_\mu\right)_{\text{emission}}}{\left(u^\mu k_\mu\right)_{\text{observation}}}, \tag{1.47}$$

where u^μ is the 4-velocity of the cosmic fluid and k^μ the tangent vector to the null geodesic relating the emission and the observation (see Fig. 1.3). The redshift depends on the structure of the past light-cone and thus on the symmetries of the spacetime. It reduces to the simple expression (1.2) only for a Roberston–Walker spacetime. Indeed, it is almost impossible to prove that a given observational relation, such as the magnitude–redshift relation, is not compatible with another spacetime geometry.

While isotropy around us seems well established observationally (see e.g. Ruiz-Lapuente, 2007), homogeneity is more difficult to test. The possibility that we may be living close to the center of a large under-dense region has sparked

considerable interest, because such models can successfully match the magnitude–
redshift relation of type Ia supernovae without the need to modify general relativity
or add dark energy.

In particular, the low-redshift (background) observations such as the magnitude–
redshift relation can be matched (Célérier, 2000; Tomita, 2001; Iguchi *et al.*, 2002;
Ellis, 2008) by a non-homogeneous spacetime of the Lemaître–Tolman–Bondi
(LTB) family, i.e. a spherically symmetric solution of Einstein equations sourced
by pressureless matter and no cosmological constant.

The geometry of a LTB spacetime (Lemaître, 1933; Tolman, 1934; Bondi, 1947)
is described by the metric

$$ds^2 = -dt^2 + S^2(r,t)dr^2 + R^2(r,t)d\Omega^2,$$

where $S(r,t) = R'/\sqrt{1+2E(r)}$ and $\dot{R}^2 = 2M(r)/R(r,t) + 2E(r)$, using a dot
and prime to refer to derivatives with respect to t and r respectively. The Einstein
equations can be solved parametrically as

$$\{R(r,\eta), t(r,\eta)\} = \left\{\frac{M(r)}{\mathcal{E}(r)}\Phi'(\eta), T_0(r) + \frac{M(r)}{[\mathcal{E}(r)]^{3/2}}\Phi(\eta)\right\}, \qquad (1.48)$$

where Φ is defined by $\Phi(\eta) = (\sinh\eta - \eta, \eta^3/6, \eta - \sin\eta)$, and $\mathcal{E}(r) = (2E,$
$2, -2E)$ according to whether E is positive, null or negative.

This solution depends on three arbitrary functions of r only, $E(r)$, $M(r)$,
and $T_0(r)$. Their choice determines the model completely. For instance
$(E, M, T_0) = (-K_0 r^2, M_0 r^3, 0)$ corresponds to a Robertson–Walker spacetime.
One can further use the freedom in the choice of the radial coordinate to fix one of
the three functions at will so that one effectively has only two arbitrary independent
functions.

Let us sketch the reconstruction and use r as the integration coordinate, instead
of z. Our past light-cone is defined as $t = \hat{t}(r)$ and we set $\mathcal{R}(r) \equiv R[\hat{t}(r), r]$.
The time derivative of R is given by $\dot{R}[\hat{t}(r), r] \equiv \mathcal{R}_1 = \sqrt{2M_0 r^3/\mathcal{R}(r) + 2E(r)}$.
Then we get $R'[\hat{t}(r), r] \equiv \mathcal{R}_2(r) = -[\mathcal{R}(r) - 3(\hat{t}(r) - T_0(r))\mathcal{R}_1(r)/2]E'/E -$
$\mathcal{R}_1(r)T_0'(r) + \mathcal{R}(r)/r$. Finally, more algebra leads to $\dot{R}'[\hat{t}(r), r] \equiv \mathcal{R}_3(r) = [\mathcal{R}_1(r)$
$-3M_0 r^3(\hat{t}(r) - T_0(r))/\mathcal{R}^2(r)]E'(r)/2E(r) + M_0 r^3 T_0'(r)/\mathcal{R}^2 + \mathcal{R}_1(r)/r$. Thus,
$\dot{R}, R'$ and $\dot{R}'$ evaluated on the light-cone are just functions of $\mathcal{R}(r)$, $E(r)$, $T_0(r)$
and their first derivatives. Now, the null geodesic equation gives that

$$\frac{d\hat{t}}{dr} = -\frac{\mathcal{R}_2(r)}{\sqrt{1+2E(r)}}, \qquad \frac{dz}{dr} = \frac{1+z}{\sqrt{1+2E(r)}}\mathcal{R}_3(r),$$

and

$$\frac{d\mathcal{R}}{dr} = \left[1 - \frac{\mathcal{R}_1(r)}{\sqrt{1+2E(r)}}\right]\mathcal{R}_2(r).$$

These are three first-order differential equations relating five functions $\mathcal{R}(r)$, $\hat{t}(r)$, $z(r)$, $E(r)$, and $T_0(r)$. To reconstruct the free functions we thus need two observational relations. The reconstruction from background data alone is under-determined and one must fix one function by hand. The angular distance–redshift relation, $\mathcal{R}(z) = D_A(z)$, is the obvious choice. Does this explain why the magnitude–redshift relation can be matched (Célérier, 2000; Tomita, 2001; Iguchi *et al.*, 2002) by a LTB geometry? Indeed the geometry is not fully reconstructed.

It follows that many issues are left open. First, can we use more observational data to close the reconstruction of the LTB geometry? Indeed, the knowledge of the growth rate of the large-scale structure could be used, as for the reconstruction of the two arbitrary functions of a scalar–tensor theory, but no full investigation of the perturbation theory around a LTB spacetime has been performed (Zibin, 2008; Dunsby and Uzan, 2008). Second, can we construct the model-independent test of the Copernican principle avoiding the necessity to restrict to a given geometry, since we may have to consider more complex spacetimes than the LTB one? Third, we would have to understand how these models reproduce the predictions of the standard cosmological model on large scales and at early times, e.g. how are the cosmic microwave background anisotropies and the big-bang nucleosynthesis dependent on these spacetime structures.

1.2.5 Conclusions

This section has investigated two different ways to modify our reference cosmological model either by extending the description of the laws of nature or by extending the complexity of the geometry of our spacetime by relaxing the Copernican principle.

Whatever the choice, we see that many possibilities are left open. All of them introduce new degrees of freedom, either as physical fields or new geometrical freedom, and free functions. They also contain the standard ΛCDM as a continuous limit (e.g. the potential can become flat, the arbitrary functions of a LTB can reduce to their FLRW form, etc.) These extensions are thus almost non-excludable by cosmological observations alone and, as we have seen, they can reduce to pure descriptions of the data. Again, we must be guided by some principles.

The advantages of the model-dependent approaches is that we know whether we are dealing with well-defined theories or spacetime structures. All cosmological observables can be consistently computed so that these models can be safely compared to observations to quantify how close to a pure ΛCDM the model of our universe should be. They can also forecast the ability of coming surveys to constrain them.

The drawback is that we cannot test all the possibilities, which are too numerous. An alternative is to design parameterizations that have the advantage, we hope, to encompass many models. The problem is then the physical interpretation of the new parameters that are measured from the observations.

Another route, which we shall now investigate, is to design null tests of the ΛCDM model in order to indicate what kind of modifications, if any, are required by the observations.

1.3 Testing the underlying hypotheses

Let us first clarify what we mean by a *null test*. Once the physical theory and the properties of its cosmological solution have been fixed, there exist rigidities between different observable quantities. They reflect the set of assumptions of our reference cosmological models. By testing these rigidities we can strengthen our confidence in the principles on which our model lies. If we can prove that some of them are violated, it will give us just a hint of how to extend our cosmological model and of which principle has to be questioned.

Let us take a few examples that will be developed below.

- The equation of state of the dark energy must be $w_{de} = -1$ and constant in time.
- The luminosity and angular distances must be related by the distance duality relation stating that $D_L(z) = (1+z)^2 D_A(z)$.
- On sub-Hubble scales, the gravitational potential and the perturbation of the matter energy density must be related by the Poisson equation, $\Delta \Phi = 4\pi G \rho_m a^2 \delta_m$, which derives from the Einstein equation in the weak field limit.
- On sub-Hubble scales, the background dynamics and the growth of structure are not independent.
- The constants of nature must be strictly constant.

These rigidities are related to different hypotheses, such as the validity of general relativity or the Maxwell theory. We shall now describe them and see how they can be implemented with cosmological data.

1.3.1 Testing the Copernican principle

The main difficulty in testing the Copernican principle, as discussed in Section 1.2.4, lies in the fact that all observations are located on our past light-cone and that many four-dimensional spacetimes may be compatible with the same three-dimensional light-like slice (Ellis, 1975).

Recently, it was realized that cosmological observations may however provide a test of the Copernican principle (Uzan *et al.*, 2008b). This test exploits the time drift of the redshift that occurs in any expanding spacetime, as first pointed out in the particular case of Robertson–Walker spacetimes for which it takes the form (Sandage, 1962; McVittie, 1962)

$$\dot{z} = (1 + z)H_0 - H(z). \tag{1.49}$$

Such an observation gives information on the dynamics outside the past light-cone since it compares the redshift of a given source at two times and thus on two infinitely close past light-cones (see Fig. 1.3, right). It follows that it contains information about the spacetime structure along the worldlines of the observed sources that must be compatible with that derived from the data along the past light-cone.

For instance, in a spherically symmetric spacetime, the expression (1.49) depends on the shear, $\sigma(z)$, of the congruence of the worldlines of the comoving observers evaluated along our past light-cone,

$$\dot{z} = (1 + z)H_0 - H(z) - \frac{1}{\sqrt{3}}\sigma(z).$$

It follows that, when combined with other distance data, it allows us to determine the shear on our past light-cone and we can check whether it is compatible with zero, as expected for any Robertson–Walker spacetime.

In a RW universe, we can go further and determine a consistency relation between several observables. From the metric (1.1), one deduces that $H^{-1}(z) = D'(z)\left[1 + \Omega_{K0}H_0^2 D^2(z)\right]^{-1/2}$, where a prime stands for ∂_z and $D(z) = D_L(z)/(1 + z)$; this relation being independent of the Friedmann equations. It follows that in any Robertson–Walker spacetime the *consistency relation*,

$$1 + \Omega_{K0}H_0^2 \left(\frac{D_L(z)}{1+z}\right)^2 - [H_0(1+z) - \dot{z}(z)]^2 \left[\frac{\mathrm{d}}{\mathrm{d}z}\left(\frac{D_L(z)}{1+z}\right)\right]^2 = 0,$$

between observables must hold whatever the matter content and the field equations, since it derives from pure kinematical relations that do not rely on the dynamics (a similar analysis is provided in Clarkson *et al.*, 2008). The measurement of $\dot{z}(z)$ will also allow (Uzan *et al.*, 2008b) us to close the reconstruction of the local geometry of such an under-dense region (as discussed in Section 1.2.4).

$\dot{z}(z)$ has a typical amplitude of order $\delta z \sim -5 \times 10^{-10}$ on a time scale of $\delta t = 10$ yr, for a source at redshift $z = 4$. This measurement is challenging, and impossible with present-day facilities. However, it was recently revisited in the context of the Extremely Large Telescopes (ELT), arguing that they could measure velocity shifts of order $\delta v \sim 1$–$10\,\mathrm{cm\,s^{-1}}$ over a 10-year period from observation of the Lyman-α forest. It is one of the science drivers in the design of the

CODEX spectrograph (Pasquini *et al.*, 2005) for the future European ELT. Indeed, many effects, such as proper motion of the sources, local gravitational potential, or acceleration of the Sun, may contribute to the time drift of the redshift. It has been shown (Liske *et al.*, 2008; Uzan *et al.*, 2008a), however, that these contributions can be brought to a 0.1% level so that the cosmological redshift is actually measured.

Let us also stress that another idea was proposed recently (Goodman, 1995; Caldwell and Stebbins, 2008). It is based on the distortion of the Planck spectrum of the cosmic microwave background.

1.3.2 Testing general relativity on astrophysical scales

1.3.2.1 Test of local position invariance

The local position invariance is one aspect of the Einstein equivalence principle that is at the basis of the hypothesis of metric coupling. It implies that all constants of nature must be strictly constant. The indication that the numerical value of any constant has drifted during cosmological evolution would be a sign in favor of models of classes C and D.

The test of the constancy of the fundamental constants has seen very intense activity in the past decade. In particular, the observations from quasar absorption spectra have relaunched a debate on the possible variation of the fine structure constant. Recently it was also argued (Coc *et al.*, 2007) that a time variation of the Yukawa couplings may allow us to solve the lithium-7 problem, which, at the moment, has no other physical explanation.

Constraints can be obtained from many physical systems such as atomic clocks ($z = 0$), the Oklo phenomenon ($z \sim 0.14$), the lifetime of unstable nuclei and meteorite data ($z \sim 0.2$), quasar absorption spectra ($z = 0.2 - 3$), cosmic microwave background ($z \sim 10^3$), and primordial nucleosynthesis ($z \sim 10^8$). The time variation of fundamental constants is also deeply related to the universality of free-fall. We refer to Uzan (2003) and (2004) for extensive reviews on the methods and the constraints, which are summarized in Fig. 1.4.

In conclusion, we have no compelling evidence for any time variation of a constant, which sets strong constraints on the couplings between the dark energy degrees of freedom and ordinary matter. We can conclude that the local position invariance holds in our observable universe and that metric couplings are favored.

1.3.2.2 Test of the Poisson equation

Extracting constraints on deviations from GR is difficult because large-scale structures entangle the properties of matter and gravity. On sub-Hubble scales, one can, however, construct tests reproducing those in the solar system. For instance, light

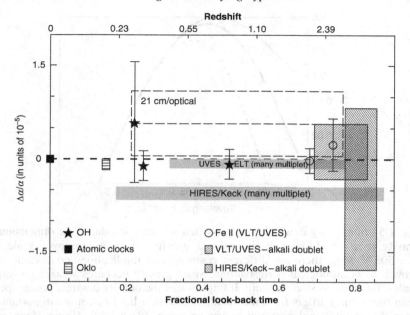

Fig. 1.4. Constraints on the time variation of the fine structure constant α from the observations of quasar absorption spectra.

deflection is a test of GR because we can measure independently the deflection angle and the mass of the Sun.

On sub-Hubble scales, relevant for the study of the large-scale structure, the Einstein equations reduce to the Poisson equation,

$$\Delta \Psi = 4\pi G \rho_m a^2 \delta_m = \frac{3}{2} \Omega_m H^2 a^2 \delta_m, \tag{1.50}$$

relating the gravitational potential and the matter density contrast.

As first pointed out by Uzan and Bernardeau (2001), this relation can be tested on astrophysical scales, since the gravitational potential and the matter density perturbation can be measured independently from the use of cosmic shear measurements and galaxy catalogs. The test was recently implemented with the CFHTLS-weak lensing data and the SDSS data to conclude that the Poisson equation holds observationally to about 10 Mpc (Doré *et al.*, 2007).

As an example, Fig. 1.5 depicts the expected modifications of the matter power spectrum and of the gravitational potential power spectrum in the case of a theory in which gravity switches from a standard four-dimensional gravity to a DGP-like five-dimensional gravity above a crossover scale of $r_s = 50h^{-1}$ Mpc. Since gravity becomes weaker on large scales, density fluctuations stop growing, exactly

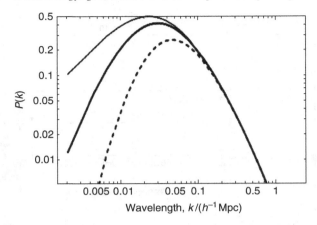

Fig. 1.5. In a theory in which gravity switches from a standard four-dimensional gravity to a DGP-like five-dimensional gravity above a crossover scale of $r_s = 50h^{-1}$ Mpc, there are different cosmological implications concerning the growth of cosmological perturbations. Since gravity becomes weaker on large scales, fluctuations stop growing. It implies that the density contrast power spectrum (thick line) differs from the standard one (thin line) but, more importantly, from the gravitational potential power spectrum (dash line). (From Uzan and Bernardeau, 2001.)

as the cosmological constant starts dominating. It implies that the density contrast power spectrum differs from the standard one but, more importantly, from the gravitational potential power spectrum.

Let us emphasize that the deviation from the standard behavior of the matter power spectrum is model dependent (it depends in particular on the cosmological parameters), but that the discrepancy between the matter and gravitational potential Laplacian power spectra is a direct signature of a modification of general relativity.

The main limitation in the applicability of this test is due to the biasing mechanisms (i.e. the fact that galaxies do not necessarily trace the matter field faithfully) even if it is thought to have no significant scale dependence at such scales.

1.3.2.3 Toward a post-ΛCDM formalism

The former test of the Poisson equation exploits one rigidity of the field equations on sub-Hubble scales. It can be improved by considering the full set of equations.

Assuming that the metric of spacetime takes the form

$$ds^2 = -(1 + 2\Phi)dt^2 + (1 - 2\Psi)a^2\gamma_{ij}\,dx^i\,dx^j \tag{1.51}$$

on sub-Hubble scales, the equation of evolution reduces to the continuity equation

$$\delta'_m + \theta_m = 0, \tag{1.52}$$

where θ is the divergence of the velocity perturbation and a prime denotes a derivative with respect to the conformal time, the Euler equation

$$\theta'_m + \mathcal{H}\theta_m = -\Delta\Phi, \tag{1.53}$$

where $\mathcal{H}$ is the comoving Hubble parameter, the Poisson equation (1.50) and

$$\Phi = \Psi. \tag{1.54}$$

These equations imply many relations between the cosmological observables. For instance, decomposing δ_m as $D(t)\epsilon(x)$ where ϵ encodes the initial conditions, the growth rate $D(t)$ evolves as

$$\ddot{D} + 2H\dot{D} - 4\pi G\rho_m D = 0.$$

This equation can be rewritten in terms of $p = \ln a$ as time variable (Peter and Uzan, 2005) and considered not as a second-order equation for $D(t)$ but as a first-order equation for $H^2(a)$,

$$(H^2)' + 2\left(\frac{3}{a} + \frac{D''}{D'}\right)H^2 = 3\frac{\Omega_{m0}H_0^2 D}{a^2 D'},$$

where a prime denotes a derivative with respect to p. It can be integrated as (Chiba and Nakamura, 2007)

$$\frac{H^2(z)}{H_0^2} = 3\Omega_{m0}\left(\frac{1+z}{D'(z)}\right)^2 \int \frac{D}{1+z}(-D')\,\mathrm{d}z. \tag{1.55}$$

This exhibits a rigidity between the growth function and the Hubble parameter. In particular the Hubble parameter determined from background data and from perturbation data using Eq. (1.55) must agree. This was used in the analysis of Wang *et al.* (2007).

Another relation exists between θ_m and δ_m. The Euler equation implies that

$$\theta_m = -\beta(\Omega_{m0}, \Omega_{\Lambda 0})\delta_m, \tag{1.56}$$

with

$$\beta(\Omega_{m0}, \Omega_{\Lambda 0}) \equiv \frac{\mathrm{d}\ln D(a)}{\mathrm{d}\ln a}. \tag{1.57}$$

We conclude that the perturbation variables are not independent and the relations between them are inherited from some assumptions on the dark energy. Phenomenologically, we can generalize Eqs. (1.52)–(1.54) to

$$\delta'_m + \theta_m = 0, \tag{1.58}$$

$$\theta'_m + \mathcal{H}\theta_m = -\Delta\Phi + S_{\text{de}}, \tag{1.59}$$

$$-k^2\Phi = 4\pi G F(k, H)\delta_m + \Delta_{\text{de}}, \tag{1.60}$$

$$\Delta(\Phi - \Psi) = \pi_{\text{de}}. \tag{1.61}$$

We assume that there is no production of baryonic matter so that the continuity equation is left unchanged. S_{de} describes the interaction between dark energy and standard matter. Δ_{de} characterizes the clustering of dark energy, F accounts for a scale dependence of the gravitational interaction and π_{de} is an effective anisotropic stress. It is clear that the ΛCDM corresponds to $(F, \pi_{\text{de}}, \Delta_{\text{de}}, S_{\text{de}}) = (1, 0, 0, 0)$. The expression of $(F, \pi_{\text{de}}, \Delta_{\text{de}}, S_{\text{de}})$ for quintessence, scalar–tensor, $f(R)$, and DGP models and more generally for models of the classes A–D can be found in Uzan (2007).

From an observational point of view, a weak lensing survey gives access to $\Phi + \Psi$, galaxy maps allow us to reconstruct $\delta_g = b\delta_m$ where b is the bias, velocity fields give access to θ. In a ΛCDM, the correlations between these observables are not independent since, for instance, $\langle\delta_g\delta_g\rangle = b^2\langle\delta_m^2\rangle$, $\langle\delta_g\theta_m\rangle = -b\beta\langle\delta_m^2\rangle$ and $\langle\delta_g\kappa\rangle = 8\pi G\rho_m a^2 b\langle\delta_m^2\rangle$.

Various ways of combining these observables have been proposed; construction of efficient estimators and forecasts for possible future space missions designed to make these tests as well as the possible limitations (arising e.g. from non-linear bias, the effect of massive neutrinos or the dependence on the initial conditions) are now being extensively studied (Zhang *et al.*, 2007; Amendola *et al.*, 2008; Jain and Zhang, 2008; Song and Koyama, 2008).

To finish let us also mention that the analysis of the weakly non-linear dynamics allows us to develop complementary tests of the Poisson equation (Bernardeau, 2004) but no full investigation in the framework presented here has been performed yet.

1.3.3 Other possible tests

1.3.3.1 Distance duality

As long as photons travel along null geodesics and the geodesic deviation equation holds, the source angular distance, r_s, and the observer area distance, r_o, must be related by the *reciprocity relation* (Ellis, 1971), $r_s^2 = r_o^2(1 + z)^2$ regardless of the metric and matter content of the spacetime.

Indeed, the solid angle from the source cannot be measured so that r_s is not an observable quantity. But, it can be shown that, if the number of photons is conserved, the source angular distance is related to the luminosity distance, D_L, by the relation $D_L = r_s(1 + z)$. It follows that there exists a *distance duality relation* between the luminosity and angular distances,

$$D_L = D_A(1 + z)^2. \tag{1.62}$$

This distance duality relation must hold if the reciprocity relation is valid and if the number of photons is conserved. In fact, one can show that in a metric theory of gravitation, if Maxwell equations are valid, then both the reciprocity relation and the area law are satisfied and so is the distance duality relation.

There are many possibilities for one of these conditions to be violated. For instance, the non-conservation of the number of photons can arise from absorption by dust, but more exotic models involving photon–axion oscillation in an external magnetic field (Csaki *et al.*, 2002; Deffayet *et al.*, 2002) (class B) can also be a source of violation. More drastic violations would arise from theories in which gravity is not described by a metric theory and in which photons do not follow the null geodesic.

A test of this distance duality relies on the X-ray observations and Sunyaev–Zel'dovich (SZ) effect of galaxy clusters (Uzan *et al.*, 2004a).

Galaxy clusters are known as the largest gravitationally bound systems in the universe. They contain large quantities of hot and ionized gas at temperatures of typically 10^{7-8} K. The spectral properties of intra-cluster gas show that it radiates through bremsstrahlung in the X-ray domain. Therefore, this gas can modify the cosmic microwave background spectral energy distribution through inverse Compton interaction of photons with free electrons. This is the so-called SZ effect. It induces a decrement in the cosmic microwave background brightness at low frequencies and an increment at high frequencies.

In brief, the method is based on the fact that the cosmic microwave background temperature (i.e. brightness) decrement due to the SZ effect is given by $\Delta T_{SZ} \sim L\overline{n_e T_e}$, where the bar refers to an average over the line of sight and L is the typical size of the line of sight in the cluster. T_e is the electron temperature and n_e the electron density. Also, the total X-ray surface brightness is given by $S_X \sim \frac{V}{4\pi D_L^2}\overline{n_e n_p T_e^{1/2}}$, where the volume V of the cluster is given in terms of its angular diameter by $V = D_A^2 \theta^2 L$. It follows that

$$S_X \sim \frac{\theta^2}{4\pi}\frac{D_A^2}{D_L^2}\overline{L n_e n_p T_e^{1/2}}. \tag{1.63}$$

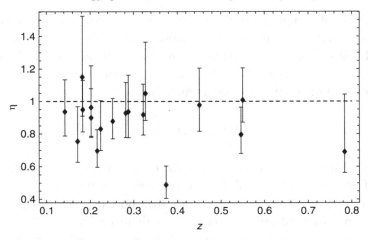

Fig. 1.6. Test of distance duality. The constraint of η defined in Eq. (1.64) at different redshifts is obtained by combining SZ and X-ray measurements from 18 clusters. No sign of violation of the distance duality relation is seen. From Uzan *et al.* (2004a).

The usual approach is to assume the distance duality relation so that forming the ratio $\Delta T_{SZ}^2 / S_X$ eliminates n_e.

We can however use these observation to measure

$$\eta(z) = \frac{D_L(z)}{(1+z)^2 D_A(z)} \tag{1.64}$$

and thus test whether $\eta = 1$. Figure 1.6 summarizes the constraints that have been obtained from the analysis of 18 clusters. No sign of violation of the distance duality relation is seen, contrary to an early claim by Bassett and Kunz (2004).

1.3.3.2 Gravity waves

In models involving two metrics, gravitons and standard matter are coupled to different metrics. It follows that the propagation of gravity waves and light may be different. As a consequence the arrival times of gravity waves and light should not be equal.

An estimation (Kahya and Woodard, 2007) in the case of the TeVeS theory for the supernovae 1987A indicates that light should arrive days before the gravity waves, which should be easily detectable.

We also emphasize that models in which gravity waves propagate slower than electromagnetic waves are also very constrained by the observations of cosmic rays (Moore and Nelson, 2001) because particles propagating faster than the gravity waves emit gravi-Cerenkov radiation.

These two examples highlight that the cosmological tests of general relativity *do not reduce* to the study of the large-scale structures.

1.3.3.3 A word on topology

The debate concerning the topology of our universe is a continuation of the historically long-discussed question on whether our universe is finite or infinite.

The hypotheses on the global topology do not influence local physics and leave most of the theoretical and observational conclusions unchanged. Local geometry has, however, a deep impact since it sets the topologies that are acceptable. In our cosmological model, the Copernican principle implies that we are dealing with three-dimensional spaces of constant curvature. Also, almost spatial flatness limits those topologies that would be detectable (Weeks *et al.*, 2003).

A non-trivial topology would violate global isotropy and allow signatures mainly on the statistical isotropy of the cosmic microwave background anisotropies (Gaussman *et al.*, 2001; Luminet *et al.*, 2003; Riazuelo *et al.*, 2004a,b; Uzan *et al.*, 2004b). The current constraints imply (Shapiro *et al.*, 2007) that the size of the universe has to be larger than 0.91 times the diameter of the last scattering surface, that is 24 Gpc.

Even though the cosmological constant can be related to a characteristic size of the order of $\Lambda^{-1/2} \sim H_0^{-1}$, no mechanism relating the size of the universe and the cosmological constant has been constructed. (Calder and Lahav (2008) however suggest a possible relation to the Mach principle.)

1.4 Conclusion

The acceleration of the cosmic expansion and the understanding of its origin drives us to reconsider the construction of our reference cosmological models.

Three possibilities seem open to us.

- *Stick to the ΛCDM.* The model is well defined, does not require extending the low-energy version of the law of nature, and is compatible with all existing data. However, in order to make sense of the cosmological constant and avoid the cosmological constant problem one needs to invoke a very large structure (Weinberg, 1989; Garriga and Vilenkin, 2004; Carr and Ellis, 2008), the multiverse, a collection of universes in which the value of the cosmological constant, as well as those of other physical constants, is randomized in different regions. Such a structure, while advocated on the basis of the string landscape (Suskind, 2006), has no clear mathematical definition (Ellis *et al.*, 2004) but it aims at suppressing the contingency of our physical models (such as their symmetry groups, value of constants, etc. that, by construction, cannot be explained by these models) at the price of an anthropic approach which may appear as half-way between pure anthropocentrism, fixing us at the center of the universe, and the cosmological principle, stating that no place can be favored in any way.

In such a situation, it is clear that the Copernican principle holds on the size of the observable universe and even much beyond. However, on the scales of the multiverse, it has to be abandoned since, according to this view, we can only live in regions of the multiverse where the value of the cosmological constant is small enough for observers to exist (see Fig. 1.7).

The alternative would be to better understand the computation of the energy density of the vacuum.

- *Assume* $\Lambda = 0$ and then
  - *Assume no new physics.* In such a case, we must abandon the Copernican principle on the size of the observable universe. This leads us to consider more involved solutions of known and established physical theories. Indeed, the main objection would be to understand why we should live in such a particular place.

    Note, however, that the Copernican principle can be restored on much larger scales (i.e. super-Hubble but without the need to invoke a structure like the multiverse). On these scales, one can argue that there will exist a distribution of over- and under-dense regions of all sizes and density profiles. In this sense, we are just living in one of them, in the same sense that stars are more likely to be in galaxies (see Fig. 1.7) and the Copernican principle seems to be violated on Hubble scales, just because we live in such a structure that happens to have a size comparable to that of the observable universe.
  - *Invoke new physics.* This can be achieved in numerous ways. The main constraint is to construct a well-defined theory. In such a case the Copernican principle can hold both on the size of the observable universe but also on much larger scales.

At the moment, none of these three possibilities is satisfactory, mainly because they force us to speculate on scales much beyond those of the observable universe. A further possibility, that was alluded to in Section 1.2.2.1, is the possibility that the acceleration is induced by the backreaction of the large-scale structures, but this still needs in-depth investigation. We have argued that future cosmological observations can shed some light on the way to modify our reference cosmological model and extend the tests of the fundamental laws of physics, such as general relativity, as well as some extra hypotheses such as the Copernican principle and the topology of space. In this sense, we follow the most standard physical approach in which any null test that can be done must be done in order to extend our understanding of the domain of validity of the description of the physical laws we are using.

From an observational point of view, demonstrating a violation of the Copernican principle on the size of the observable universe will indicate that the second solution is the most likely, but nothing forces us to accept the associated larger spacetime described in Fig. 1.7. If any of the tests presented here, or any other to be designed, is positive then we will have an indication that the dark energy is not the cosmological constant and on the kind of extension required. The question of

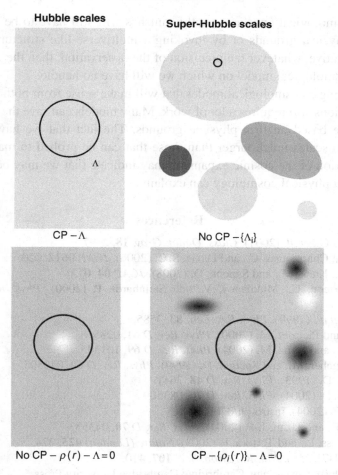

Fig. 1.7. On the scale of the observable universe (circle), the acceleration of the universe can be explained by a cosmological constant (or more generally a dark energy component) in which case the construction of the cosmological model relies on the Copernican principle (upper-left). To make sense of a cosmological constant, one introduces a large structure known as the multiverse (upper-right), which can be seen as a collection of universes of all sizes and in which the values of the cosmological constant, as well as other constants, are randomized. The anthropic principle then states that we observe only those universes where the values of these constants are such that observers can exist. In this sense we have to abandon the Copernican principle on the scales of the multiverse. An alternative is to assume that there is no need for a cosmological constant or new physics, in which case we have to abandon the Copernican principle and assume e.g. that we are living in an under-dense region (lower-left). However, we may recover the Copernican principle on larger scales if there exists a distribution of over- and under-dense regions of all sizes and densities on super-Hubble scales, without the need for a multiverse. In such a view, the Copernican principle will be violated on Hubble scale, just because we live in such a structure which happens to have a size comparable to that of the observable universe.

why the cosmological constant strictly vanishes will still have to be understood, either on physical grounds or by invoking a multiverse-like structure. If all the tests are negative, whatever the precision of the observation, then the ΛCDM will remain a cosmological model on which we will have no handle.

Constructing a cosmological model that will make sense from both physics and the observations still requires a lot of work. Many models can save the phenomena but none are based on firm physical grounds. The fact that we have to invoke structures on scales much larger than those that can be probed to make sense of the acceleration of the cosmic expansion may indicate that we may be reaching a limit of what physical cosmology can explain.

References

Adelberger, E. G., *et al.* (2001). *Class. Quant. Grav.* **18**, 2397.

Amendola, L., Charmousis, C., and Davis, S. C. (2006). *JCAP* **0612**, 020.

Amendola, L., Kunz, M., and Sapone, D. (2008). *JCAP* **04**, 013.

Armendariz-Picon, C., Mukhanov, V., and Steinhardt, P. (2000). *Phys. Rev. Lett.* **85**, 103510.

Baessler, S., *et al.* (1999). *Phys. Rev. Lett.* **83**, 3585.

Bartolo, N., and Pietroni, M. (2000). *Phys. Rev. D* **61**, 023518.

Bassett, B. A., and Kunz, M. (2004). *Phys. Rev. D* **69**, 101305.

Battye, R., Bucher, M., and Spergel, D. (1999). *Phys. Rev. D* **60**, 043505.

Bekenstein, J. D. (1993). *Phys. Rev. D* **48**, 3641.

Bekenstein, J. D. (2004). *Phys. Rev. D* **70**, 083509.

Bernardeau, F. (2004). [astro-ph/0409224].

Bernardeau, F., and Uzan, J.-P. (2004). *Phys. Rev. D* **70**, 043533.

Bertotti, B., Iess, L., and Tortora, P. (2003). *Nature (London)* **425**, 374.

Bondi, H. (1947). *Mon. Not. R. Astron. Soc.* **107**, 410.

Bondi, H. (1960). *Cosmology*. Cambridge: Cambridge Univerity Press.

Bruneton, J.-P., (2007). *Phys. Rev. D* **75**, 085013.

Calder, L., and Lahav O. (2008). *News Rev. Astron. Geophys.*, **49**, 1.13–1.18.

Caldwell, R., and Stebbins, A. (2008). *Phys. Rev. Lett.* **100**, 191302.

Carr, B., and Ellis G. F. R. (2008). *News Rev. Astron. Geophys.*, **49**, 2.29–2.37.

Carter, B., *et al.* (2001). *Class. Quant. Grav.* **18**, 4871.

Célérier, M.-N. (2000). *Astron. Astrophys.* **353**, 63.

Chiba, T., and Nakamura, T. (2007). *Prog. Theor. Phys.* **118**, 815.

Chiba, T., Okabe, T., and Yagamuchi, M. (2000). *Phys. Rev. D* **62**, 023511.

Chupp, T. E., *et al.* (1989). *Phys. Rev. Lett.* **63**, 1581.

Clarkson, C., Basset, B., and Lu, T. (2008). *Phys. Rev. Lett.* **101**, 011301.

Coc, A., *et al.* (2006). *Phys. Rev. D.* **73**, 083525.

Coc, A., *et al.* (2007). *Phys. Rev. D.* **76**, 023511.

Copeland, E., Sami M., and Tsujikawa S. (2006). *Int. J. Mod. Phys. D* **15**, 1753.

Csaki, C., Kaloper, N., and Terning, J. (2002). *Phys. Rev. Lett.* **88**, 161302.

Damour, T., and Esposito-Farèse, G. (1992). *Class. Quant. Grav.* **9**, 2093.

Damour, T., and Esposito-Farèse, G. (1998). *Phys. Rev. D* **58**, 042001.

Damour, T., and Lilley, M. (2008). In *Les Houches Summer School in Theoretical Physics: Session 87, String Theory and the Real World*, Amsterdam: Elsevier.

Damour, T., and, Nordtvedt, K. (1993). *Phys. Rev. Lett.* **70**, 2217.

Damour, T., and Polyakov, A. M. (1994). *Nucl. Phys. B* **423**, 532.

Deffayet, C. (2005). *Phys. Rev. D* **71**, 103501.

Deffayet, C., *et al.* (2002). *Phys. Rev. D* **66**, 043517.

Doré, O., *et al.* (2007). [arXiv:0712.1599].

Duhem, P. (1908). *Sozein ta phainomena: Essai sur la notion de théorie physique de Platon à Galilée*, Paris: Vrin; translated as *Sozein ta phainomena: An Essay on the Idea of Physical Theory from Plato to Galileo*, Chicago: University of Chicago Press.

Duhem, P. (1913–1917). *Le Système du Monde: Histoire des doctrines cosmologiques de Platon à Copernic*, Paris: Hermann.

Dunsby, P., and Uzan, J.-P. (2008). *In preparation.*

Dvali, G., Gabadadze, G., and Porati, M. (2000). *Phys. Lett. B* **485**, 208.

Ellis, G. F. R. (1971). In *Relativity and Cosmology*, R. K. Sachs, Ed., New York: Academic Press.

Ellis, G. F. R. (1975). *Q. J. Astron. Soc.* **16**, 245.

Ellis, G. F. R. (2008). *Nature (London)* **452**, 158.

Ellis, G. F. R., and Buchert T. (2005). *Phys. Lett. A* **347**, 38.

Ellis, G. F. R., Kirchner, U., and Stoeger, W. R. (2004). *Mon. Not. R. Astron. Soc.* **347**, 921–936.

Esposito-Farèse, G. (2005). *eConf C0507252 SLAC-R-819*, T025.

Esposito-Farèse, G., and Bruneton, J.-P. (2007). *Phys. Rev. D* **76**, 124012.

Esposito-Farèse, G., and Polarski D. (2001). *Phys. Rev. D* **63**, 063504.

Flanagan, É. É. (2004). *Phys. Rev. Lett.* **92**, 071101.

Garriga, J., and Vilenkin A. (2001). *Phys. Rev. D* **64**, 023517.

Gaussman, E., *et al.* (2001). *Class. Quant. Grav.* **18**, 5155.

Goodman, J. (1995). *Phys. Rev. D* **52**, 1821–1827.

Gottlöber, S., Schmidt, H. J., and Starobinsky, A. A. (1990). *Class. Quant. Grav.* **7**, 893.

Gregory, R., Rubakov, V., and Sibiryakov, S. (2000). *Phys. Rev. Lett.* **84**, 4690.

Gross, D., and Sloan, J. H. (1987). *Nucl. Phys. B* **291**, 41.

Hindawi, A., Ovrut, B., and Waldram, D. (1996). *Phys. Rev. D* **53**, 5583.

Hoyle, C. D., *et al.* (2004). *Phys. Rev. D* **70**, 042004.

Iguchi, H., Nakamura, T., and Nakao, K.-I. (2002). *Prog. Theor. Phys.* **108**, 809.

Jain, B., and Zhang, P. (2008). *Phys. Rev. D* **78**, 063503.

Kahya, E. O., and Woodard, R. P. (2007). *Phys. Lett. B* **652**, 213.

Kamenshchik, A. Y., Moschella, U., and Pasquier, V. (2001). *Phys. Lett. B* **511**, 265.

Khoury, J., and Weltman, A. (2004). *Phys. Rev. Lett.* **92**, 171104.

Kogan, I., *et al.* (2000). *Nucl. Phys. B* **584**, 313.

Lamoreaux, S. K., *et al.* (1986). *Phys. Rev. Lett.* **57**, 3125.

Lemaître, G. (1933). *Ann. Soc. Sci. Bruxelles A* **53**, 51.

Liske, J., *et al.* (2008). *Mon. Not. R. Astron. Soc.* **386**, 1192–1218.

Luminet, J.-P., *et al.* (2003). *Nature (London)* **425**, 593.

Mangano, G., and Sokolowski, M. (1994). *Phys. Rev. D* **50**, 5039.

Martin, J., Schimd C., and Uzan J.-P. (2006). *Phys. Rev. Lett.* **96**, 061303.

McVittie, G. (1962). *Astrophys. J.* **136**, 334.

Milgrom, M. (1983). *Astrophys. J.* **270**, 365.

Moore, G. D., and Nelson, A. E. (2001). *JHEP* **0109**, 023.

North, J. D. (1965). *The Measure of the Universe*, Oxford: Oxford University Press.

Ostrogradski, M. (1850). *Mem. Ac. St. Petersbourgh* **VI-4**, 385.

Pasquini, L., *et al.* (2005). *The Messenger* **122**, 10.

Peebles, P. J. E., and Ratra, B. (2003). *Rev. Mod. Phys.* **75**, 559.

Pereira, T., *et al.* (2007). *JCAP* **09**, 006.

Peter, P., and Uzan, J.-P. (2005). *Cosmologie Primordiale*, Paris: Belin. Translated as *Primordial Cosmology*, Oxford: Oxford University Press, (2009).

Pitrou, C., *et al.* (2008). *JCAP* **04**, 004.

Prestage, J.D., *et al.* (1985). *Phys. Rev. Lett.* **54**, 2387.

Psaltis, D. (2008). *Living Rev. Relat.*, [arXiv:0806.1531].

Ratra, B., and Peebles, P. J. E. (1988). *Phys. Rev. D* **37**, 321.

Riazuelo, A., and Uzan J.-P. (2002). *Phys. Rev. D* **65**, 043525.

Riazuelo, A., *et al.* (2004a). *Phys. Rev. D* **69**, 103514.

Riazuelo, A., *et al.* (2004b). *Phys. Rev. D* **69**, 103518.

Ruiz-Lapuente, P. (2007). *Class. Quant. Grav.* **24**, R91.

Sandage, A. (1962). *Astrophys. J.* **136**, 319.

Sanders, R. H. (1997). *Astrophys. J.* **480**, 492.

Schimd, C., Uzan J.-P., and Riazuelo, A. (2005). *Phys. Rev. D* **71**, 083512.

Sen, A. (1999). *JHEP* **9910**, 008.

Shapiro, I. I., *et al.* (1990). In *General Relativity and Gravitation 12*, Cambridge: Cambridge University Press, p. 313.

Shapiro, S. S., *et al.* (2004). *Phys. Rev. Lett.* **92**, 121101.

Shapiro, J., *et al.* (2007). *Phys. Rev. D* **75**, 084034.

Song, Y.-S., and Koyama, K. (2008). [arXiv:0802.3897].

Stelle, K. (1978). *Gen. Relat. Grav.* **9**, 353.

Suskind, L. (2006). In *Universe or Multiverse*, B. J. Carr, Ed., Cambridge: Cambridge University Press.

Teyssandier, P., and Tourrenc, P. (1993). *J. Math. Phys.* **24**, 2793.

Tolman, R. (1934). *Proc. Natl. Acad. Sci. U.S.A.* **20**, 169.

Tomboulis, E. T. (1996). *Phys. Lett. B* **389**, 225.

Tomita, K. (2001). *Mon. Not. R. Astron. Soc.* **326**, 287.

Turyshev, S. G. (2008). [arXiv:0806.1731].

Uzan, J.-P. (1999). *Phys. Rev. D* **59**, 123510.

Uzan, J.-P. (2003). *Rev. Mod. Phys.* **75**, 403.

Uzan, J.-P. (2004). *AIP Conf. Proc.* **736**, 3.

Uzan, J.-P. (2007). *Gen. Relat. Grav.* **39**, 307.

Uzan, J.-P., and Bernardeau F. (2001). *Phys. Rev. D.* **64**, 083004.

Uzan, J.-P., and Leclercq, B. (2008). *The Natural Laws of the Universe: Understanding Fundamental Constants*, Praxis.

Uzan, J.-P., Kirchner, U., and Ellis, G. F. R. (2003). *Mon. Not. R. Astron. Soc.* **344**, L65.

Uzan, J.-P., Aghanim N., and Mellier Y. (2004a). *Phys. Rev. D* **70**, 083533.

Uzan, J.-P., *et al.* (2004b). *Phys. Rev. D* **69**, 043003.

Uzan, J.-P., Bernardeau, B., and Mellier, Y. (2008a). *Phys. Rev. D* **70**, 021301.

Uzan, J.-P., Clarkson, C., and Ellis, G. F. R. (2008b). *Phys. Rev. Lett.* **100**, 191303.

van Dam, H., and Veltman, M. J. (1970). *Nucl. Phys. B* **22**, 397.
Vessot, R. F. C., and Levine, M. W. (1978). *Gen. Relat. Grav.* **10**, 181.
Wands, D. (1994). *Class. Quant. Grav.* **5**, 269.
Wang, S., *et al.* (2007). *Phys. Rev. D* **76**, 063503.
Weeks, J., Lehoucq R., and Uzan, J.-P. (2003). *Class. Quant. Grav.* **20**, 1529.
Weinberg, S. (1989). *Rev. Mod. Phys.* **61**, 1.
Wetterich, C. (1988). *Nucl. Phys. B* **302**, 668.
Will, C. M. (1981). *Theory and Experiment in Gravitational Physics*, Cambridge: Cambridge University Press.
Williams, J. G., *et al.* (2004). *Phys. Rev. Lett.* **93**, 21101.
Woodard, R. P. (2006). [astro-ph/0601672].
Zakharov, V. I. (1970). *Sov. Phys. JETP Lett.* **12**, 312.
Zel'dovich, Y. B. (1988). *Sov. Phys. Usp.* **11**, 381.
Zhang, P., *et al.* (2007). *Phys. Rev. Lett.* **99**, 141302.
Zibin, J.-P. (2008). *Phys. Rev. D* **78**, 043504.

2

Dark energy and modified gravity

ROY MAARTENS AND RUTH DURRER

2.1 Introduction

The current 'standard model' of cosmology is the inflationary cold dark matter model with cosmological constant Λ, usually called ΛCDM, which is based on general relativity and particle physics (i.e., the Standard Model and its minimal supersymmetric extensions). This model provides an excellent fit to the wealth of high-precision observational data, on the basis of a remarkably small number of cosmological parameters (Dunkley *et al.*, 2009). In particular, independent data sets from cosmic microwave background (CMB) anisotropies, galaxy surveys and supernova luminosities lead to a consistent set of best-fit model parameters (see Fig. 2.1) – which represents a triumph for ΛCDM.

The standard model is remarkably successful, but we know that its theoretical foundation, general relativity, breaks down at high enough energies, usually taken to be at the Planck scale,

$$E \gtrsim M_{\rm p} \sim 10^{16}\,{\rm TeV}. \tag{2.1}$$

The ΛCDM model can only provide limited insight into the very early universe. Indeed, the crucial role played by inflation belies the fact that inflation remains an effective theory without yet a basis in fundamental theory. A quantum gravity theory will be able to probe higher energies and earlier times, and should provide a consistent basis for inflation, or an alternative that replaces inflation within the standard cosmological model.

An even bigger theoretical problem than inflation is that of the late-time acceleration in the expansion of the universe (see, e.g., Bludman, 2006; Copeland, Sami

Dark Energy: Observational and Theoretical Approaches, ed. Pilar Ruiz-Lapuente. Published by Cambridge University Press. © Cambridge University Press 2010.

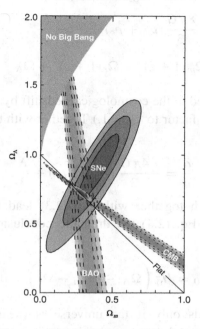

Fig. 2.1. Observational constraints in the $(\Omega_m, \Omega_\Lambda)$ plane: joint constraints from supernovae (SNe), baryon acoustic oscillations (BAO) and CMB. (From Kowalski *et al.*, 2008.)

and Tsujikawa, 2006; Padmanabhan, 2006a; Perivolaropoulos, 2006; Polarski, 2006; Straumann, 2006; Nojiri and Odintsov, 2007; Ruiz-Lapuente, 2007; Uzan, 2007; Ellis *et al.*, 2008). In terms of the fundamental energy density parameters, the data indicate that the present cosmic energy budget is given by (see Fig. 2.1)

$$\Omega_\Lambda \equiv \frac{\Lambda}{3H_0^2} \approx 0.75, \quad \Omega_m \equiv \frac{8\pi G \rho_{m0}}{3H_0^2} \approx 0.25,$$

$$\Omega_K \equiv \frac{-K}{H_0^2} \approx 0, \qquad \Omega_r \equiv \frac{8\pi G \rho_{r0}}{3H_0^2} \approx 8 \times 10^{-5}. \tag{2.2}$$

Here $H_0 = 100h$ km(s Mpc)$^{-1}$ is the present value of the Hubble parameter, Λ is the cosmological constant, K is spatial curvature, ρ_{m0} is the present matter density and ρ_{r0} is the present radiation density. Newton's constant is related to the Planck mass by $G = M_p^{-2}$ (we use units where the speed of light, $c = 1$ and the Planck constant $\hbar = 1$).

The Friedmann equation governs the evolution of the scale factor $a(t)$. It is given by

$$\left(\frac{\dot{a}}{a}\right)^2 \equiv H^2 = \frac{8\pi G}{3}(\rho_m + \rho_r) + \frac{\Lambda}{3} - \frac{K}{a^2}$$

$$= H_0^2 \left[\Omega_m(1+z)^3 + \Omega_r(1+z)^4 + \Omega_\Lambda + \Omega_K(1+z)^2\right]. \qquad (2.3)$$

The scale factor is related to the cosmological redshift by $z = a^{-1} - 1$. (We normalize the present scale factor to $a_0 = 1$.) Together with the energy conservation equation this implies

$$\frac{\ddot{a}}{a} = -\frac{4\pi G}{3}(\rho_m + 2\rho_r) + \frac{\Lambda}{3}. \qquad (2.4)$$

The observations, which together with Eq. (2.3) lead to the values given in Eq. (2.2), produce via Eq. (2.4) the dramatic conclusion that the universe is currently accelerating,

$$\ddot{a}_0 = H_0^2 \left(\Omega_\Lambda - \frac{1}{2}\Omega_m - \Omega_r\right) > 0. \qquad (2.5)$$

This conclusion holds only if the universe is (nearly) homogeneous and isotropic, i.e., a Friedmann–Lemaître model. In this case the distance to a given redshift, z, and the time elapsed since that redshift are tightly related via the only free function of this geometry, $a(t)$. If the universe instead is isotropic around us but not homogeneous, i.e., if it resembles a Tolman–Bondi–Lemaître solution with our galaxy cluster at the centre, then this tight relation between distance and time for a given redshift would be lost and present data would not necessarily imply acceleration – or the data may imply acceleration without dark energy. This remains a controversial and unresolved issue (see e.g. Paranjape and Singh, 2006; Vanderveld, Flanagan and Wasserman, 2006, 2008; Enqvist, 2008; Zibin, Moss and Scott, 2008; Garcia-Bellido and Haugboelle, 2009).

Of course isotropy without homogeneity violates the Copernican Principle as it puts us in the centre of the universe. However, it has to be stressed that up to now observations of homogeneity are very limited, unlike isotropy, which is firmly established. Homogeneity is usually inferred from isotropy together with the Copernican Principle. With future data, it will in principle be possible to distinguish observationally an isotropic but inhomogeneous universe from an isotropic and homogeneous universe (see e.g. Goodman, 1995; Clarkson, Bassett and Lu, 2008; Clifton, Ferreira and Land, 2008; Uzan, Clarkson and Ellis, 2008). Testing the Copernican Principle is a crucial aspect of testing the standard cosmological model. But in the following, we will assume that the Copernican Principle applies.

The data also indicate that the universe is currently (nearly) spatially flat,

$$|\Omega_K| \ll 1. \qquad (2.6)$$

It is common to assume that this implies $K=0$ and to use inflation as a motivation. However, inflation does *not* imply $K=0$, but only $\Omega_K \to 0$. Even if this distinction may be negligible in the present universe, a non-zero curvature can have significant implications for the onset of inflation (see e.g. Ellis and Maartens, 2004; Lasenby and Doran, 2005). In fact, if the present curvature is small but non-vanishing, neglecting it in the analysis of supernova data can sometimes induce surprisingly large errors (Clarkson, Cortes and Bassett, 2007).

The simplest way to produce acceleration is probably a cosmological constant, i.e., the ΛCDM model. Even though the cosmological constant can be considered as simply an additional gravitational constant (in addition to Newton's constant), it enters the Einstein equations in exactly the same way as a contribution from the vacuum energy, i.e., via a Lorentz-invariant energy–momentum tensor $T_{\mu\nu}^{\text{vac}} = -(\Lambda/8\pi G)g_{\mu\nu}$. The only observable signature of both a cosmological constant and vacuum energy is their effect on spacetime – and so a vacuum energy and a classical cosmological constant cannot be distinguished by observation. Therefore the 'classical' notion of the cosmological constant is effectively physically indistinguishable from quantum vacuum energy.

Even though the absolute value of vacuum energy cannot be calculated within quantum field theory, *changes* in the vacuum energy (e.g. during a phase transition) can be calculated, and they do have a physical effect – for example, on the energy levels of atoms (Lamb shift), which is well known and well measured. Furthermore, differences of vacuum energy in different locations, e.g. between or on one side of two large metallic plates, have been calculated and their effect, the Casimir force, is well measured (Bordag, Mohideen and Mostepanenko, 2001; Bressi *et al.*, 2002). Hence, there is no doubt about the reality of vacuum energy. For a field theory with cutoff energy scale E, the vacuum energy density scales with the cutoff as $\rho_{\text{vac}} \sim E^4$, corresponding to a cosmological constant $\Lambda_{\text{vac}} = 8\pi G\rho_{\text{vac}}$. If $E = M_{\text{p}}$, this yields a naïve contribution to the 'cosmological constant' of about $\Lambda_{\text{vac}} \sim 10^{38}\ \text{GeV}^2$, whereas the measured effective cosmological constant is the sum of the 'bare' cosmological constant and the contribution from the cutoff scale,

$$\Lambda_{\text{eff}} = \Lambda_{\text{vac}} + \Lambda \simeq 10^{-83}\ \text{GeV}^2. \tag{2.7}$$

Hence a cancellation of about 120 orders of magnitude is required. This is called the *fine-tuning* or *size* problem of dark energy: a cancellation is needed to arrive at a result that is many orders of magnitude smaller than each of the terms.[1] It is possible that the quantum vacuum energy is much smaller than the Planck scale.

---

[1] In quantum field theory we actually have to add to the cutoff term $\Lambda_{\text{vac}} \simeq E_c^4/M_{\text{pl}}^2$ the unmeasurable 'bare' cosmological constant. In this sense, the cosmological constant problem is a fine tuning between the unobservable 'bare' cosmological constant and the term coming from the cutoff scale.

But even if we set it to the lowest possible SUSY scale, $E_{susy} \sim 1\,\mathrm{TeV}$, arguing that at higher energies vacuum energy exactly cancels due to supersymmetry, the required cancellation is still about 60 orders of magnitude.

A reasonable attitude towards this open problem is the hope that quantum gravity will explain this cancellation. But then it is much more likely that we shall obtain directly $\Lambda_{vac} + \Lambda = 0$ and not $\Lambda_{vac} + \Lambda \simeq 24\pi\,G\rho_m(t_0)$. This unexpected observational result leads to a second problem, *the coincidence problem*: given that

$$\rho_\Lambda = \frac{\Lambda_{eff}}{8\pi\,G} = \text{constant}, \quad \text{while} \quad \rho_m \propto (1+z)^3, \tag{2.8}$$

why is ρ_Λ of the order of the *present* matter density $\rho_m(t_0)$? It was completely negligible in most of the past and will entirely dominate in the future.

These problems prompted cosmologists to look for other explanations of the observed accelerated expansion. Instead of a cosmological constant, one may introduce a scalar field or some other contribution to the energy–momentum tensor that has an equation of state $w < -1/3$. Such a component is called 'dark energy'. So far, no consistent model of dark energy has been proposed that can yield a convincing or natural explanation of either of these problems (see, e.g. Linder, 2008).

Alternatively, it is possible that there is no dark energy field, but instead the late-time acceleration is a signal of a *gravitational* effect. Within the framework of general relativity, this requires that the impact of inhomogeneities somehow acts to produce acceleration, or the appearance of acceleration (within a Friedman–Lemaître interpretation). A non-Copernican possibility is the Tolman–Bondi–Lemaître model (Paranjape and Singh, 2006; Vanderveld, Flanagan and Wasserman, 2006, 2008; Enqvist, 2008; Zibin, Moss and Scott, 2008; Garcia-Bellido and Haugboelle, 2009). Another (Copernican) possibility is that the 'back-reaction' of inhomogeneities on the background, treated via nonlinear averaging, produces effective acceleration (Buchert, 2008).

A more radical version is the 'dark gravity' approach, the idea that gravity itself is weakened on large scales, i.e., that there is an 'infrared' modification to general relativity that accounts for the late-time acceleration. The classes of modified gravity models that have been most widely investigated are scalar–tensor models (Capozziello and Francaviglia, 2008) and brane-world models (Koyama, 2008).

Schematically, we are modifying the geometric side of the field equations,

$$G_{\mu\nu} + G_{\mu\nu}^{dark} = 8\pi\,G T_{\mu\nu}, \tag{2.9}$$

rather than the matter side,

$$G_{\mu\nu} = 8\pi\,G \left(T_{\mu\nu} + T_{\mu\nu}^{dark} \right), \tag{2.10}$$

as in the general relativity approach. Modified gravity represents an intriguing possibility for resolving the theoretical crisis posed by late-time acceleration. However, it turns out to be extremely difficult to modify general relativity at low energies in cosmology, without violating observational constraints – from cosmological and solar system data, or without introducing ghosts and other instabilities into the theory. Up to now, there is no convincing alternative to the general relativity dark energy models – which themselves are not convincing.

The plan of the remainder of this chapter is as follows. In Section 2.2 we discuss constraints that one may formulate for a dark energy or modified gravity theory from basic theoretical requirements. In Section 2.3 we briefly discuss models that address the dark energy problem within general relativity. In Section 2.4 we present modified gravity models. In Section 2.5 we conclude. This chapter is based on a previous review published in Durrer and Maartens (2008).

2.2 Constraining effective theories

Theories of both dark matter and dark energy often have very unusual Lagrangians that cannot be quantized in the usual way, e.g. because they have non-standard kinetic terms. We then simply call them 'effective low energy theories' of some unspecified high energy theory that we do not elaborate. In this section, we want to point out a few properties that we nevertheless can require of low energy effective theories. We first enumerate the properties we require from a good basic physical theory at the classical and at the quantum level. We then discuss which of these requirements are inherited by low energy effective descriptions.

2.2.1 Fundamental physical theories

Here we give a minimal list of properties that we require from a fundamental physical theory. Of course, all the points enumerated below are open for discussion, but at least we should be aware of what we lose when we let go of them.

In our list we start with very basic requirements that become more strict as we go on. Even though some theorists would be able to live without one or several of the criteria discussed here, we think they are all very well founded. Furthermore, all known current physical theories, including string and M-theory, do respect them.

(i) **A physical theory allows a mathematical description**
This is the basic idea of theoretical physics.

(ii) **A fundamental physical theory allows a Lagrangian formulation**
This requirement is of course much stronger than the previous one. But it has been extremely successful in the past and was the guiding principle for the entire development of quantum field theory and string theory in the twentieth century.

Some 'varying speed of light theories' without Lagrangian formulation leave us more or less free to specify the evolution of the speed of light during the expansion history of the universe. However, if we introduce a Lagrangian formulation, we realize that most of these theories are simply some variant of scalar–tensor theories of gravity, which are well defined and have been studied in great detail.

If we want to keep deep physical insights such as Nöther's theorem, which relates symmetries to conservation laws, we need to require a Lagrangian formulation for a physical theory. A basic ingredient of a Lagrangian physical theory is that every physical degree of freedom has a kinetic term that consists (usually) of first-order time derivatives and may also have a 'potential term' that does not involve derivatives. In the Lagrangian formulation of a fundamental physical theory, we do not allow for external, arbitrarily given functions. Every function has to be a degree of freedom of the theory so that its evolution is determined self-consistently via the Lagrangian equations of motion, which are of first or second order. It is possible that the Lagrangian contains also higher than first-order derivatives, but such theories are strongly constrained by the problem of ghosts, which we mention below, and by the fact that the corresponding equations of motion are usually described by an unbounded Hamiltonian, i.e. the system is unstable (Ostrogradski's theorem: Ostrogradski, 1850; Woodard, 2007).

(iii) **Lorentz invariance**

We also want to require that the theory be Lorentz invariant. Note that this requirement is much stronger than demanding simply 'covariance'. It requires that there be no 'absolute element' in the theory apart from true constants. Lorentz covariance can always be achieved by rewriting the equations. As an example, consider a Lagrangian given in flat space by $(\partial_t \phi)^2 - (\partial_x \phi)^2$. This is clearly not Lorentz invariant. However, we can trivially write this term in the covariant form $\alpha^{\mu\nu} \partial_\nu \partial_\mu \phi$, by setting $(\alpha^{\mu\nu}) = \text{diag}(1, -1, 0, 0)$. Something like this should of course not be allowed in a fundamental theory. A term of the form $\alpha^{\mu\nu} \partial_\nu \partial_\mu \phi$ is only allowed if $\alpha^{\mu\nu}$ is itself a dynamical field of the theory. This is what we mean by requiring that the theory is not allowed to contain 'absolute elements', i.e. it is Lorentz invariant and not simply covariant.

(iv) **Ghosts**

Ghosts are fields whose kinetic term has the wrong sign. Such a field, instead of slowing down when it climbs up a potential, is speeding up. This unstable situation leads to severe problems when we want to quantize it, and it is generally accepted that one cannot make sense of such a theory, at least not at the quantum level. This is not surprising, since quantization usually is understood as defining excitations above some ground state, and a theory with a ghost has no well-defined ground state. Its kinetic energy has the wrong sign and the larger $\dot{\phi}^2$ is, the lower is the energy.

(v) **Tachyons**

These are degrees of freedom that have a negative mass squared, $m^2 < 0$. Using again the simple scalar field example, this means that the second derivative of the potential about the 'vacuum value' ($\phi = 0$ with $\partial_\phi V(0) = 0$) is negative, $\partial_\phi^2 V(0) < 0$. In

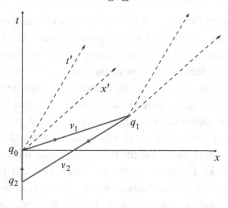

Fig. 2.2. We assume a signal that can propagate at speeds v_1, $v_2 > 1$. The frame R' with coordinates (t', x') moves with speed $v < 1$ in the x-direction. The speed v is chosen such that both, v_1, $v_2 > 1/v$. A signal is sent with velocity v_1 from q_0 to q_1 in the frame R. Since $v_1 > 1/v$, this signal travels backward in time with respect to frame R'. Then a signal is sent with speed v_2 from q_1 to q_2. Since $|v_2| > 1/v$, this signal, which is sent forward in time in frame R', travels backward in time with respect to R and can arrive at an event q_2 with $t_2 < 0$. The loop generated in this way is not 'causal' since both the trajectory from q_0 to q_1 and the one from q_1 to q_2 are spacelike. So we cannot speak of the formation of closed causal loops, but it is nevertheless a closed loop along which a signal can propagate and which therefore enables the construction of a time machine, leading to the usual problems with causality and entropy. (From Bonvin, Caprini and Durrer, 2007.)

general, this need not mean that the theory makes no sense, but rather that $\phi = 0$ is a bad choice for expanding around, since it is a maximum rather than a minimum of the potential and therefore an unstable equilibrium.

This means also that the theory cannot be quantized around the classical solution $\phi = 0$, but it may become a good quantum theory by a simple shift, $\phi \to \phi - \phi_0$, where ϕ_0 is the minimum of the potential. If the potential of a fundamental scalar field has no minimum but only a maximum, the situation is more severe. Then the theory is truly unstable.

The last two problems, together with the Ostrogradski instability that appears in theories with higher derivatives, can be summarized in the requirement that a meaningful theory needs to have an energy functional that is bounded from below.

(vi) **Superluminal motion and causality**
A fundamental physical theory that does respect Lorentz invariance must not allow for superluminal motions. If this condition is not satisfied, we can construct closed curves along which a signal can propagate (Bonvin, Caprini and Durrer, 2007). (See Fig. 2.2.)

At first sight one might think that a Lorentz invariant Lagrangian will automatically forbid superluminal motions. But the situation is not so simple. Generic Lorentz invariant higher spin theories, $s \geq 1$, lead to superluminal motion

(Velo and Zwanzinger, 1969a,b). While the equations are manifestly Lorentz invariant, their characteristics in general do not coincide with the light cone and can very well be spacelike. There are exceptions, among which are Yang Mills theories for spin 1 and the linearized Einstein equations for spin 2.

One may object to this restriction, on the grounds that general relativity, which is certainly a theory that is acceptable (at least at the classical level), can lead to closed causal curves, even though it does not admit superluminal motion (Morris, Thorne and Yurtsever, 1988; Gott, 1991; Bonnor and Steadman, 2005; Ori, 2007).

The situation is somewhat different if superluminal motion is only possible in a background that breaks Lorentz invariance. Then one has in principle a preferred frame and one can specify that perturbations should always propagate with the Green's function that corresponds to the retarded Green's function in this frame (Babichev, Mukhanov and Vikman, 2008). Nevertheless, one has to accept that there will be boosted frames relative to which the Cauchy problem for the superluminal modes is not well defined. The physics experienced by an observer in such a frame is most unusual (to say the least).

Causality of a theory is intimately related to the analyticity properties of the S-matrix of scattering, without which perturbative quantum theory does not make sense. Furthermore, we require the S-matrix to be unitary. Important consequences of these basic requirements are the Kramers–Kronig dispersion relations, which are a result of the analyticity properties and hence of causality, and the optical theorem, which is a result of unitarity. The analyticity properties have many further important consequences, such as the Froissart bound, which implies that the total cross section converges at high energy (Froissart, 1961; for an overview, see Itzykson and Zuber, 1980).

2.2.2 Low energy effective theories

The concept of low energy effective theories is extremely useful in physics. As one of the most prominent examples, consider superconductivity. It would be impossible to describe this phenomenon by using full quantum electrodynamics with a typical energy scale of MeV, where the energy scale of superconductivity is meV and less. However, many aspects of superconductivity can be successfully described with the Ginzburg–Landau theory of a complex scalar field. Microscopically, this scalar field is to be identified with a Cooper pair of two electrons, but this is irrelevant for many aspects of superconductivity.

Another example is weak interaction and four-Fermi theory. The latter is a good approximation to weak interactions at energy scales far below the Z-boson mass. Most physicists also regard the standard model of particle physics as a low energy effective theory that is valid below some high energy scale beyond which new degrees of freedom become relevant, be this supersymmetry, GUT or string theory.

We now want to investigate which of the properties in the previous subsection may be lost if we 'integrate out' high energy excitations and consider only processes that take place at energies below some cutoff scale E_c. We cannot completely ignore all particles with masses above E_c, since in the low energy quantum theory they can still be produced 'virtually', i.e., for a time shorter than $1/E_c$. This is not relevant for the initial and final states of a scattering process, but plays a role in the interaction.

Coming back to our list in the previous subsection, we certainly want to keep the first point – a mathematical description. But the Lagrangian formulation will also survive if we proceed in a consistent way by simply integrating out the high energy degrees of freedom.

What about higher order derivatives in the Lagrangian? The problem is that, in general, there is no Hamiltonian that is bounded from below if the Lagrangian contains higher derivatives, i.e. the system is unstable (Woodard, 2007). Of course it is possible to find well-behaved solutions of this system, since for a given solution energy is conserved. But as soon as the system is interacting, with other degrees of freedom, it will lower its energy and produce more and more modes of these other degrees of freedom. This is especially serious when one quantizes the system. The vacuum is exponentially unstable to simultaneous production of modes of positive and negative energy. Of course one cannot simply 'cut away' the negative energy solutions without violating unitarity. And even if the theory under consideration is only a low energy effective theory, it should at least be 'unitary at low energy'. Introducing even higher derivatives only worsens the situation, since the Hamiltonian acquires more unstable directions.

For this argument, it does not matter whether the degrees of freedom we are discussing are fundamental or only low energy effective degrees of freedom. Even if we modify the Hamiltonian at high energies, the instability, which is a *low energy problem*, will not disappear. There are only two ways out of the Ostrogradski instability: firstly, if the necessary condition that the Lagrangian be non-degenerate is not satisfied; secondly, via constraints, whereby one might be able to eliminate the unstable directions. In practice, this has to be studied on a case by case basis. An important example for the dark energy problem, which avoids the Ostrogradski instability via constraints, is modified gravity Lagrangians of the form $f(R)$, discussed below.

If the Ostradgradski theorem does not apply, we have still no guarantee that the theory has no ghosts or that the potential energy is bounded from below (no 'serious' tachyon). The limitation from the Ostragradski theorem, and also the ghost and tachyon problem, can be cast in the requirement that the theory needs to have an energy functional that is bounded from below. This condition can certainly not disappear in a consistent low energy version of a fundamental theory that satisfies it.

The high energy cutoff will be given by some mass scale, i.e., some Lorentz invariant energy scale of the theory, and therefore the effective low energy theory should also admit a Lorentz invariant Lagrangian. Lorentz invariance is not a high energy phenomenon that can simply be lost at low energies.

What about superluminal motion and causality? We do not want to require certain properties of the S-matrix of the low energy theory, since the latter may not have a meaningful perturbative quantum theory; like the 4-Fermi theory, it may not be renormalizable. Furthermore, one can argue that in cosmology we do have a preferred frame, the cosmological frame, hence Lorentz invariance is broken and we can simply demand that all superluminal modes of a field propagate forward in cosmic time. Then no closed signal curves are possible.

But this last argument is very dangerous. Clearly, most solutions of a Lagrangian theory do break several or most of the symmetries of the Lagrangian spontaneously. But when applying a Lorentz transformation to a solution, we produce a new solution that, from the point of view of the Lagrangian, has the same right of existence. If some modes of a field propagate with superluminal speed, this means that their characteristics are spacelike. The condition that the mode has to travel forward in time with respect to a certain frame implies that one has to use the retarded Green's function in this frame. Since spacelike distances have no frame-independent chronology, for spacelike characteristics this is a frame-dependent statement. Depending on the frame of reference, a given mode can represent a normal propagating degree of freedom, or it can satisfy an elliptic equation, a constraint.

Furthermore, to make sure that the mode propagates forward with respect to one fixed reference frame, one would have to use sometimes the retarded, sometimes the advanced and sometimes a mixture of both functions, depending on the frame of reference. In a cosmological setting this can be done in a consistent way, but it is far from clear that such a prescription can be unambiguously implemented for generic low energy solutions. Indeed in Adams *et al.* (2006) a solution is sketched that would not allow this, so that closed signal curves are again possible.

Therefore, we feel that Lorentz invariant low energy effective Lagrangians, which allow for superluminal propagation of certain modes, have to be rejected. Nevertheless, this case is not clear-cut and there are opposing opinions in the literature, e.g. Babichev, Mukhanov and Vikman (2008).

With the advent of the 'landscape' (Polchinski, 2006; Bousso, 2006, 2008), physicists have begun to consider anthropic arguments to justify their theory, whenever it fits the data. Even though the existence of life on Earth is an experimental fact, we consider this argument weak, nearly tantamount to giving up physics: 'Things are like they are since otherwise we would not be here'. We nevertheless find it important to inquire, also from a purely theoretical point of view, whether really 'anything goes' for effective theories. In the following sections we shall come back to the basic requirements we have outlined in this section.

2.3 General relativistic approaches

We give a very brief overview of models for the late-time acceleration within general relativity, before moving on to the main topic of modified gravity.

The 'standard' general relativistic interpretation of dark energy is based on the cosmological constant as vacuum energy:

$$G_{\mu\nu} = 8\pi G \left[T_{\mu\nu} + T_{\mu\nu}^{\text{vac}} \right], \qquad T_{\mu\nu}^{\text{vac}} = -\frac{\Lambda_{\text{eff}}}{8\pi G} g_{\mu\nu}, \tag{2.11}$$

where the vacuum energy–momentum tensor is Lorentz invariant. This approach faces the problem of accounting for the incredibly small and highly fine-tuned value of the vacuum energy, as summarized in Eq. (2.7).

String theory provides a tantalizing possibility in the form of the 'landscape' of vacua (Polchinski, 2006; Bousso, 2006, 2008). There appears to be a vast number of vacua admitted by string theory, with a broad range of vacuum energies above and below zero. The idea is that our observable region of the universe corresponds to a particular small positive vacuum energy, whereas other regions with greatly different vacuum energies will look entirely different. This multitude of regions forms in some sense a 'multiverse'. This is an interesting idea, but it is highly speculative, and it is not clear how much of it will survive the further development of string theory and cosmology.

An alternative view of ΛCDM is the interpretation of Λ as a classical geometric constant (Padmanabhan, 2006b), on a par with Newton's constant G. Thus the field equations are interpreted in the geometrical way,

$$G_{\mu\nu} + \Lambda g_{\mu\nu} = 8\pi G T_{\mu\nu}. \tag{2.12}$$

In this approach, the small and fine-tuned value of Λ is no more of a mystery than the host of other fine-tunings in the constants of nature. For example, more than a 2% change in the strength of the strong interaction means that no atoms beyond hydrogen can form, so that stars and galaxies would not emerge. But it is not evident whether this distinction between Λ and ρ_{vac} is really a physical statement, or a purely theoretical statement that cannot be tested by any experiments. Furthermore, this classical approach to Λ does not evade the vacuum energy problem – it simply shifts that problem to 'why does the vacuum not gravitate?' The idea is that particle physics and quantum gravity will somehow discover a cancellation or symmetry mechanism to explain why $\rho_{\text{vac}} = 0$. This would be a simpler solution than that indicated by the string landscape approach, and would evade the disturbing anthropic aspects of that approach.

Within general relativity, various alternatives to ΛCDM have been investigated, in an attempt to address the coincidence problem.

2.3.1 Dynamical dark energy: quintessence

Here we replace the constant $\Lambda/8\pi G$ by the energy density of a scalar field φ, with Lagrangian

$$L_\varphi = -\frac{1}{2}g^{\mu\nu}\partial_\mu\varphi\partial_\nu\varphi - V(\varphi), \tag{2.13}$$

so that in a cosmological setting,

$$\rho_\varphi = \frac{1}{2}\dot{\varphi}^2 + V(\varphi), \qquad p_\varphi = \frac{1}{2}\dot{\varphi}^2 - V(\varphi), \tag{2.14}$$

$$\ddot{\varphi} + 3H\dot{\varphi} + V'(\varphi) = 0, \tag{2.15}$$

$$H^2 + \frac{K}{a^2} = \frac{8\pi G}{3}\left(\rho_r + \rho_m + \rho_\varphi\right). \tag{2.16}$$

The field rolls down its potential and the dark energy density varies through the history of the universe. 'Tracker' potentials have been found for which the field energy density follows that of the dominant matter component. This offers the possibility of solving or alleviating the fine-tuning problem of the resulting cosmological constant. Although these models are insensitive to initial conditions, they do require a strong fine-tuning of the *parameters of the Lagrangian* to secure recent dominance of the field, and hence do not evade the coincidence problem. An attempt to address the coincidence problem is proposed in Wetterich (2007), where the transition from the tracker behaviour to dark energy domination is tied to the neutrino mass.

More generally, the quintessence potential, somewhat like the inflaton potential, remains arbitrary, until and unless fundamental physics selects a potential. There is currently no natural choice of potential.

In conclusion, there is no compelling reason as yet to choose quintessence above the ΛCDM model of dark energy. Quintessence models do not seem more natural, better motivated or less contrived than ΛCDM. Nevertheless, they are a viable possibility and computations are straightforward. Therefore, they remain an interesting target for observations to shoot at (Linder, 2008).

2.3.2 Dynamical dark energy: more general models

It is possible to couple quintessence to cold dark matter without violating current constraints from fifth force experiments. This could lead to a new approach to the coincidence problem, since a coupling may provide a less unnatural way to explain why acceleration kicks in when $\rho_m \sim \rho_{de}$. In the presence of coupling, the energy conservation equations in the background become

$$\dot{\varphi}\left[\ddot{\varphi} + 3H\dot{\varphi} + V'(\varphi)\right] = Q, \qquad (2.17)$$

$$\dot{\rho}_{dm} + 3H\rho_{dm} = -Q, \qquad (2.18)$$

where Q is the rate of energy exchange. It is relatively simple to match the geometric data on the background expansion history (Amendola, Camargo Campos and Rosenfeld, 2007; Guo, Ohta and Tsujikawa, 2007; Quartin *et al.*, 2008). The perturbations show that there is a momentum transfer as well as an energy transfer. Analysis of the perturbations typically leads to more stringent constraints, with some forms of coupling being ruled out by instabilities (Koivisto, 2005; Bean, Flanagan and Trodden, 2008; Valiviita, Majerotto and Maartens, 2008).

Another possibility is a scalar field with non-standard kinetic term in the Lagrangian, for example,

$$L_\varphi = F(\varphi, X) - V(\varphi), \quad \text{where } X \equiv -\frac{1}{2}g^{\mu\nu}\partial_\mu\varphi\partial_\nu\varphi. \qquad (2.19)$$

The standard Lagrangian has $F(\varphi, X) = X$. Some of the non-standard F models may be ruled out on theoretical grounds. An example is provided by 'phantom' fields with negative kinetic energy density (ghosts), $F(\varphi, X) = -X$. They have $w < -1$, so that their energy density *grows* with expansion. This bizarre behaviour is reflected in the instability of the quantum vacuum for phantom fields.

Another example is 'k-essence' fields (Armendariz-Picon, Mukhanov and Steinhardt, 2000), which have $F(\varphi, X) = \varphi^{-2}f(X)$. These theories have no ghosts, and they can produce late-time acceleration. The sound speed of the field fluctuations for the Lagrangian in Eq. (2.19) is

$$c_s^2 = \frac{F_{,X}}{F_{,X} + 2XF_{,XX}}. \qquad (2.20)$$

For a standard Lagrangian, $c_s^2 = 1$. But for the class of F that produce accelerating k-essence models, it turns out that there is always an epoch during which $c_s^2 > 1$, so that these models may be ruled out according to our causality requirement. They violate standard causality (Bonvin, Caprini and Durrer, 2006; Ellis, Maartens and McCallum, 2007).

For models not ruled out on theoretical grounds, there is the same general problem as with quintessence, i.e. that no model is better motivated than ΛCDM, none is selected by fundamental physics and any choice of model is more or less arbitrary. Quintessence then appears to at least have the advantage of simplicity – although ΛCDM has the same advantage over quintessence.

When investigating generic dark energy models we always have to keep in mind that since both dark energy and dark matter are only detected gravitationally, we can only measure the total energy–momentum tensor of the dark component,

$$T_{\mu\nu}^{\text{dark}} = T_{\mu\nu}^{\text{de}} + T_{\mu\nu}^{\text{dm}}. \tag{2.21}$$

Hence, if we have no information on the equation of state of dark energy, there is a degeneracy between the dark energy equation of state $w(t)$ and Ω_{dm}. Without additional assumptions, we cannot measure either of them by purely gravitational observations (Kunz, 2007). This degeneracy becomes even worse if we allow for interactions between dark matter and dark energy.

2.3.3 Dark energy as a nonlinear effect from structure

As structure forms and the matter density perturbation becomes nonlinear, there are two questions that are posed: (1) what is the back-reaction effect of this nonlinear process on the background cosmology? (2) how do we perform a covariant and gauge-invariant averaging over the inhomogeneous universe to arrive at the correct FRW background? The simplistic answers to these questions are: (1) the effect is negligible since it occurs on scales too small to be cosmologically relevant; (2) in light of this, the background is independent of structure formation, i.e., it is the same as in the linear regime. A quantitative analysis is needed to fully resolve both issues. However, this is very complicated because it involves the nonlinear features of general relativity in an essential way.

There have been claims that these simplistic answers are wrong, and that, on the contrary, the effects are large enough to mimic an accelerating universe. This would indeed be a dramatic and satisfying resolution of the coincidence problem, without the need for any dark energy field. This issue is discussed in Buchert (2008). Of course, the problem of why the vacuum does not gravitate would remain.

However, these claims have been disputed, and it is fair to say that there is as yet no convincing demonstration that acceleration could emerge naturally from nonlinear effects of structure formation (see e.g. Coley, Pelvas and Zalaletdinav, 2005; Flanagan, 2005; Geshnizjani, Chung and Afshordi 2005; Giovannini, 2005; Hirata and Seljak, 2005; Kolb *et al.*, 2005; Martineau and Brandenberger, 2005; Nambu and Tanimoto, 2005; Buchert, 2006; Ishibashi and Wald, 2006;

Mansouri, 2006; Moffat, 2006; Rasanen, 2006; Alnes, Amarzguioui and Gron, 2007; Vanderveld, Flanagan and Wasserman, 2007; Rosenthal and Flanagan, 2008; Wiltshire, 2008). We should however note that back-reaction/averaging effects could significantly affect our estimations of cosmological parameters, even if they do not lead to acceleration (Li and Schwarz, 2007).

It is in principle also possible that the universe around us resembles more a spherically symmetric but inhomogeneous solution of Einstein's equation, a Tolman–Bondi–Lemaître universe, than a Friedmann–Lemaître universe. In this case, what appears as cosmic acceleration to us could perhaps be explained within simple matter models that only contain dust (Paranjape and Singh, 2006; Vanderveld, Flanagan and Wasserman, 2006, 2008; Enqvist, 2008; Garcia-Bellido and Haugboelle, 2009; Zibin, Moss and Scott, 2008). However, this would imply that we are situated very close to the centre of a huge (nearly) spherical struc-ture. Apart from violating the Copernican principle, this poses another fine-tuning problem, and it also is not clear to us whether these models are consistent with all observations – not just supernova, but baryon acoustic oscillations, CMB anisotropies, and weak lensing.

2.4 The modified gravity approach: dark gravity

Late-time acceleration from nonlinear effects of structure formation is an attempt, within general relativity, to solve the coincidence problem without a dark energy field. The modified gravity approach shares the assumption that there is no dark energy field, but generates the acceleration via 'dark gravity', i.e. a weakening of gravity on the largest scales, due to a modification of general relativity itself.

Could the late-time acceleration of the universe be a gravitational effect? (Note that in general also this does not remove the problem of why vacuum energy does not gravitate or is very small.) A historical precedent is provided by attempts to explain the anomalous precession of Mercury's perihelion by a 'dark planet', named Vulcan. In the end, it was discovered that a modification to Newtonian gravity was needed.

As we have argued in Section 2.2, a consistent modification of general relativity requires a covariant formulation of the field equations in the general case, i.e., including inhomogeneities and anisotropies. It is not sufficient to propose ad hoc modifications of the Friedmann equation, of the form

$$f(H^2) = \frac{8\pi G}{3}\rho \quad \text{or} \quad H^2 = \frac{8\pi G}{3}g(\rho), \tag{2.22}$$

for some functions f or g. Apart from the fundamental problems outlined in Section 2.2, such a relation allows us to compute the supernova distance/redshift relation using this equation – but we *cannot* compute the density perturbations without knowing the covariant parent theory that leads to such a modified Friedmann equation. And we also cannot compute the solar system predictions.

It is very difficult to produce infrared corrections to general relativity that meet all the minimum requirements:

- Theoretical consistency in the sense discussed in Section 2.2.
- Late-time acceleration consistent with supernova luminosity distances, baryon acoustic oscillations and other data that constrain the expansion history.
- A matter-dominated era with an evolution of the scale factor $a(t)$ that is consistent with the requirements of structure formation.
- Density perturbations that are consistent with the observed growth factor, matter power spectrum, peculiar velocities, CMB anisotropies and weak lensing power spectrum.
- Stable static spherical solutions for stars, and consistency with terrestrial and solar system observational constraints.
- Consistency with binary pulsar period data.

One of the major challenges is to compute the cosmological perturbations for structure formation in a modified gravity theory. In general relativity, the perturbations are well understood. The perturbed metric in Newtonian gauge is

$$ds^2 = -(1 + 2\Psi)dt^2 + a^2(1 + 2\Phi)d\vec{x}^{\,2}, \tag{2.23}$$

and the metric potentials define two important combinations:

$$\Phi_+ = \frac{1}{2}(\Phi + \Psi), \qquad \Phi_- = \frac{1}{2}(\Phi - \Psi). \tag{2.24}$$

In the Newtonian limit $\Psi = -\Phi = -\Phi_-$ is the ordinary Newtonian potential and $\Phi_+ = 0$. The potential Φ_+ is sourced by anisotropic stresses. It vanishes if the gravitational field is entirely due to non-relativistic matter or a perfect fluid. The (comoving) matter density perturbation $\Delta = \delta - 3aHv$ obeys the Poisson and evolution equations on sub-Hubble scales:

$$k^2\Phi = 4\pi Ga^2\rho\Delta, \tag{2.25}$$

$$\ddot{\Delta} + 2H\dot{\Delta} - 4\pi G\rho\Delta = 0. \tag{2.26}$$

These equations are exact on all scales, if perturbations are purely matter ($w = 0$) and there are no anisotropic stresses. On super-Hubble scales (and for adiabatic perturbations, but in the presence of anisotropic stresses), the evolution of the perturbations is entirely determined by the background (Bertschinger, 2006) (and the anisotropic stresses that relate the potentials Ψ and Φ)

$$\Phi'' - \Psi'' - \frac{H''}{H'}\Phi' - \left(\frac{H'}{H} - \frac{H''}{H'}\right)\Psi = 0, \tag{2.27}$$

where a prime denotes $d/d\ln a$.

The large-angle anisotropies in the CMB temperature encode a signature of the formation of structure. They are determined by the propagation of photons along the geodesics of the perturbed geometry. For adiabatic perturbations one obtains on large scales the following expression (Durrer, 2008) for the temperature fluctuation in direction $\mathbf{n}$:

$$\frac{\delta T}{T}(\mathbf{n})\bigg|_{\text{SW}} = \left[\frac{1}{3}\Psi - \frac{2}{3}H\dot{\Phi} + \frac{1}{4}\Delta_r + \mathbf{V}_b \cdot \mathbf{n}\right](\mathbf{x}_{\text{dec}}, t_{\text{dec}})$$

$$-2\int a\dot{\Phi}_-(\mathbf{x}(v), t(v))\,dv. \tag{2.28}$$

The integral is along the (unperturbed) trajectory of the light ray with affine parameter v, from last scattering to today. The position at decoupling, $\mathbf{x}_{\text{dec}}$, depends on $\mathbf{n}$. The same integral also determines the weak lensing signal, since the deflection angle is given by (see, e.g. Durrer, 2008)

$$\vec{\alpha} = 2\int \vec{\nabla}_\perp \Phi_-\,dv, \tag{2.29}$$

where $\nabla_\perp$ is the gradient operator in the plane normal to $\mathbf{n}$.

The first term in the square brackets of Eq. (2.28) is called the ordinary Sachs–Wolfe effect (OSW), the second term is usually small since at the time of decoupling the universe is matter dominated and this term vanishes in a purely matter dominated universe. The third term is responsible for the acoustic peaks in the CMB anisotropy spectrum and the fourth term is the Doppler term, due to the motion of the emitting electrons, $\mathbf{V}_b$ is the baryon velocity field. The integral is the integrated Sachs–Wolfe effect (ISW). It comes from the fact that the photons are blueshifted when they fall into a gravitational potential and redshifted when they climb out of it. Hence if the potential varies during this time, they acquire a net energy shift.

In a modified gravity theory, which we assume to be a metric theory obeying energy–momentum conservation, Eq. (2.23) still holds, and so does the super-Hubble evolution equation (2.27), and the SW and lensing relations (2.28) and (2.29). But in general

$$\Phi_+ \neq 0, \tag{2.30}$$

even in the absence of matter anisotropic stress – the modified-gravity effects produce a 'dark' anisotropic stress. In addition, the Poisson equation and the evolution of density perturbations will be modified.

2.4.1 $f(R)$ and scalar–tensor theories

General relativity has a unique status as a theory where gravity is mediated by a massless spin-2 particle, and the field equations are second order. Consider modifications to the Einstein–Hilbert action of the general form

$$-\int d^4x \sqrt{-g}\, R \;\rightarrow\; -\int d^4x \sqrt{-g}\, f(R, R_{\mu\nu}R^{\mu\nu}, C_{\mu\nu\alpha\beta}C^{\mu\nu\alpha\beta}), \quad (2.31)$$

where $R_{\mu\nu}$ is the Ricci tensor, $C_{\mu\nu\alpha\beta}$ is the Weyl tensor and $f(x_1, x_2, x_3)$ is an arbitrary (at least three times differentiable) function. Since the curvature tensors contain second derivatives of the metric, the resulting equations of motion will in general be fourth order, and gravity is carried also by massless spin-0 and spin-1 fields. However Ostrogradski's theorem applies: the usual Hamiltonian formulation of general relativity leads to six independent metric components g_{ij}, which all acquire higher derivative terms. There is actually only one way out, which is the case $\partial_2 f = \partial_3 f = 0$, i.e., f may only depend on the Ricci scalar.[2] The reason is that in the Ricci scalar R, only a single component of the metric contains second derivatives. In this case, the consequent new degree of freedom can be fixed completely by the g_{00} constraint, so that the only instability in $f(R)$ theories is the usual one associated with gravitational collapse (Woodard, 2007).

Therefore, the only acceptable low energy generalizations of the Einstein–Hilbert action of general relativity are $f(R)$ theories, with $f''(R) \neq 0$. The field equations are

$$f'(R)R_{\mu\nu} - \frac{1}{2}f(R)g_{\mu\nu} - \left[\nabla_\mu\nabla_\nu - g_{\mu\nu}\nabla^\alpha\nabla_\alpha\right]f'(R) = 8\pi G T_{\mu\nu}, \quad (2.32)$$

and standard energy–momentum conservation holds:

$$\nabla_\nu T^{\mu\nu} = 0. \quad (2.33)$$

The trace of the field equations is a wave-like equation for f', with source term $T = T_\mu{}^\mu$:

$$3\nabla^\alpha\nabla_\alpha f'(R) + Rf'(R) - 2f(R) = 8\pi G T. \quad (2.34)$$

This equation is important for investigating issues of stability in the theory, and it also implies that Birkhoff's theorem does not hold.

---

[2] Another possibility is the addition of a Gauss–Bonnet term, $\sqrt{-g}f(L_{GB})$, where $L_{GB} = R^2 - 4R_{\mu\nu}R^{\mu\nu} + R_{\mu\nu\sigma\rho}R^{\mu\nu\sigma\rho}$. In four dimensions $\sqrt{-g}L_{GB}$ contributes only a surface term and does not enter the equations of motion. However, $\sqrt{-g}f(L_{GB})$ is non-trivial. Such a term also becomes interesting in scalar–tensor theories of gravity where one may consider a contribution of the form $\sqrt{-g}\phi L_{GB}$ to the Lagrangian.

There has been a revival of interest in $f(R)$ theories due to their ability to produce late-time acceleration (Sotiriou and Faraoni, 2008). However, it turns out
to be extremely difficult for this simplified class of modified theories to pass
the observational and theoretical tests. A simple example of an $f(R)$ model is
(Capozziello, Carloni and Trosi, 2003)

$$f(R) = R - \frac{\mu}{R}. \tag{2.35}$$

For $|\mu| \sim H_0^4$, this model successfully achieves late-time acceleration as the μ/R
term starts to dominate. But the model strongly violates solar system constraints,
can have a strongly non-standard matter era before the late-time acceleration, and
suffers from nonlinear matter instabilities (see, e.g. Dolgov and Kawasaki, 2003;
Amendola *et al.*, 2007; Bean *et al.*, 2007; Chiba, Smith and Erickcek, 2007).

In $f(R)$ theories, the additional degree of freedom can be interpreted as a
scalar field, and in this sense $f(R)$ theories are mathematically equivalent to
scalar–tensor theories via

$$\psi \equiv f'(R), \qquad U(\psi) \equiv -\psi R(\psi) + f(R(\psi)), \tag{2.36}$$

$$L = -\frac{1}{16\pi G}\sqrt{-g}\,[\psi R + U(\psi)]. \tag{2.37}$$

This Lagrangian is the Jordan-frame representation of $f(R)$. It can be conformally
transformed to the Einstein frame, via the transformation

$$\tilde{g}_{\mu\nu} = \psi g_{\mu\nu}, \qquad \varphi = \sqrt{\frac{3}{4\pi G}}\ln\psi. \tag{2.38}$$

In terms of $\tilde{g}_{\mu\nu}$ and φ the Lagrangian then becomes a standard scalar field
Lagrangian,

$$L = \frac{-1}{16\pi G}\sqrt{-\tilde{g}}\left[\tilde{R} + \frac{1}{2}\tilde{g}^{\mu\nu}\partial_\mu\varphi\partial_\nu\varphi + V(\varphi)\right], \tag{2.39}$$

where

$$V(\varphi) = \frac{U(\psi(\varphi))}{\psi(\varphi)^2}. \tag{2.40}$$

This example shows that modifying gravity (dark gravity) or modifying the
energy–momentum tensor (dark energy) can be seen as a different description of
the same physics. Only the coupling of the scalar field φ to ordinary matter shows

that this theory originates from a scalar–tensor theory of gravity – and this non-standard coupling reflects the fact that gravity is also mediated by a spin-0 degree of freedom, in contrast to general relativity with a standard scalar field.

The spin-0 field is precisely the cause of the problem with solar system constraints in most $f(R)$ models, since the requirement of late-time acceleration leads to a very light mass for the scalar. The modification to the growth of large-scale structure due to this light scalar may be kept within observational limits. But on solar system scales, the coupling of the light scalar to the Sun and planets induces strong deviations from the weak-field Newtonian limit of general relativity, in obvious violation of observations. In terms of the Lagrangian (2.39), this scalar has an associated Brans–Dicke parameter that vanishes, $\omega_{BD} = 0$, whereas solar system and binary pulsar data currently require $\omega_{BD} > 40\,000$.

The only way to evade this problem is to increase the mass of the scalar near massive objects like the Sun, so that the Newtonian limit can be recovered, while preserving the ultralight mass on cosmological scales. This 'chameleon' mechanism can be used to construct models that evade solar system/binary pulsar constraints (Hu and Sawicki, 2007a; Nojiri and Odintsov, 2007b; Starobinsky, 2007). However, the price to pay is that additional parameters must be introduced, and the chosen $f(R)$ tends to look unnatural and strongly fine-tuned. An example is

$$f(R) = R + \lambda R_0 \left[\left(1 + \frac{R^2}{R_0^2} \right)^{-n} - 1 \right], \tag{2.41}$$

where λ, R_0, n are positive parameters.

Cosmological perturbations in $f(R)$ theory are well understood (Song, Hu and Sawicki, 2007). The modification to general relativity produces a dark anisotropic stress

$$\Phi_+ \propto \frac{f''(R)}{f'(R)}, \tag{2.42}$$

and deviations from general relativity are conveniently characterized by the dimensionless parameter

$$B = \frac{dR/d\ln a}{d\ln H/d\ln a} \frac{f''(R)}{f'(R)}. \tag{2.43}$$

If we invoke a chameleon mechanism, then it is possible for these models to match the observed large-angle CMB anisotropies (see Fig. 2.3) and linear matter power

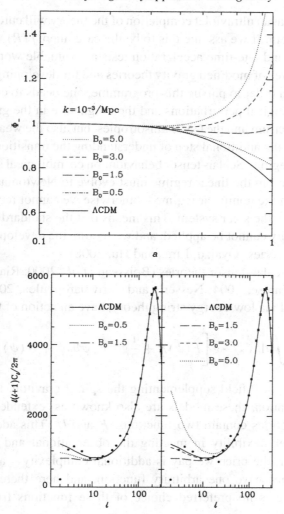

Fig. 2.3. *Top*: The ISW potential, $(\Phi - \Psi)/2$, for $f(R)$ models, where the parameter B_0 indicates the strength of deviation from general relativity (see Eq. (2.43)). *Bottom*: The large-angle CMB anisotropies for the models of the top figure. (For more details see Song, Peiris and Hu, 2007, where this figure is taken from.)

spectrum (Song, Peiris and Hu, 2007). However, there may also be fatal problems with singularities in the strong gravity regime, which would be incompatible with the existence of neutron stars (Kobayashi and Maeda, 2008; Frolov, 2008). These problems appear to arise in the successful chameleon models, and they are another unintended, and unexpected, consequence of the scalar degree of freedom, this time at high energies.

It is possible that an ultraviolet completion of the theory will cure the high energy singularity problem. If we assume this to be the case, then $f(R)$ models that pass the solar system and late-time acceleration tests are valuable working models for probing the features of modified gravity theories and for developing tests of general relativity itself. In order to pursue this programme, one needs to compute not only the linear cosmological perturbations and their signature in the growth factor, the matter power spectrum and the CMB anisotropies, but also the weak lensing signal. For this, we need the additional step of understanding the transition from the linear to the nonlinear regime. Scalar–tensor behaviour on cosmological scales relevant to structure formation in the linear regime must evolve to Newtonian-like behaviour on small scales in the nonlinear regime – otherwise we cannot recover the general relativistic limit in the solar system. This means that the standard fitting functions in general relativity cannot be applied, and we require the development of N-body codes in $f(R)$ theories (Oyaizu, Lima and Hu, 2008).

More general scalar–tensor theories (Boisseau *et al.*, 2000; Riazuelo and Uzan, 2002; Esposito-Farese, 2004; Nesseris and Perivolaropoulos, 2007), which may also be motivated via low-energy string theory, have an action of the form

$$-\int d^4x \sqrt{-g}\left[F(\psi)R + \frac{1}{2}g^{\mu\nu}\partial_\mu\psi\partial_\nu\psi + U(\psi)\right], \qquad (2.44)$$

where ψ is the spin-0 field supplementing the spin-2 graviton. In the context of late-time acceleration, these models are also known as 'extended quintessence'. Scalar–tensor theories contain two functions, F and U. This additional freedom allows for greater flexibility in meeting the observational and theoretical constraints. However, the price we pay is additional complexity – and arbitrariness. The $f(R)$ theories have one arbitrary function, and here there are two, $F(\psi)$ and $U(\psi)$. There is no preferred choice of these functions from fundamental theory.

Modifications of the Einstein–Hilbert action, which lead to fourth-order field equations, either struggle to meet the minimum requirements in the simplest cases, or contain more complexity and arbitrary choices than quintessence models in general relativity. Therefore, none of these models appears to be a serious competitor to quintessence in general relativity.

2.4.2 Brane-world models

Modifications to general relativity within the framework of quantum gravity are typically ultraviolet corrections that must arise at high energies in the very early universe or during collapse to a black hole. The leading candidate for a quantum gravity theory, string theory, is able to remove the infinities of quantum field theory

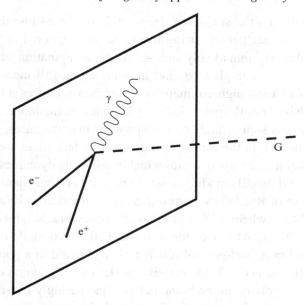

Fig. 2.4. The confinement of matter to the brane, while gravity propagates in the bulk. (From Cavaglia, 2003.)

and unify the fundamental interactions, including gravity. But there is a price – the theory is only consistent in nine space dimensions. Branes are extended objects of higher dimension than strings, and play a fundamental role in the theory, especially D-branes, on which open strings can end. Roughly speaking, the endpoints of open strings, which describe the standard model particles like fermions and gauge bosons, are attached to branes, while the closed strings of the gravitational sector can move freely in the higher-dimensional 'bulk' spacetime. Classically, this is realized via the localization of matter and radiation fields on the brane, with gravity propagating in the bulk (see Fig. 2.4).

The implementation of string theory in cosmology is extremely difficult, given the complexity of the theory. This motivates the development of phenomenological models, as an intermediary between observations and fundamental theory. Brane-world cosmological models inherit some aspects of string theory, but do not attempt to impose the full machinery of the theory. Instead, simplifications are introduced in order to be able to construct cosmological models that can be used to compute observational predictions (see Brax, van de Bruck and Davis, 2004; Maartens, 2004; Durrer, 2005; Sahni, 2005; Langlois, 2006; Lue, 2006; Wands, 2006, for reviews in this spirit). Cosmological data can then be used to constrain the brane-world models, and hopefully provide constraints on string theory, as well as pointers for the further development of string theory.

It turns out that even the simplest (5D, we effectively assume that five of the extra dimensions in the 'parent' string theory may be ignored at low energies) brane-world models are remarkably rich – and the computation of their cosmological perturbations is complicated, and in many cases still incomplete. A key reason for this is that the higher-dimensional graviton produces a tower of four-dimensional massive spin-0, spin-1 and spin-2 modes on the brane, in addition to the standard massless spin-2 mode on the brane (or in some cases, instead of the massless spin-2 mode). In the case of some brane models, there are in addition a massless gravi-scalar and a gravi-vector which modify the dynamics.

Most brane-world models modify general relativity at high energies. The main examples are those of Randall–Sundrum (RS) type (Randall and Sundrum, 1999; Binetruy *et al.*, 2000), where a FRW brane is embedded in a 5D anti de Sitter bulk, with curvature radius ℓ. At low energies $H\ell \ll 1$, the zero-mode of the graviton dominates on the brane, and general relativity is recovered to a good approximation. At high energies, $H\ell \gg 1$, the massive modes of the graviton dominate over the zero-mode, and gravity on the brane behaves increasingly five-dimensionally. On the brane, the standard conservation equation holds,

$$\dot{\rho} + 3H(\rho + p) = 0, \tag{2.45}$$

but the Friedmann equation is modified by an ultraviolet correction:

$$H^2 + \frac{K}{a^2} = \frac{8\pi G}{3}\rho\left(1 + \frac{2\pi G\ell^2}{3}\rho\right) + \frac{\Lambda}{3}. \tag{2.46}$$

The ρ^2 term is the ultraviolet correction. At low energies, this term is negligible, and we recover $H^2 + K/a^2 \propto \rho + \Lambda/8\pi G$. At high energies, gravity 'leaks' off the brane and $H^2 \propto \rho^2$. This 5D behaviour means that a given energy density produces a greater rate of expansion than it would in general relativity. As a consequence, inflation in the early universe is modified in interesting ways (Brax, van de Bruck and Davis, 2004; Maartens, 2004; Sahni, 2005; Durrer, 2005; Langlois, 2005; Lue, 2006; Wands, 2006).

By contrast, the brane-world model of Dvali–Gabadadze–Porrati (Dvali *et al.*, 2000) (DGP), which was generalized to cosmology by Deffayet (2001), modifies general relativity at *low* energies. This model produces 'self-acceleration' of the late-time universe due to a weakening of gravity at low energies. Like the RS model, the DGP model is a 5D model with infinite extra dimension.

The action is given by

$$\frac{-1}{16\pi G}\left[\frac{1}{r_{\rm c}}\int_{\rm bulk} {\rm d}^5x\,\sqrt{-g^{(5)}}\,R^{(5)} + \int_{\rm brane} {\rm d}^4x\,\sqrt{-g}\,R\right]. \tag{2.47}$$

The bulk is assumed to be 5D Minkowski spacetime. Unlike the AdS bulk of the RS model, the Minkowski bulk has infinite volume. Consequently, there is no normalizable zero-mode of the 4D graviton in the DGP brane-world. Gravity leaks off the 4D brane into the bulk at large scales, $r \gg r_c$, where the first term in the sum (2.47) dominates. On small scales, gravity is effectively bound to the brane and 4D dynamics is recovered to a good approximation, as the second term dominates. The transition from 4D to 5D behaviour is governed by the crossover scale r_c. For a Minkowski brane, the weak-field gravitational potential behaves as

$$\Psi \propto \begin{cases} r^{-1} & \text{for } r \ll r_c, \\ r^{-2} & \text{for } r \gg r_c. \end{cases} \tag{2.48}$$

On a Friedmann brane, gravity leakage at late times in the cosmological evolution can initiate acceleration – not due to any negative pressure field, but due to the weakening of gravity on the brane. 4D gravity is recovered at high energy via the lightest massive modes of the 5D graviton, effectively via an ultralight metastable graviton.

The energy conservation equation remains the same as in general relativity, but the Friedmann equation is modified:

$$\dot{\rho} + 3H(\rho + p) = 0, \tag{2.49}$$

$$H^2 + \frac{K}{a^2} - \frac{1}{r_c}\sqrt{H^2 + \frac{K}{a^2}} = \frac{8\pi G}{3}\rho. \tag{2.50}$$

To arrive at Eq. (2.50) we have to take a square root, which implies a choice of sign. As we shall see, the above choice has the advantage of leading to acceleration but the disadvantage of the presence of a 'ghost' in this background. It is not clear whether these facts are related. We shall discuss the 'normal' DGP model, where the opposite sign of the square root is chosen, in the next section.

From Eq. (2.50) we infer that at early times, i.e., $Hr_c \gg 1$, the general relativistic Friedmann equation is recovered. By contrast, at late times in an expanding CDM universe, with $\rho \propto a^{-3} \to 0$, we have

$$H \to H_\infty = \frac{1}{r_c}, \tag{2.51}$$

so that expansion accelerates and is asymptotically de Sitter. The above equations imply

$$\dot{H} - \frac{K}{a^2} = -4\pi G\rho \left[1 + \frac{1}{\sqrt{1 + 32\pi Gr_c^2\rho/3}} \right]. \tag{2.52}$$

In order to achieve self-acceleration at late times, we require

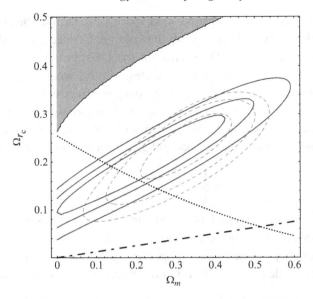

Fig. 2.5. The confidence contours for supernova data in the DGP density param-eter plane. The solid contours are for SNLS data, and the dashed contours are for the Gold data. The dotted curve defines the flat models, the dot-dashed curve defines zero acceleration today, and the shaded region contains models without a big bang. (From Maartens and Majerotto, 2006.)

$$r_c \gtrsim H_0^{-1}, \tag{2.53}$$

since $H_0 \lesssim H_\infty$. This is confirmed by fitting supernova observations, as shown in Fig. 2.5. The dimensionless crossover parameter is defined as

$$\Omega_{r_c} = \frac{1}{4(H_0 r_c)^2}, \tag{2.54}$$

and the ΛCDM relation,

$$\Omega_m + \Omega_\Lambda + \Omega_K = 1, \tag{2.55}$$

is modified to

$$\Omega_m + 2\sqrt{\Omega_{r_c}}\sqrt{1 - \Omega_K} + \Omega_K = 1. \tag{2.56}$$

ΛCDM and DGP can both account for the supernova observations, with the fine-tuned values $\Lambda \sim H_0^2$ and $r_c \sim H_0^{-1}$ respectively. When we add further constraints on the expansion history from the baryon acoustic oscillation peak at $z = 0.35$ and the CMB shift parameter, the flat DGP models are in strong tension with data,

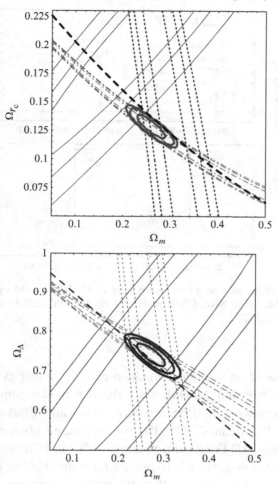

Fig. 2.6. Joint constraints (thick solid) from the SNLS data (thin solid), the BAO peak at $z = 0.35$ (dotted) and the CMB shift parameter from WMAP3 (dot-dashed). The upper plot show DGP models, the lower plot shows ΛCDM. The thick dashed line represents the flat models, $\Omega_K = 0$. (From Maartens and Majerotto, 2006.)

whereas ΛCDM models provide a consistent fit. This is evident in Fig. 2.6. The open DGP models provide a somewhat better fit to the geometric data – essentially because the lower value of Ω_m favoured by supernovae reduces the distance to last scattering and an open geometry is able to extend that distance. For a combination of SNe, CMB shift and Hubble Key Project data, the best-fit open DGP also performs better than the flat DGP (Song, Sawicki and Hu, 2007), as shown in Fig. 2.7.

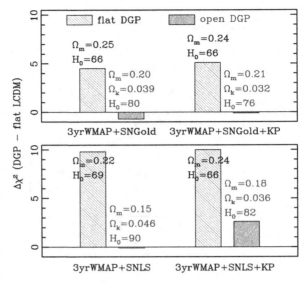

Fig. 2.7. The difference in χ^2 between best-fit DGP (flat and open) and best-fit (flat) ΛCDM, using SNe, CMB shift and H_0 Key Project data. (From Song, Sawicki and Hu, 2007.)

Observations based on structure formation provide further evidence of the difference between DGP and ΛCDM, since the two models suppress the growth of density perturbations in different ways (Lue and Starkman, 2003; Lue, Scoccimarro and Starkman, 2004). The distance-based observations draw only upon the background 4D Friedmann equation (2.50) in DGP models – and therefore there are quintessence models in general relativity that can produce precisely the same supernova distances as DGP. By contrast, structure formation observations require the 5D perturbations in DGP, and one cannot find equivalent quintessence models (Koyama and Maartens, 2006). One can find 4D general relativity models, with dark energy anisotropic stress and variable sound speed, that can in principle mimic DGP (Kunz and Sapone, 2007). However, these models are highly unphysical and can be discounted on grounds of theoretical consistency.

For ΛCDM, the analysis of density perturbations is well understood. For DGP the perturbations are much more subtle and complicated (Koyama, 2008). Although matter is confined to the 4D brane, gravity is fundamentally 5D, and the 5D bulk gravitational field responds to and back-reacts on 4D density perturbations. The evolution of density perturbations requires an analysis based on the 5D nature of gravity. In particular, the 5D gravitational field produces an effective 'dark' anisotropic stress on the 4D universe. If one neglects this stress and other 5D effects, and simply treats the perturbations as 4D perturbations with a modified

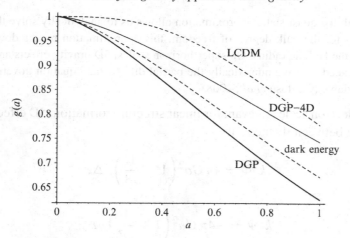

Fig. 2.8. The growth factor $g(a) = \Delta(a)/a$ for ΛCDM (long dashed) and DGP (solid, thick), as well as for a dark energy model with the same expansion history as DGP (short dashed). DGP-4D (solid, thin) shows the incorrect result in which the 5D effects are set to zero. (From Koyama and Maartens, 2006.)

background Hubble rate – then as a consequence, the 4D Bianchi identity on the brane is violated, i.e., $\nabla^{\nu} G_{\mu\nu} \neq 0$, and the results are inconsistent. When the 5D effects are incorporated (Koyama and Maartens, 2006; Cardoso *et al.*, 2008), the 4D Bianchi identity is automatically satisfied. (See Fig. 2.8.)

There are three regimes governing structure formation in DGP models:

(i) On small scales, below the so-called Vainshtein radius (which for cosmological purposes is roughly the scale of clusters), the spin-0 scalar degree of freedom becomes strongly coupled, so that the general relativistic limit is recovered (Koyama and Silva, 2007).

(ii) On scales relevant for structure formation, i.e., between cluster scales and the Hubble radius, the spin-0 scalar degree of freedom produces a scalar–tensor behaviour. A quasi-static approximation to the 5D perturbations shows that DGP gravity is like a Brans–Dicke theory with parameter (Koyama and Maartens, 2006)

$$\omega_{\mathrm{BD}} = \frac{3}{2}(\beta - 1), \tag{2.57}$$

where

$$\beta = 1 + 2H^2 r_{\mathrm{c}} \left(H^2 + \frac{K}{a^2} \right)^{-1/2} \left[1 + \frac{\dot{H}}{3H^2} + \frac{2K}{3a^2 H^2} \right]. \tag{2.58}$$

At late times in an expanding universe, when $Hr_{\mathrm{c}} \gtrsim 1$, it follows that $\beta < 1$, so that $\omega_{\mathrm{BD}} < 0$. (This signals a pathology in DGP which is discussed below.)

(iii) Although the quasi-static approximation allows us to analytically solve the 5D wave equation for the bulk degree of freedom, this approximation breaks down near and beyond the Hubble radius. On super-horizon scales, 5D gravity effects are dominant, and we need to solve numerically the partial differential equation governing the 5D bulk variable (Cardoso *et al.*, 2008).

On sub-horizon scales relevant for linear structure formation, 5D effects produce a difference between Φ and $-\Psi$:

$$k^2 \Phi = 4\pi G a^2 \left(1 - \frac{1}{3\beta}\right) \rho \Delta, \tag{2.59}$$

$$k^2 \Psi = -4\pi G a^2 \left(1 + \frac{1}{3\beta}\right) \rho \Delta, \tag{2.60}$$

so that there is an effective dark anisotropic stress on the brane:

$$k^2 (\Phi + \Psi) = -\frac{8\pi G a^2}{3\beta^2} \rho \Delta. \tag{2.61}$$

The density perturbations evolve as

$$\ddot{\Delta} + 2H\dot{\Delta} - 4\pi G \left(1 - \frac{1}{3\beta}\right) \rho \Delta = 0. \tag{2.62}$$

The linear growth factor, $g(a) = \Delta(a)/a$ (i.e., normalized to the flat CDM case, $\Delta \propto a$), is shown in Fig. 2.8. This shows the dramatic suppression of growth in DGP relative to ΛCDM – from both the background expansion and the metric perturbations. If we parameterize the growth factor in the usual way, we can quantify the deviation from general relativity with smooth dark energy (Linder, 2005):

$$f := \frac{d \ln \Delta}{d \ln a} = \Omega_m(a)^\gamma,$$

$$\gamma \approx \begin{cases} 0.55 + 0.05[1 + w(z = 1)] & \text{GR, smooth DE,} \\ 0.68 & \text{DGP.} \end{cases} \tag{2.63}$$

Observational data on the growth factor (Guzzo *et al.*, 2008) are not yet precise enough to provide meaningful constraints on the DGP model. Instead, we can look at the large-angle anisotropies of the CMB, i.e., the ISW effect. This requires a treatment of perturbations near and beyond the horizon scale. The full numerical solution has been given by Cardoso *et al.* (2008), and is illustrated in Fig. 2.9. The CMB anisotropies are also shown in Fig. 2.9, as computed in Fang *et al.*

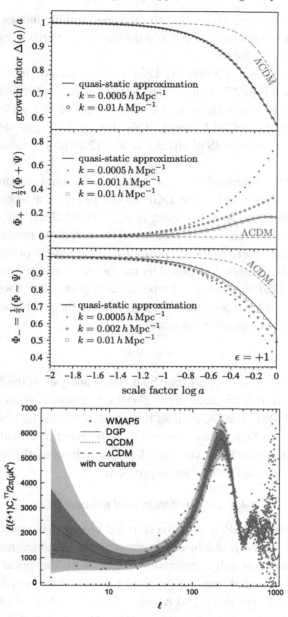

Fig. 2.9. *Top*: Numerical solutions for DGP density and metric perturbations, showing also the quasi-static solution, which is an increasingly poor approximation as the scale is increased. (From Cardoso *et al.*, 2008.) *Bottom*: Constraints on DGP (the open model that provides a best fit to geometric data) from CMB anisotropies (WMAP5). The DGP model is the solid curve, QCDM (short-dashed curve) is the quintessence model with the same background expansion history as the DGP model, and ΛCDM is the dashed curve (a slightly closed model that gives the best fit to WMAP5, HST and SNLS data). (From Fang *et al.*, 2008.)

(2008) using a scaling approximation to the super-Hubble modes (Sawicki, Song and Hu, 2007) (the accuracy of the scaling ansatz is verified by the numerical results: Cardoso *et al.*, 2008).

It is evident from Fig. 2.9 that the DGP model that provides a best fit to the geometric data (see Fig. 2.7) is in serious tension with the WMAP5 data on large scales. The problem arises from the large deviation of $\Phi_- = (\Phi - \Psi)/2$ in the DGP model from the ΛCDM model. This deviation, i.e., a stronger decay of Φ_-, leads to an over-strong ISW effect (see Eq. (2.28)), in tension with WMAP5 observations.

As a result of the combined observations of background expansion history and large-angle CMB anisotropies, the DGP model provides a worse fit to the data than ΛCDM at about the 5σ level (Fang *et al.*, 2008). Effectively, the DGP model is ruled out by observations in comparison with the ΛCDM model.

In addition to the severe problems posed by cosmological observations, a problem of theoretical consistency is posed by the fact that the late-time asymptotic de Sitter solution in DGP cosmological models has a ghost. The ghost is signaled by the negative Brans–Dicke parameter in the effective theory that approximates the DGP on cosmological sub-horizon scales:

$$\omega_{BD} < 0. \tag{2.64}$$

The existence of the ghost is confirmed by detailed analysis of the 5D perturbations in the de Sitter limit (Charmousis *et al.*, 2006; Gorbunov, Koyama and Sibiryakov 2006; Koyama, 2007). The DGP ghost is a ghost mode in the scalar sector of the gravitational field – which is more serious than the ghost in a phantom scalar field. It effectively rules out the DGP, since it is hard to see how an ultraviolet completion of the DGP can cure the *infrared* ghost problem.

2.4.3 Degravitation and normal DGP

The self-accelerating DGP is effectively ruled out as a cosmological model by observations and by the problem of the ghost in the gravitational sector. Indeed, it may be the case that self-acceleration comes with the price of ghost states. An alternative idea is that massive-graviton theories (like the DGP) may lead to *degravitation* (Dvali, Hofmann and Khoury, 2007; de Rham *et al.*, 2008a,b), i.e., the feature that the vacuum energy (cosmological constant) does not gravitate at the level expected (as in Eq. (2.7)), and possibly not at all.

To achieve a reduction of gravitation on very large scales, degravitation, Newton's constant is promoted to a 'high-pass filter' and Einstein's equations are modified to

$$G^{-1}(L^2(\nabla_\mu \nabla^\mu))G_{\mu\nu} = 8\pi T_{\mu\nu}. \tag{2.65}$$

We want $G(L^2(\nabla_\mu \nabla^\mu))$ to act as a high-pass filter: for scales smaller than L it is constant while scales much larger than L are filtered out, degravitated. For this to work, G^{-1} must contain inverse powers of $(\nabla_\mu \nabla^\mu)$, hence it must be non-local. Furthermore, this equation cannot describe a massless spin-2 graviton with only two degrees of freedom, but it leads, at the linear level to massive gravitons with mass $1/L$ or a superposition (spectral density) of massive gravitons. These are known to carry three additional polarizations: two of helicity 1 and one helicity 0 state. The latter couples to the trace of the energy momentum tensor and remains present also in the zero-mass limit, the well-known van Dam–Veltman–Zakharov discontinuity of massive gravity (van Dam and Veltman, 1970; Zakharov, 1970). This problem might be solved on small scales, where the extra polarizations become strongly coupled due to nonlinear self interactions (Vainshtein, 1972). One can show that in regions where the curvature exceeds L^{-2}, the extra polarizations are suppressed by powers of L and we recover ordinary spin-2 gravity.

Contrary to the models discussed so far, these theories can in principle address the cosmological constant problem: the cosmological constant is not necessarily small, but we cannot see it in gravitational experiments since it is (nearly) degravitated. On the other hand, the problem of the present cosmological acceleration is not addressed.

Apart from a simple massive graviton, the simplest example of degravitation is provided by the so-called 'normal' (i.e., non-self-accelerating and ghost-free) branch of the DGP (Sahni and Shtanov, 2003; Lue and Starkman, 2004), which arises from a different embedding of the DGP brane in the Minkowski bulk (see Fig. 2.10). In the background dynamics, this amounts to a replacement $r_c \to -r_c$ in Eq. (2.50) – and there is no longer late-time self-acceleration. It is therefore necessary to include a Λ term in order to accelerate the late universe:

$$H^2 + \frac{K}{a^2} + \frac{1}{r_c}\sqrt{H^2 + \frac{K}{a^2}} = \frac{8\pi G}{3}\rho + \frac{\Lambda}{3}. \tag{2.66}$$

Normal DGP models with a quintessence field have also been investigated (Chimento *et al.*, 2006). Using the dimensionless crossover parameter defined in Eq. (2.54), the densities are related at the present time by

$$\sqrt{1 - \Omega_K} = -\sqrt{\Omega_{r_c}} + \sqrt{\Omega_{r_c} + \Omega_m + \Omega_\Lambda}, \tag{2.67}$$

which can be compared with the self-accelerating DGP relation (2.56).

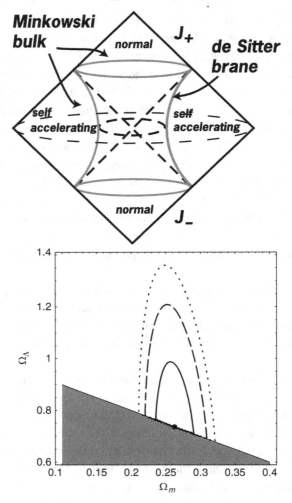

Fig. 2.10. *Top*: The embedding of the self-accelerating and normal branches of the DGP brane in a Minkowski bulk. (From Charmousis *et al.*, 2006.) *Bottom*: Joint constraints on normal DGP (flat, $K = 0$) from SNLS, CMB shift (WMAP3) and BAO ($z = 0.35$) data. The best fit is the solid point, and is indistinguishable from the ΛCDM limit. The shaded region is unphysical and its upper boundary represents flat ΛCDM models. (From Lazkoz, Maartens and Majerotto, 2006.)

The degravitation feature of normal DGP is that Λ is effectively screened by 5D gravity effects. This follows from rewriting the modified Friedmann equation (2.66) in standard general relativistic form, with

$$\Lambda_{\text{eff}} = \Lambda - \frac{3}{r_c}\sqrt{H^2 + \frac{K}{a^2}} < \Lambda. \tag{2.68}$$

Thus 5D gravity in normal DGP can in principle reduce the bare vacuum energy significantly. However, Fig. 2.10 shows that best-fit flat models, using geometric data, only admit insignificant screening (Lazkoz, Maartens and Majerotto, 2006). The closed models provide a better fit to the data (Giannantonio, Song and Koyama, 2008), and can allow a bare vacuum energy term with $\Omega_\Lambda > 1$, as shown in Fig. 2.11. This does not address the fundamental problem of the smallness of Ω_Λ, but it is nevertheless an interesting feature. We can define an effective equation of state parameter via

$$\dot{\Lambda}_{\text{eff}} + 3H(1 + w_{\text{eff}})\Lambda_{\text{eff}} = 0. \qquad (2.69)$$

At the present time (setting $K = 0$ for simplicity),

$$w_{\text{eff},0} = -1 - \frac{(\Omega_m + \Omega_\Lambda - 1)\Omega_m}{(1 - \Omega_m)(\Omega_m + \Omega_\Lambda + 1)} < -1, \qquad (2.70)$$

where the inequality holds since $\Omega_m < 1$. This reveals another important property of the normal DGP model: effective phantom behaviour of the recent expansion history. This is achieved without any pathological phantom field (similar to what can be done in scalar–tensor theories (Boisseau *et al.*, 2000; Riazuelo and Uzan, 2002; Esposito-Farese, 2004; Nesseris and Perivolaropoulos, 2007). Furthermore, there is no 'big rip' singularity in the future associated with this phantom acceleration, unlike the situation that typically arises with phantom fields. The phantom behaviour in the normal DGP model is also not associated with any ghost problem – indeed, the normal DGP branch is free of the ghost that plagues the self-accelerating DGP (Charmousis *et al.*, 2006).

Perturbations in the normal branch have the same structure as those in the self-accelerating branch, with the same regimes – i.e., below the Vainshtein radius (recovering a GR limit), up to the Hubble radius (Brans–Dicke behaviour), and beyond the Hubble radius (strongly 5D behaviour). The quasi-static approximation and the numerical integrations can be simply repeated with the replacement $r_c \to -r_c$ (and the addition of Λ to the background). In the sub-Hubble regime, the effective Brans–Dicke parameter is still given by Eqs. (2.57) and (2.58), but now we have $\omega_{\text{BD}} > 0$ – and this is consistent with the absence of a ghost. Furthermore, a positive Brans–Dicke parameter signals an extra positive contribution to structure formation from the scalar degree of freedom, so that there is *less* suppression of structure formation than in ΛCDM – the reverse of what happens in the self-accelerating DGP. This is confirmed by computations, as illustrated in Fig. 2.11.

The closed normal DGP models fit the background expansion data reasonably well, as shown in Fig. 2.11. The key remaining question is how well do these

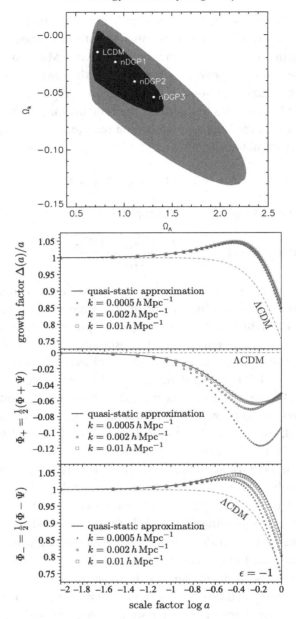

Fig. 2.11. *Top*: Joint constraints on normal DGP from SNe Gold, CMB shift (WMAP3) and H_0 data in the projected curvature-Λ plane, after marginalizing over other parameters. The best fits are the solid points, corresponding to different values of Ω_m. (From Giannantonio, Song and Koyama, 2008.) *Bottom*: Numerical solutions for the normal DGP density and metric perturbations, showing also the quasi-static solution, which is an increasingly poor approximation as the scale is increased. Compare with the self-accelerating DGP case in Fig. 2.9. (From Cardoso *et al.*, 2008.)

models fit the large-angle CMB anisotropies, which is yet to be computed at the time of writing. The derivative of the ISW potential $\dot{\Phi}_-$ can be seen in Fig. 2.11, and it is evident that the ISW contribution is negative relative to ΛCDM at high redshifts, and goes through zero at some redshift before becoming positive. This distinctive behaviour may be contrasted with the behaviour in $f(R)$ models (see Fig. 2.3): both types of model lead to less suppression of structure than ΛCDM, but they produce different ISW effects. However, in the limit $r_c \to \infty$, normal DGP tends to ordinary ΛCDM, hence observations which fit ΛCDM will always just provide a lower limit for r_c.

2.5 Conclusion

The evidence for a late-time acceleration of the universe continues to mount, as the number of experiments and the quality of data grow. This revolutionary discovery by observational cosmology confronts theoretical cosmology with a major crisis – how to explain the origin of the acceleration. The core of this problem may be 'handed over' to particle physics, since we require, at the most fundamental level, an explanation for why the vacuum energy either has an incredibly small and fine-tuned value or is exactly zero. Both options violently disagree with naïve estimates of the vacuum energy.

If one accepts that the vacuum energy is indeed non-zero, then the dark energy is described by Λ, and the ΛCDM model is the best current model. The cosmological model requires completion via developments in particle physics that will explain the value of the vacuum energy. In many ways, this is the best that we can do currently, since the alternatives to ΛCDM, within and beyond general relativity, do not resolve the vacuum energy crisis, and furthermore have no convincing theoretical motivation. None of the contenders so far appears any better than ΛCDM, and it is fair to say that, at the theoretical level, there is as yet no serious challenger to ΛCDM. One consequence of this is the need to develop better observational tests of ΛCDM, which could in principle rule it out, e.g. by showing, to some acceptable level of statistical confidence, that $w \neq -1$. However, observations are still quite far from the necessary precision for this.

It remains necessary and worthwhile to continue investigating alternative dark energy and dark gravity models, in order better to understand the space of possibilities, the variety of cosmological properties, and the observational strategies needed to distinguish them. The lack of any consistent and compelling theoretical model means that we need to keep exploring alternatives – and also to keep challenging the validity of general relativity itself on cosmological scales.

We have focused in this chapter on two of the simplest infrared-modified gravity models: the $f(R)$ models (the simplest scalar–tensor models), and the DGP models (the simplest brane-world models). In both types of model, the new scalar degree of freedom introduces severe difficulties at theoretical and observational levels. Strictly speaking, the $f(R)$ models are probably ruled out by the presence of singularities that exclude neutron stars (even if they can match all cosmological observations, including weak lensing). And the DGP models are likely ruled out by the appearance of a ghost in the asymptotic de Sitter state – as well as by a combination of geometric and structure-formation data.

Nevertheless, the intensive investigation of $f(R)$ and DGP models has left an important legacy – in a deeper understanding of

- the interplay between gravity and expansion history and structure;
- the relation between cosmological and local observational constraints;
- the special properties of general relativity itself;
- the techniques needed to distinguish different candidate models, and the limitations and degeneracies within those techniques;
- the development of tests that can probe the validity of general relativity itself on cosmological scales, independent of any particular alternative model.

The last point is one of the most important by-products of the investigation of modified gravity models. It involves a careful analysis of the web of consistency relations that link the background expansion to the evolution of perturbations (Song, 2005, 2008; Ishak, Upadhye and Spergel, 2006; Caldwell, Cooray and Melchiorri, 2007; Dore *et al.*, 2007; Hu and Sawicki, 2007a,b; Huterer and Linder, 2007; Linder and Cahn, 2007; Wang *et al.*, 2007; Zhang *et al.*, 2007; Amendola, Kunz and Sapone, 2008; Daniel *et al.*, 2008; Di Porto and Amendola, 2008; Hu, 2008; Jain and Zhang, 2008; Nesseris and Perivolaropoulos, 2008; Thomas, Abdalla and Weller, 2008; Zhang *et al.*, 2008; Mortonson, Hu and Huterer, 2009; Zhao *et al.*, 2009) and opens up the real prospect of testing general relativity well beyond the solar system and its neighbourhood.

Acknowledgments

We thank Camille Bonvin, Chiara Caprini, Kazuya Koyama, Martin Kunz, Sanjeev Seahra and Norbert Straumann for stimulating and illuminating discussions. This work is supported by the Swiss National Science Foundation and the UK STFC.

References

Adams, A., Arkani-Hamed, N., Dubovsky, S., Nicolis, A., and Rattazzi, R. (2006). *JHEP* **0610**, 014 [hep-th/0602178].

Alnes, H., Amarzguioui, M., and Gron, O. (2007). *JCAP* **0701**, 007 [astro-ph/0506449].

Amendola, L., Camargo Campos, G., and Rosenfeld, R. (2007). *Phys. Rev. D* **75**, 083506 [astro-ph/0610806].

Amendola, L., Gannouji, R., Polarski, D., and Tsujikawa, S. (2007). *Phys. Rev. D* **75**, 083504 [gr-qc/0612180].

Amendola, L., Kunz, M., and Sapone, D. (2008). *JCAP* **0804**, 013 [arXiv:0704.2421].

Armendariz-Picon, C., Mukhanov, V., and Steinhardt, P.J. (2000). *Phys. Rev. Lett.* **85**, 4438 [astro-ph/0004134].

Babichev, E., Mukhanov, V., and Vikman, A. (2008). *JHEP* **0802**, 101 [arXiv:0708.0561].

Bean, R., Bernat, D., Pogosian, L., Silvestri, A., and Trodden, M. (2007). *Phys. Rev. D* **75**, 064020 [astro-ph/0611321].

Bean, R., Flanagan, E.E., and Trodden, M. (2008). *New J. Phys.* **10**, 033006 [arXiv:0709.1124].

Bertschinger, E. (2006). *Astrophys. J.* **648**, 797 [astro-ph/0604485].

Binetruy, P., Deffayet, C., Ellwanger, U., and Langlois, D. (2000). *Phys. Lett. B* **477**, 285 [hep-th/9910219].

Bludman, S. (2006). [astro-ph/0605198].

Boisseau, B., Esposito-Farese, G., Polarski, D., and Starobinsky, A.A. (2000). *Phys. Rev. Lett.* **85**, 2236 [gr-qc/0001066].

Bonnor, W.B., and Steadman, B.R. (2005). *Gen. Relat. Grav.* **37**, 1833.

Bonvin, C., Caprini, C., and Durrer, R. (2006). *Phys. Rev. Lett.* **97**, 081303 [astro-ph/0606584].

Bonvin, C., Caprini, C., and Durrer, R. (2007). [arXiv:0706.1538].

Bordag, M., Mohideen, U., and Mostepanenko, V.M. (2001). *Phys. Rep.* **353**, 1 [quant-ph/0106045].

Bousso, R. (2006). [hep-th/0610211].

Bousso, R. (2008). In *Special Issue on Dark Energy*, G.F.R. Ellis *et al.*, Eds. *Gen. Relat. Grav.* **40**, Nos. 2–3.

Brax, P., van de Bruck, C., and Davis, A.C. (2004). *Rep. Prog. Phys.* **67**, 2183 [hep-th/0404011].

Bressi, G., Carugno, G., Onofrio, R., and Ruoso, G. (2002). *Phys. Rev. Lett.* **88**, 041804 (2002).

Buchert, T. (2006). *Class. Quant. Grav.* **23**, 817 [gr-qc/0509124].

Buchert, T. (2008). In *Special Issue on Dark Energy*, G.F.R. Ellis *et al.*, Eds. *Gen. Relat. Grav.* **40**, Nos. 2–3.

Caldwell, R., Cooray, A., and Melchiorri, A. (2007). *Phys. Rev. D* **76**, 023507 [astro-ph/0703375].

Capozziello, S., Carloni, S., and Troisi, A. (2003). *Recent Res. Dev. Astron. Astrophys.* **1**, 625 [astro-ph/0303041].

Capozziello, S., and Francaviglia, M. (2008). In *Special Issue on Dark Energy*, G.F.R. Ellis *et al.*, Eds. *Gen. Rel. Grav.* **40**, Nos. 2–3.

Cardoso, A., Koyama, K., Seahra S. S., and Silva, F. P. (2008). *Phys. Rev. D* **77**, 083512 [arXiv:0711.2563].

Cavaglia, M. (2003). *Int. J. Mod. Phys. A* **18**, 1843 [hep-ph/0210296].

Charmousis, G., Gregory, R., Kaloper, N., and Padilla, A. (2006). *JHEP* **0610**, 066 [hep-th/0604086].

Chiba, T., Smith, T. L., and Erickcek, A. L. (2007). *Phys. Rev. D* **75**, 124014 [astro-ph/0611867].

Chimento, L. P., Lazkoz, R., Maartens, R., and Quiros, I. (2006). *JCAP* **0609**, 004 [astro-ph/0605450].

Clarkson, C., Bassett, B., and Lu, T. C. (2008). *Phys. Rev. Lett.* **101**, 011301 [arXiv:0712.3457].

Clarkson, C., Cortes, M., and Bassett, B. A. (2007). *JCAP* **0708**, 011 [astro-ph/0702670].

Clifton, T., Ferreira, P. G., and Land, K. (2008). *Phys. Rev. Lett.* **101**, 131302 [arXiv:0807.1443].

Coley, A. A., Pelavas, N., and Zalaletdinov, R. M. (2005). *Phys. Rev. Lett.* **95**, 151102 [gr-qc/0504115].

Copeland, E. J., Sami, M., and Tsujikawa, S. (2006). *Int. J. Mod. Phys. D* **15**, 1753 [arXiv:hep-th/0603057].

Daniel, S. F., Caldwell, R. R., Cooray A., and Melchiorri, A. (2008). *Phys. Rev. D* **77**, 103513 [arXiv:0802.1068].

Deffayet, C. (2001). *Phys. Lett. B* **502**, 199 [hep-th/0010186].

de Rham, C., Dvali, G., Hofmann, S., *et al.* (2008a). *Phys. Rev. Lett.* **100**, 251603 [arXiv:0711.2072].

de Rham, C., Hofmann, S., Khoury, J., and Tolley, A. J. (2008b). *JCAP* **0802**, 011 [arXiv:0712.2821].

Di Porto, C., and Amendola, L. (2008). *Phys. Rev. D* **77**, 083508 [arXiv:0707.2686].

Dolgov, A. D., and Kawasaki, M. (2003). *Phys. Lett. B* **573**, 1 [astro-ph/0307285].

Dore, O., *et al.* (2007). [arXiv:0712.1599].

Dunkley, J., *et al.* (WMAP Collaboration) (2009). *Astrophys. J. Suppl.* **180**, 306 [arXiv:0803.0586].

Durrer, R. (2005). *AIP Conf. Proc.* **782**, 202 [hep-th/0507006].

Durrer, R. (2008). *The Cosmic Microwave Background*, Cambridge: Cambridge University Press.

Durrer, R., and Maartens, R. (2008). In *Special Issue on Dark Energy*, G. F. R. Ellis *et al.*, eds. *Gen. Relat. Grav.* **40**, Nos. 2–3.

Dvali, G. R., Gabadadze, G., and Porrati, M. (2000). *Phys. Lett. B* **485**, 208 [hep-th/0005016].

Dvali, G. R., Hofmann, S., and Khoury, J. (2007). *Phys. Rev. D* **76**, 084006 [hep-th/0703027].

Ellis, G. F. R., and Maartens, R. (2004). *Class. Quant. Grav.* **21**, 223 [gr-qc/0211082].

Ellis, G., Maartens, R., and MacCallum, M. A. H. (2007). *Gen. Relat. Grav.* **39**, 1651 [gr-qc/0703121].

Ellis, G. F. R., Nicolai, H., Durrer, R., and Maartens, R. (Eds.) (2008). *Special Issue on Dark Energy*, *Gen. Relat. Grav.* **40**, Nos. 2–3.

Enqvist, K. (2008). In *Special Issue on Dark Energy*, G. F. R. Ellis *et al.*, Eds. *Gen. Relat. Grav.* **40**, Nos. 2–3.

Esposito-Farese, G. (2004). *AIP Conf. Proc.* **736**, 35 [gr-qc/0409081].

Fang, W., Wang, S., Hu, W., Haiman, Z., Hui, L., and May, M. (2008). *Phys. Rev. D* **78**, 103509 [arXiv:0808.2208].

Flanagan, E. E. (2005). *Phys. Rev. D* **71**, 103521 [hep-th/0503202].

Froissart, M. (1961). *Phys. Rev.* **123**, 1053.

Frolov, A. V. (2008). *Phys. Rev. Lett.* **101**, 061103 [arXiv:0803.2500].

Garcia-Bellido, J., and Haugboelle, T. (2009). *JCAP* **0909**, 028 [arXiv:0810.4939].

Geshnizjani, G., Chung, D. J. H., and Afshordi, N. (2005). *Phys. Rev. D* **72**, 023517 [astro-ph/0503553].

Giannantonio, T., Song, Y. S., and Koyama, K. (2008). *Phys. Rev. D* **78**, 044017 [arXiv:0803.2238].

Giovannini, M. (2005). *JCAP* **0509**, 009 [astro-ph/0506715].

Goodman, J. (1995). *Phys. Rev. D* **52**, 1821 [astro-ph/9506068].

Gorbunov, D., Koyama, K., and Sibiryakov, S. (2006). *Phys. Rev. D* **73**, 044016 [hep-th/0512097].

Gott, J. R. (1991). *Phys. Rev. Lett.* **66**, 1126.

Guo, Z. K, Ohta, N., and Tsujikawa, S. (2007). *Phys. Rev. D* **76**, 023508 [astro-ph/0702015].

Guzzo, L., *et al.* (2008). *Nature* **451**, 541 [arXiv:0802.1944].

Hirata, C. M., and Seljak, U. (2005). *Phys. Rev. D* **72**, 083501 [astro-ph/0503582].

Hu, W. (2008). *Phys. Rev. D* **77**, 103524 [arXiv:0801.2433].

Hu, W., and Sawicki, I. (2007a). *Phys. Rev. D* **76**, 064004 [arXiv:0705.1158].

Hu, W., and Sawicki, I. (2007b). *Phys. Rev. D* **76**, 104043 [arXiv:0708.1190].

Huterer, D., and Linder, E. V. (2007). *Phys. Rev. D* **75**, 023519 [astro-ph/0608681].

Ishak, M., Upadhye, A., and Spergel, D. N. (2006). *Phys. Rev. D* **74**, 043513 [astro-ph/0507184].

Ishibashi, A., and Wald, R. M. (2006). *Class. Quant. Grav.* **23**, 235 [gr-qc/0509108].

Itzykson, C., and Zuber, J. B. (1980). *Quantum Field Theory*, New York: McGraw Hill, Chapter 5.

Jain, B., and Zhang, P. (2008). *Phys. Rev. D* **78**, 063503 [arXiv:0709.2375].

Kobayashi, T., and Maeda, K. I. (2008). *Phys. Rev. D* **78**, 064019 [arXiv:0807.2503].

Kolb, E. W., Matarrese, S., Notari, A., and Riotto, R. (2005). [hep-th/0503117].

Koivisto, T. (2005). *Phys. Rev. D* **72**, 043516 [astro-ph/0504571].

Kowalski, M., *et al.* (the Supernova Cosmology Project) (2008). [arXiv:0804.4142].

Koyama, K. (2007). *Class. Quant. Grav.* **24**, R231 [arXiv:0709.2399].

Koyama, K. (2008). In *Special Issue on Dark Energy*, G. F. R. Ellis *et al.*, Eds. *Gen. Relat. Grav.* **40**, Nos. 2–3.

Koyama, K., and Maartens, R. (2006). *JCAP* **0610**, 016 [astro-ph/0511634].

Koyama, K., and Silva, F. P. (2007). *Phys. Rev. D* **75**, 084040 [hep-th/0702169].

Kunz, M. (2007). [astro-ph/0702615].

Kunz, M., and Sapone, D. (2007). *Phys. Rev. Lett.* **98**, 121301 [astro-ph/0612452].

Langlois, D. (2006). *Prog. Theor. Phys. Suppl.* **163**, 258 [hep-th/0509231].

Lasenby, A., and Doran, C. (2005). *Phys. Rev. D* **71**, 063502 [astro-ph/0307311].

Lazkoz, R., Maartens, R., and Majerotto, E. (2006). *Phys. Rev. D* **74**, 083510 [astro-ph/0605701].

Li, N., and Schwarz, D. J. (2007). *Phys. Rev. D* **76**, 083011 [gr-qc/0702043].

Linder, E. V. (2005). *Phys. Rev. D* **72**, 043529 [astro-ph/0507263].

Linder, E. (2008). In *Special Issue on Dark Energy*, G. F. R. Ellis *et al.*, Eds. *Gen. Relat. Grav.* **40**, Nos. 2–3.

Linder, E. V., and Cahn, R. N. (2007). *Astropart. Phys.* **28**, 481 [astro-ph/0701317].

Lue, A. (2006). *Phys. Rep.* **423**, 1 [astro-ph/0510068].

Lue, A., and Starkman, G. (2003). *Phys. Rev. D* **67**, 064002 [astro-ph/0212083].

Lue, A., and Starkman, G. D. (2004). *Phys. Rev. D* **70**, 101501 [astro-ph/0408246].

Lue, A., Scoccimarro, R., and Starkman, G. D. (2004). *Phys. Rev. D* **69**, 124015 [astro-ph/0401515].

Maartens, R. (2004). *Living Rev. Relat.* **7**, 7 [gr-qc/0312059].

Maartens, R., and Majerotto, E. (2006). *Phys. Rev. D* **74**, 023004 [astro-ph/0603353].

Mansouri, R. (2006). [astro-ph/0601699].

Martineau, P., and Brandenberger, R. (2005). [astro-ph/0510523].

Moffat, J. W. (2006). [astro-ph/0603777].

Morris, M. S., Thorne, K. S., and Yurtsever, U. (1988). *Phys. Rev. Lett.* **61**, 1446.

Mortonson, M. J., Hu, W., and Huterer, D. (2009). *Phys. Rev. D* **79**, 023004 [arXiv:0810.1744].

Nambu, Y., and Tanimoto, M. (2005). [gr-qc/0507057].

Nesseris, S., and Perivolaropoulos, L. (2007). *Phys. Rev. D* **75**, 023517 [astro-ph/0611238].

Nesseris, S., and Perivolaropoulos, L. (2008). *Phys. Rev. D* **77**, 023504 [arXiv:0710.1092].

Nojiri, S., and Odintsov, S. D. (2007). *Int. J. Geom. Meth. Math. Phys.* **4**, 115 [hep-th/0601213].

Nojiri, S., and Odintsov, S. D. (2007). *Phys. Lett. B* **657**, 238 [arXiv:0707.1941].

Ori, A. (2007). *Phys. Rev. D* **76**, 044002 [gr-qc/0701024].

Ostrogradski, M. (1850). *Mem. Acad. St. Petersbourg, Ser. VI* **4**, 385.

Oyaizu, H., Lima, M., and Hu, W. (2008). *Phys. Rev. D* **78**, 123524 [arXiv:0807.2462].

Padmanabhan, T. (2006). *AIP Conf. Proc.* **861**, 179 [astro-ph/0603114].

Padmanabhan, T. (2006). *Int. J. Mod. Phys. D* **15**, 2029 [gr-qc/0609012].

Paranjape, A., and Singh, T. P. (2006). *Class. Quant. Grav.* **23**, 6955 [astro-ph/0605195].

Perivolaropoulos, L. (2006). [astro-ph/0601014].

Polarski, D. (2006). *AIP Conf. Proc.* **861**, 1013 [astro-ph/0605532].

Polchinski, J. (2006). [hep-th/0603249].

Quartin, M., Calvao, O., Joras, S. E., Reis, R. R. R., and Waga, I. (2008). *JCAP* **0805**, 007 [arXiv:0802.0546].

Randall, L., and Sundrum, R. (1999). *Phys. Rev. Lett.* **83**, 4690 [hep-th/9906064].

Rasanen, S. (2006). *Class. Quant. Grav.* **23**, 1823

Riazuelo, A., and Uzan, J. P. (2002). *Phys. Rev. D* **66**, 023525 [astro-ph/0107386], [astro-ph/0504005].

Rosenthal, E., and Flanagan, E. E. (2008). [arXiv:0809.2107].

Ruiz-Lapuente, P. (2007). *Class. Quant. Grav.* **24**, R91 [arXiv:0704.1058].

Sahni, V. (2005). [astro-ph/0502032].

Sahni, V., and Shtanov, Y. (2003). *JCAP* **0311**, 014 [astro-ph/0202346].

Sawicki, I., Song, Y. S., and Hu, W. (2007). *Phys. Rev. D* **75**, 064002 [astro-ph/0606285].

Song, Y. S. (2005). *Phys. Rev. D* **71**, 024026 [astro-ph/0407489].

Song, Y. S. (2008). *Proceedings of the 14th Workshop on General Relativity and Gravitation* (JGRG14), 95 [astro-ph/0602598].

Song, Y. S., and Koyama, K. (2008). [arXiv:0802.3897].

Song, Y. S., Hu, W., and Sawicki, I. (2007). *Phys. Rev. D* **75**, 044004 [astro-ph/0610532].

Song, Y. S., Peiris, H., and Hu, W. (2007). *Phys. Rev. D* **76**, 063517 [arXiv:0706.2399].

Song, Y. S., Sawicki, I., and Hu, W. (2007). *Phys. Rev. D* **75**, 064003 [astro-ph/0606286].

Sotiriou, T. P., and Faraoni, V. (2008). [arXiv:0805.1726].

Starobinsky, A. A. (2007). *JETP Lett.* **86**, 157 [arXiv:0706.2041v2].

Straumann, N. (2006). *Mod. Phys. Lett. A* **21**, 1083 [hep-ph/0604231].

Thomas, S. A., Abdalla, F. B., and Weller, J. (2008). [arXiv:0810.4863].

Uzan, J. P. (2007). *Gen. Rel. Grav.* **39**, 307 [astro-ph/0605313].

Uzan, J. P., Clarkson, C., and Ellis, G. F. R (2008). *Phys. Rev. Lett.* **100**, 191303 [arXiv:0801.0068].

Vainshtein, A. I. (1972). *Phys. Lett. B* **39**, 393.

Valiviita, J., Majerotto, E., and Maartens, R. (2008). *JCAP* **0807**, 020 [arXiv:0804.0232].

van Dam, H., and Veltman, M. J. G. (1970). *Nucl. Phys. B* **22**, 397.

Vanderveld, R. A., Flanagan, E. E., and Wasserman, I. (2006). *Phys. Rev. D* **74**, 023506 [astro-ph/0602476].

Vanderveld, R. A., Flanagan, E. E., and Wasserman, I. (2007). *Phys. Rev. D* **76**, 083504 [arXiv:0706.1931].

Vanderveld, R. A., Flanagan, E. E., and Wasserman, I. (2008). *Phys. Rev. D* **78**, 083511 [arXiv:0808.1080]

Velo, G., and Zwanzinger, D. (1969a). *Phys. Rev.* **186**, 1337.

Velo, G., and Zwanzinger, D. (1969b). *Phys. Rev.* **188**, 2218.

Wands, D. (2006). [gr-qc/0601078].

Wang, S., Hui, L., May, M., and Haiman, Z. (2007). *Phys. Rev. D* **76**, 063503 [arXiv:0705.0165].

Wetterich, C. (2007). *Phys. Lett. B* **655**, 201 [arXiv:0706.4477].

Wiltshire, D. L. (2008). *Phys. Rev. D* **78**, 084032 [arXiv:0809.1183].

Woodard, R. P. (2007). *Lect. Notes Phys.* **720**, 403 [astro-ph/0601672].

Zakharov, V. I. (1970). *JETP Lett.* **12**, 312.

Zhang, P., Liguori, M., Bean, R., and Dodelson, S. (2007). *Phys. Rev. Lett.* **99**, 141302 [arXiv:0704.1932].

Zhang, P., Bean, R., Liguori, M., and Dodelson, S. (2008). [arXiv:0809.2836].

Zhao, G. B., Pogosian, L., Silvestri, A., and Zylberberg, J. (2009). *Phys. Rev. D* **79**, 083513 [arXiv:0809.3791].

Zibin, J. P., Moss, A., and Scott, D. (2008). *Phys. Rev. Lett.* **101**, 251303 [arXiv:0809.3761].

3

Some views on dark energy

DAVID POLARSKI

3.1 Introduction

The late-time accelerated expansion of our universe (Perlmutter *et al.*, 1997, 1998, 1999; Riess *et al.*, 1998; Astier *et al.*, 2006) is certainly a major challenge for cosmologists. It will durably affect the way we look at our universe and its future. It is interesting to recall the scientific context in which this accelerated expansion was discovered and promoted to a pillar of the present paradigm. Big Bang cosmology was spectacularly confirmed by the discovery of a remarkably homogeneous cosmic microwave background (CMB) possessing a perfect blackbody spectrum. Nucleosynthesis of the light elements is another of its successes. Important shortcomings of Big Bang cosmology were cured by the introduction of an inflationary stage in the early universe. Inflationary models are constrained by the primordial perturbations they produce, which leave their imprint on the CMB and eventually lead to the formation of cosmic structures through gravitational instability. The inflationary scenario found spectacular support in the detection of the tiny CMB angular anisotropies. These are in agreement with the simplest (single-field slow-roll) inflationary models. In particular, these anisotropies are in agreement with a spatially flat universe, a generic key prediction of inflationary models.

Intererestingly, earlier observations, such as the measurement of cosmic peculiar velocity fields made at the end of the eighties, pointed to a rather low content of dustlike matter, whether dark or baryonic, with $\Omega_{m,0} \leq 0.3$. At that time, this observation was often interpreted as putting the inflationary scenario, despite its beauty and simplicity, in a delicate situation. Indeed, if there is nothing else in our universe, putting aside a negligible amount of relativistic species today, then our universe is (very) open, at odds with inflation. The discovery of the present

Dark Energy: Observational and Theoretical Approaches, ed. Pilar Ruiz-Lapuente. Published by Cambridge University Press. © Cambridge University Press 2010.

accelerated expansion of the universe, together with the later confirmation of the spatial flatness of our universe (see the latest exquisite constraints on the spatial curvature released recently by WMAP (Wilkinson Microwave Anisotropy Probe) under certain assumptions like a constant dark energy equation of state: Komatsu *et al.*, 2008) was a welcome surprise: while it opened the door to dark energy, it also reconciled the inflationary scenario with the present low content of dustlike matter.

So the paradigm that has formed in the last decade is that the universe underwent two phases of accelerated expansion: the inflationary stage in the very early universe, and a late-time acceleration in which our universe entered only recently. Models trying to explain this late-time acceleration are dubbed dark energy (DE) models (Sahni and Starobinsky, 2000, 2006; Padmanabhan, 2003; Copeland, Sami and Tsujikawa, 2006; Ruiz-Lapuente, 2007; Durrer and Maartens, 2008). It is not the aim of this chapter to give an exhaustive review of all DE models. Rather, we would like to emphasize all the concepts that have emerged during the quest for the 'true' DE model.

3.2 Dark energy

In principle, the simplest way to account for late-time accelerated expansion, or for any non-standard expansion, of our universe is to introduce a new smooth component and we call this component dark energy (DE). The gravitational effect of DE can be extracted from the gravitational equations connecting the contents of the universe to its expansion. So in this first approach, our starting point is the standard Friedmann equations for a homogeneous and isotropic universe, as they appear in the context of general relativity, namely (we take $c = 1$)

$$\left(\frac{\dot{a}}{a}\right)^2 = \frac{8\pi G}{3} \sum_i \rho_i - \frac{k}{a^2}, \tag{3.1}$$

$$\frac{\ddot{a}}{a} = -\frac{4\pi G}{3} \sum_i (\rho_i + 3p_i), \tag{3.2}$$

where a dot stands for a derivative with respect to (cosmic) time t. Here the different components labelled i are all isotropic perfect fluids. The important point from (3.2) is that a component i can induce accelerated expansion provided $\rho_i + 3p_i < 0$, a necessary but not yet sufficient condition. Hence the lesson to be drawn here is that DE can be an isotropic perfect fluid provided it has a sufficiently negative pressure. We will give below an estimate of what is meant by sufficiently negative.

It is convenient to introduce the following dimensionless quantities

$$\Omega_i = \frac{\rho_i}{\rho_{\rm cr}}, \quad w_i \equiv \frac{p_i}{\rho_i}, \quad \Omega_k = -\frac{k}{a^2 H^2}, \quad k = 0, \pm 1, \tag{3.3}$$

with

$$H^2 \equiv \frac{8\pi G}{3} \rho_{\rm cr}, \tag{3.4}$$

in terms of which Eq. (3.1) becomes

$$\left(\frac{\dot{a}}{a}\right)^2 \equiv H^2 = H^2 \left(\sum_i \Omega_i + \Omega_k\right). \tag{3.5}$$

Written in this way, Eq. (3.1) appears clearly as a constraint equation, which takes the form *at all times*

$$\sum_i \Omega_i + \Omega_k = \Omega_m + \Omega_r + \Omega_{\rm de} + \Omega_k = 1, \tag{3.6}$$

if we assume three kinds of components, radiation (r), dustlike matter (m) and dark energy (de). If we introduce the deceleration parameter q, Eq. (3.2) can be recast as

$$q \equiv -\frac{\ddot{a}}{a H^2} = \frac{1}{2} \sum_{i \neq k} \Omega_i (1 + 3w_i). \tag{3.7}$$

At late times in a flat universe, Eq. (3.2) becomes

$$q \simeq \frac{1}{2}(1 + 3\, w_{\rm de}\, \Omega_{\rm de}). \tag{3.8}$$

Therefore it is straightforward to derive for accelerated (i.e. $q < 0$) expansion

$$w_{\rm de} < -\frac{1}{3}\, \Omega_{\rm de}^{-1}. \tag{3.9}$$

Observations imply that a very large part of the universe contents today is in the form of DE with $\Omega_{\rm de,0} \equiv \Omega_{\rm de}(z = 0) \simeq 0.7$. Hence, typically accelerated expansion today requires $w_{\rm de,0} < -0.5$ in a flat universe. In addition, if we have in mind that $\Omega_{\rm de}$ rapidly becomes small when going back in time, it is natural to expect from (3.9) that accelerated expansion is a late-time phenomenon. This leads to the problem of cosmic coincidence: Why is the transition to accelerated expansion taking place so close to us? Another question that arises immediately is about the identity of this mysterious DE component.

3.3 Dark energy models inside general relativity

3.3.1 Cosmological constant Λ

The simplest DE candidate is certainly a cosmological constant Λ. It is interesting that ΛCDM seems at the present time in good agreement with observational data on large cosmological scales. In this scenario we just have

$$\rho_\Lambda = \frac{\Lambda}{8\pi G} = -p_\Lambda. \tag{3.10}$$

In this case we have $w_\Lambda = -1$ and ρ_Λ is constant. As we must have $\Omega_{\Lambda,0} \simeq 0.7$, this poses immediately the problem of naturalness. Indeed, ρ_Λ is about $\rho_{cr,0}$ but this latter quantity is many orders of magnitude smaller than the energy density at the Planck scale, which is expected to be the natural order of magnitude of the vacuum energy density. One must understand why the vacuum energy density that arises from a more fundamental theory just gives us a very tiny non-vanishing piece. One can also view Λ as part of the classical gravitational action, but this seems less appealing.

3.3.2 Constant $w_{de} \neq -1$

Of course one can also consider DE with a constant equation of state, $w_{de} = $ constant, not necessarily equal to -1. This should be seen as a phenomenological approach that allows us to get insight into the effects of the introduction of DE (see e.g. Chevallier and Polarski, 2001). From (3.2), for a universe containing dustlike matter and some DE component, it is not hard to deduce that our universe is presently accelerating provided

$$w_{de} < -\frac{1}{3}\left(1 + \frac{\Omega_{m,0}}{\Omega_{de,0}}\right), \tag{3.11}$$

and in particular for a flat universe we recover the result (3.9) taken at $z = 0$.

For instance, for $\Omega_{m,0} = 0.3$, $\Omega_{de,0} = 0.7$, $w_{de} < -0.47$ is required. Therefore, present experimental evidence yielding $\Omega_{m,0} \sim 0.3$, combined with the location of the first acoustic peak in the CMB anisotropy first detected by balloon experiments and confirmed by later CMB data (see the recent WMAP data: Komatsu *et al.*, 2008) suggesting a nearly flat universe, imply that our universe would be presently accelerating for a wide range of constant values, roughly $w_{de} < -0.5$. Introducing the dimensionless quantity $x \equiv \frac{a}{a_0}$, accelerated expansion starts at x_a given by

$$x_a^{-3|w_{de}|} = (-1 + 3|w_{de}|)\frac{\Omega_{de,0}}{\Omega_{m,0}}, \tag{3.12}$$

which corresponds to redshifts

$$z_a = (-1 + 3|w_{de}|)^{\frac{1}{3|w_{de}|}} \left(\frac{\Omega_{de,0}}{\Omega_{m,0}} \right)^{\frac{1}{3|w_{de}|}} - 1. \tag{3.13}$$

For $-1 < w_{de} \rightarrow -1/3$, z_a is shifted towards smaller redshifts, for $(\Omega_{m,0}, \Omega_{de,0})$ $= (0.3, \ 0.7)$, we have $z_a = 0.671$ for a constant Λ-term, and $z_a = 0.414$ when $w_{de} = -0.6$. The fact that z_a is so close to zero is the cosmic coincidence problem mentioned above. Finally, these models can yield a significant increase, depending on w_{de}, $\Omega_{m,0}$, $\Omega_{de,0}$, of the age of the universe for given Hubble parameter H_0 compared to an Einstein–de Sitter universe, i.e. a flat universe with $\Omega_{m,0} = 1$ (Chevallier and Polarski, 2001). This was one of the early incentives to introduce a cosmological constant Λ: to reconcile a high value for H_0 with a sufficiently old universe.

3.3.3 Quintessence

Inspired by the great successes of the inflationary paradigm in which a homogeneous scalar field is so successful in implementing the inflationary stage of accelerated expansion, the DE component could well be a time-dependent minimally coupled scalar field $\phi(t)$ (Ratra and Peebles, 1988; Wetterich, 1988) called quintessence. As in inflationary models, what characterizes a quintessence model is its potential $V(\phi)$. Such a scalar field can be considered as a perfect fluid with

$$\rho_\phi = \frac{1}{2}\dot{\phi}^2 + V(\phi), \qquad p_\phi = \frac{1}{2}\dot{\phi}^2 - V(\phi), \tag{3.14}$$

and therefore the equation of state parameter w_ϕ is given by

$$w_\phi = \frac{\dot{\phi}^2 - 2V(\phi)}{\dot{\phi}^2 + 2V(\phi)}. \tag{3.15}$$

It is straightforward to see that

$$\rho_\phi + p_\phi = \dot{\phi}^2 \geq 0. \tag{3.16}$$

For $\rho_\phi \geq 0$, this implies that the equation of state must satisfy

$$w_\phi \geq -1. \tag{3.17}$$

In other words ϕ cannot be of the phantom type, where we mean by 'DE of the phantom type' a component satisfying $w_{de} < -1$. Closer inspection shows that we have

$$-1 \leq w_\phi \leq 1 \qquad V > 0, \tag{3.18}$$

$$1 \leq w_\phi < \infty \qquad V < 0. \tag{3.19}$$

Note that we have in particular in an expanding universe

$$\dot{\rho}_\phi \leq 0. \tag{3.20}$$

It is possible to have scaling solutions with $\rho_\phi \propto x^m$, $m = -3(1 + w_\phi) =$ constant (Liddle and Scherrer, 1999). However, this requires a very particular potential $V(\phi)$ for which,

$$V(\phi) = \frac{1 - w}{1 + w} \frac{\dot{\phi}^2}{2}. \tag{3.21}$$

Therefore, the most natural thing for quintessence is to have a varying equation of state; one, however, that cannot violate the weak energy condition as it must satisfy the inequality (3.17). Another interesting point is the way in which quintessence can handle the cosmic coincidence problem. It was shown that this problem is alleviated for some quintessence potentials V leading to tracking solutions such that a large range of initial conditions would all lead to the desired late time behaviour of the quintessence field (Zlatev, Wang and Steinhardt, 1999) but some fine tuning of the model is still required to solve the cosmic coincidence problem.

The models reviewed in this section, though of a different kind, are inside the framework of general relativity (GR). A further dramatic step is to consider models in which the laws of gravity are modified and to consider DE models outside the framework of GR. Before considering in detail some models of this kind, we review first the general formalism used in DE models.

3.4 General formalism

In this section we derive the basic equations related to DE and in particular to its equation of state. We assume our universe is isotropic with Friedmann–Lemaître–Robertson–Walker metric

$$ds^2 = dt^2 - a^2(t)\, d\ell^2. \tag{3.22}$$

Though we shall later apply our formalism to a flat universe, we consider in this section a non-flat universe as well. Its expansion dynamics is fully specified by the time evolution of the scale factor $a(t)$, which obeys the usual Friedmann equations (3.1, 3.2) of general relativity. The universe expansion can be probed through the measurement of the luminosity–distance $d_L(z)$ as a function of redshift z (remember that we take $c = 1$)

$$d_L(z) = (1 + z)\, H_0^{-1}\, |\Omega_{k,0}|^{-\frac{1}{2}}\, \mathcal{S}\left(|\Omega_{k,0}|^{\frac{1}{2}} \int_0^z \frac{dz'}{h(z')}\right). \tag{3.23}$$

Here $\mathcal{S}(u) = \sin u$ for a closed universe, $\mathcal{S}(u) = \sinh u$ for an open universe, while $\mathcal{S}$ is the identity for a flat universe. In Eq. (3.23), $h(z)$ stands for the (dimensionless) reduced Hubble parameter $h(z) \equiv \frac{H(z)}{H_0}$, $H_0 \equiv H(z=0)$, and neglecting radiation at small z it is given by

$$h^2(z) = \Omega_{m,0} (1+z)^3 + \Omega_{de,0} f(z) + \Omega_{k,0} (1+z)^2, \qquad (3.24)$$

with all quantities defined in (3.3). The function $f(z)$ expresses the a-priori unknown time evolution of DE

$$f(z) \equiv \frac{\rho_{de}(z)}{\rho_{de,0}}. \qquad (3.25)$$

It is directly related to the equation of state parameter $w_{de}(z) = \frac{p_{de}(z)}{\rho_{de}(z)}$ through the energy conservation equation for an isotropic perfect fluid

$$\frac{d\rho}{dt} = -3H(\rho + p). \qquad (3.26)$$

We get

$$f(z) = \exp\left[3 \int_0^z dz' \frac{1 + w_{de}(z')}{1 + z'}\right]. \qquad (3.27)$$

For constant equation of state, we recover the well-known result $f(z) = (1+z)^{3(1+w_{de})}$. We note also from (9.2) that when DE is of the phantom type ($w_{de} < -1$), its energy density decreases with increasing redshifts or increases in the course of time.

There has been enormous interest in models able to produce a phantom phase (see e.g. Feng, Wang and Zhang, 2005; Wei, Cai and Zeng, 2005) and in their general properties (Dabrowski, Stachowiak and Szydlowski, 2003). The observational signature for phantom DE at low redshifts is that the following inequality be satisfied

$$\frac{dh^2}{dz} < 3 \Omega_{m,0} (1+z)^2 + 2 \Omega_{k,0} (1+z). \qquad (3.28)$$

This condition is local: for any z where (3.28) is satisfied, we have phantom DE. As there are potential degeneracies in the ($\Omega_{k,0}, w$) plane (Gong and Zhang, 2005; Linder, 2005; Polarski and Ranquet, 2005; Clarkson, Cortes and Bassett, 2007; Ichikawa and Takahashi, 2007; Wang and Mukherjee, 2007), these can in particular falsify conclusions drawn from the data concerning the phantomness of DE.

Sometimes one defines the effective equation of state w_{eff} and we have for a flat space

$$\dot{H} = -\frac{3}{2} H^2 (1 + w_{eff}). \qquad (3.29)$$

This is as if all the right hand side is a perfect fluid defined by w_{eff}. When we have only dust and DE, we have the equality

$$w_{\text{eff}} = \Omega_{\text{de}}\, w_{\text{de}}, \tag{3.30}$$

from which we recover the condition (3.9) for accelerated expansion in a flat universe with DE and dustlike matter.

At this point we note that $w_{\text{de}}(z)$ contains all the information that is needed in order to find $h(z)$ from (3.24) so $w_{\text{de}}(z)$ is actually a central property of DE models. One essentially phenomenological approach has been to parameterize $w_{\text{de}}(z)$. In this way a function is replaced by a finite number of parameters. Of course, it is assumed that such a parameterization is a good approximation for a large class of DE models and that it is a reasonable fit for the true DE model. It is also a very useful tool to design experiments for the study of DE. The following two-parameters parameterization has been widely used (Chevallier and Polarski, 2001; Linder, 2003):

$$w_{\text{de}}(z) = (-1 + \alpha) + \beta\,(1 - x) \equiv w_0 + w_a\, \frac{z}{1+z}. \tag{3.31}$$

The same equation of state can also be represented in the following way:

$$w_{\text{de}}(z) = (-1 + \alpha_p) + \beta(x_p - x), \tag{3.32}$$

with $w_{\text{de}}(z_p) = -1 + \alpha_p$ and same parameter β. This can be more convenient if we believe there is some pivot z_p where $w_{\text{de}}(z_p)$ is best constrained observationally. Note that we have for (3.31)

$$f(z) = (1 + z)^{3(\alpha+\beta)} e^{-3\beta \frac{z}{1+z}}. \tag{3.33}$$

Amusingly this was originally introduced in Chevallier and Polarski (2001) to quantify departures from a cosmological constant, which is still very topical in view of the latest data.

If we had ideal data we could just compute $w_{\text{de}}(z)$ from the observations. We measure the apparent magnitude, or equivalently the distance modulus, of SN Ia assumed to be standard candles and from this we derive the luminosity distance $d_L(z)$ as a function of redshift. A first differentiation will give us $h(z)$:

$$h^{-1}(z) = \left(\frac{D_L(z)}{1+z}\right)' \left(1 + \left(\frac{D_L(z)}{1+z}\right)^2 \Omega_{k,0}\right)^{-\frac{1}{2}}, \tag{3.34}$$

where a prime denotes differentiation with respect to z and we have introduced the dimensionless Hubble-free luminosity distance $D_L(z) \equiv H_0\, d_L(z)$. The expression for $w_{de}(z)$ is found after a second differentiation:

$$w_{de}(z) = \frac{\dfrac{1+z}{3}\dfrac{dh^2}{dz} - h^2 + \dfrac{1}{3}\Omega_{k,0}(1+z)^2}{h^2 - \Omega_{m,0}(1+z)^3 - \Omega_{k,0}(1+z)^2}. \tag{3.35}$$

We see that $w_{de,0} \equiv w_{de}(z=0)$ is found from

$$w_{de,0} = \frac{\dfrac{1}{3}\dfrac{dh^2}{dz}|_{z=0} - 1 + \dfrac{1}{3}\Omega_{k,0}}{\Omega_{de,0}}, \tag{3.36}$$

where $\Omega_{de,0}$ can be obtained from $\Omega_{m,0}$ *and* $\Omega_{k,0}$, namely

$$\Omega_{de,0} = 1 - \Omega_{m,0} - \Omega_{k,0}. \tag{3.37}$$

Equations (3.34)–(3.36) apply to *arbitrary* equations of state and spatial curvature and allow for a determination of $w_{de}(z)$ from the Hubble diagram $h(z)$. As we shall see later, though we have derived all the equations of this section having in mind GR, they are easily extended to DE models outside GR.

3.5 Observational constraints

Before considering some specific DE models outside GR, we list here briefly some of the observations that can be used in order to constrain DE models. In order to assess the viability of models, the following observations are central: supernova data, BAO (baryonic acoustic oscillations) data and CMB data. Of course, many more observations will be efficiently used to make further progress (see e.g. Upadhye, Ishak and Steinhardt, 2005).

For the SN Ia data we can compare the observed distance modulus $\mu_{obs,i}$ for each supernova i with the theoretical prediction of a given model:

$$\mu_{th,i} = 5 \log \left((1+z_i) \int_0^{z_i} \frac{dz}{h} \right) + \mu_0, \tag{3.38}$$

where $\mu_0 = 25 + 5\ \log \left(\frac{cH_0^{-1}}{\text{Mpc}} \right)$. The distance modulus μ is the difference between the apparent magnitude m and the absolute magnitude M. For this purpose we can use a sample consisting of 192 supernovae (Davis *et al.*, 2007; Riess *et al.*, 2007; Wood-Vasey *et al.*, 2007). We get rid of the nuisance parameter H_0

using the simple way suggested by Di Pietro and Claeskens (2003), Nesseris and Perivolaropoulos (2004) and Lazkoz, Nesseris and Perivolaropoulos (2005); integrating over H_0 gives essentially the same result. However the two approaches are different in spirit (frequentist vs. Bayesian approach).

Because the baryons and the photons are coupled before recombination, baryons oscillate together with radiation before they decouple. The scale of the radiation oscillation is seen from the CMB acoustic oscillations. These oscillations were also detected in the galaxy power spectrum at low redshift. They allow us to relate the acoustic oscillations scale in the CMB at decoupling and at low redshift and therefore give us information on the expansion history. The BAO constraints can be expressed as a constraint on the quantity A:

$$A(z) = \frac{\sqrt{\Omega_{m,0}}}{z} \left[\frac{z}{h(z)} \left(\int_0^z dz' \frac{1}{h(z')} \right)^2 \right]^{\frac{1}{3}}, \tag{3.39}$$

with (Eisenstein *et al.*, 2005)

$$A = 0.469 \pm 0.017. \tag{3.40}$$

Note more recent intriguing data related to BAO oscillations (Percival *et al.*, 2007) that could even trouble the ΛCDM paradigm.

We have also a constraint from the CMB on the shift parameter R, which in a flat space reduces to the expression

$$R = \sqrt{\Omega_{m,0}} \int_0^{1089} \frac{dz}{h(z)}, \tag{3.41}$$

with (Wang and Mukherjee, 2006)

$$R = 1.70 \pm 0.03. \tag{3.42}$$

The shift parameter is perhaps the least model-dependent quantity that can be extracted from the CMB, hence its interest.

A simple and straightforward way to use these data is to perform a χ^2 analysis. If we use only the above constraints, we have to maximize the probability density function (pdf)

$$P \propto e^{-\frac{1}{2} \chi^2}, \tag{3.43}$$

where $\chi^2 = \chi_{SN}^2 + \chi_A^2 + \chi_R^2$ for a given cosmological model. The pdf P is defined on the measured quantities and depends on the parameters of the cosmological model considered. We have respectively:

$$\chi^2_{SN} = \sum_{i=1}^{192} \frac{(\mu_{th,i} - \mu_{exp,i})^2}{\sigma_i^2}, \tag{3.44}$$

$$\chi^2_A = \frac{(A(z = 0.35) - 0.469)^2}{(0.017)^2}, \tag{3.45}$$

$$\chi^2_R = \frac{(R - 1.7)^2}{(0.03)^2}. \tag{3.46}$$

As there are large degeneracies between $\Omega_{m,0}$ and the parameters entering the DE equation of state, we also need independent additional constraints on $\Omega_{m,0}h^2$ taken from 2dF and SDSS combined with the value H_0 from HST Cepheids, which gives $\Omega_{m,0} = 0.28 \pm 0.04$.

Let us say a few words about the statistical aspects of the problem (see e.g. Verde, 2007, for a practical review). We are interested in the information that can be extracted on the model parameters θ_i from the data x_j. This is the subject of statistical inference and many results and approaches were developed long ago. The observed quantities (data) x_j, for example $\mu_{i,obs}$, obey a probability density function $P(x_j, \theta_i)$ which depends on parameters θ_i, here the parameters defining our cosmological model. We want to go in the opposite way, namely to find the most likely values of the model parameters θ_j given the data. In this approach, $P(x_j, \theta_i)$ is viewed as a likelihood function $\mathcal{L}(\theta_i, x_j)$ on the model parameters θ_i. Maximizing $\mathcal{L}(\theta_i, x_j)$, we find the most likely values of θ_i: this is the maximum likelihood approach.

The likelihood function is well understood within the Bayesian approach where one introduces probabilities for the models themselves (dropping indices):

$$P(\theta|x) = \frac{P(x|\theta)\, P(\theta)}{P(x)} \tag{3.47}$$

$$= \mathcal{L}(\theta, x)\frac{P(\theta)}{P(x)}, \tag{3.48}$$

where $P(\theta|x)$, the conditional probability of a model given the data, is called the posterior distribution, and $P(\theta)$ is called the prior. So if we set $P(x) = 1$ and we assume uniform prior, maximizing the likelihood function $\mathcal{L}(\theta, x)$, we find the 'best model'. In many cases, due to the central limit theorem, $\mathcal{L}(\theta, x)$ is well approximated as a multivariate Gaussian distribution and very often this ansatz is used in practice. For example, assuming all observables are independent and Gaussian then we can write $- \ln P(x|\theta) = \frac{1}{2}\chi^2$ and maximizing $\mathcal{L}(\theta, x)$ is equivalent to minimizing χ^2. The point in parameter space maximizing $\mathcal{L}$ is called the maximum likelihood point $\bar{\theta}$. A way to find out how much information we can have on the θ_i is through the Fisher information matrix F_{ij}. With $L = -\ln \mathcal{L}$, the definition reads

$$F_{ij} \equiv \left\langle \frac{\partial^2 L}{\partial \theta_i \, \partial \theta_j} \right\rangle. \tag{3.49}$$

Assuming that $\mathcal{L}$ is a multivariate Gaussian centred around $\bar{\theta}$, the Fisher matrix is just the expectation value of the inverse of the covariance matrix computed at the maximum likelihood point $\bar{\theta}$. Fisher called his matrix the information matrix, it really tells us how much information we have on the model parameters θ_i: for an unbiased estimator, we have the Cramer–Rao inequality:

$$\sigma_{\theta_i} \geq \frac{1}{\sqrt{F_{ii}}}, \tag{3.50}$$

which gives the minimal uncertainty on θ_i, hence larger F_{ii} means more information on θ_i.

We should also mention that the analysis described above looks for the best parameters θ_i inside a given model. In order to assess whether data are able to discriminate between different models the Bayes factor can be used (see e.g. Kitching *et al.*, 2008). It is interesting that the traditional debate between the frequentist and the Bayesian approaches has become the subject of intense debates in DE studies.

Finally, we have listed here a few constraints on the background evolution. Actually, as we shall see for DE models outside GR, the evolution of the matter perturbations may well yield a decisive test and it will be considered separately.

3.6 Dark energy models outside general relativity

We now consider two classes of DE models outside GR. In some sense, considering such models is quite unavoidable even though it can be seen rightly as a drastic departure from conventional cosmology. Indeed, when we talk about dark components, and this is also true for dark matter, we mean a component whose presence is only felt gravitationally. Clearly, there are two ways to deal with such gravitational effects. Either we postulate the existence of a 'dark' component, here dark energy, without changing the theory of gravity. Another possible solution is to postulate a new theory of gravity capable of accelerating the expansion of the universe at recent times. After all, DE is felt only gravitationally. In some models, like scalar–tensor DE models (Bartolo and Pietroni, 2000; Boisseau *et al.*, 2000; Perrotta, Baccigalupi and Matarrese, 2000; Esposito-Farèse and Polarski, 2001; Fujii and Maeda, 2003; Faraoni, 2004; Gannouji *et al.*, 2006; Martin, Schimd and Uzan, 2006; Barenboim and Lykken, 2007; Capozziello *et al.*, 2007; Demianski *et al.*, 2007) the new gravity theory implies the presence of a scalar field, the scalar partner of the graviton that mediates the gravitational interaction. But in other models, as in $f(R)$ DE models, there is no additional physical degree of freedom in the form of a fundamental scalar field. Sometimes one refers to these models as modified gravity DE models. We shall discuss these two DE models, putting the emphasis on the lessons to be drawn for general modified gravity DE models.

3.6.1 Scalar–tensor DE models

We consider a model where gravity is described by a scalar–tensor theory with the following microscopic Lagrangian density in the Jordan frame (JF):

$$L = \frac{1}{2}\left(F(\Phi)\, R - Z(\Phi)\, g^{\mu\nu}\partial_\mu\Phi\,\partial_\nu\Phi\right) - U(\Phi) + L_m(g_{\mu\nu}). \tag{3.51}$$

As we do not couple L_m to Φ, the Jordan frame is the physical frame. In particular, fermion masses are constant and atomic clocks measure the proper time t in this frame. The quantity $Z(\Phi)$ can be set to either 1 or -1 by a redefinition of the field Φ, apart from the exceptional case $Z(\Phi) \equiv 0$ when the scalar–tensor theory (3.51) reduces to the higher-derivative gravity theory $R + f(R)$ that we shall consider in the next subsection. Below, we write all equations and quantities for $Z = 1$ in the Jordan frame. At low redshifts ($z \ll z_{\rm eq}$, $z_{\rm eq}$ is the equality redshift when the energy densities of non-relativistic matter and radiation are equal), L_m describes non-relativistic dustlike matter, whether baryons or cold dark matter (CDM). In such a model, the effective Newton gravitational constant for homogeneous cosmological models is given by

$$G_N = (8\pi F)^{-1}. \tag{3.52}$$

As could be expected, G_N does not have the same physical meaning as in general relativity. The effective gravitational constant $G_{\rm eff}$ for the attraction between two test masses is given by

$$G_{\rm eff} = G_N\, \frac{F + 2(\mathrm{d}F/\mathrm{d}\Phi)^2}{F + \frac{3}{2}(\mathrm{d}F/\mathrm{d}\Phi)^2} = G_N\, \frac{1 + 2\omega_{\rm BD}^{-1}}{1 + \frac{3}{2}\omega_{\rm BD}^{-1}}, \tag{3.53}$$

on all scales for which the field Φ is effectively massless. The stability conditions of the theory imply $G_{\rm eff} > 0$, $F > 0$ (see Gannouji et al., 2006 for more details).

Specializing to a spatially flat FLRW universe, the background equations are

$$3FH^2 = \rho_m + \frac{\dot\Phi^2}{2} + U - 3H\dot F, \tag{3.54}$$

$$-2F\dot H = \rho_m + \dot\Phi^2 + \ddot F - H\dot F. \tag{3.55}$$

The equation of motion of the scalar field Φ is contained in (3.54) and (3.55).

As in all DE models where the Friedmann equations are modified, a meaningful definition of energy density and pressure of the DE sector requires some care (Gannouji *et al.*, 2006; see also Torres, 2002). So we *define* the energy density ρ_{de} and the pressure p_{de} in the following way:

$$3F_0 H^2 = \rho_m + \rho_{de}, \tag{3.56}$$

$$-2F_0 \dot{H} = \rho_m + \rho_{de} + p_{de}. \tag{3.57}$$

This just corresponds to the representation of the true equation for scalar–tensor gravity in the *Einsteinian* form with the constant $G_{N,0}$:

$$R_{\mu\nu} - 12R\, g_{\mu\nu} = 8\pi\, G_{N,0} \left(T_{\mu\nu,m} + T_{\mu\nu,de} \right). \tag{3.58}$$

With these definitions, the usual conservation equation applies and we recover the general formalism derived in Section 3.4. In this way the data can be treated in the same way for all DE models whether inside or outside GR. What changes is the physics behind the symbols. For example, the condition for late-time accelerated expansion in a flat universe is again (3.9). Expressed as a function of the microscopic quantities it reads

$$\rho_m + 2\dot{\Phi}^2 - 2U + 3\ddot{F} + 3H\dot{F} < 0. \tag{3.59}$$

Note that we recover the condition for quintessence when $\dot{F} = 0$. As was first emphasized in Boisseau *et al.* (2000), scalar–tensor models allow for phantom DE and this is an important incentive to consider these models. The condition for DE to be of the phantom type, $w_{de} < -1$, can be obtained from (3.35) and it is simply condition (3.28). Expressed as a function of the quantities in the Lagrangian, it is

$$\rho_{de} + p_{de} = \dot{\Phi}^2 + \ddot{F} - H\dot{F} + 2(F - F_0)\,\dot{H} < 0. \tag{3.60}$$

Note that this condition cannot hold for quintessence ($F = F_0$).

A second point concerns the gravitational constraints. In any DE model where gravity is modified, one must make sure that at least solar system gravity constraints are satisfied. This aspect has attracted a lot of interest and is inevitably present in scalar–tensor DE models. Solar system experiments constrain the two post-Newtonian parameters γ_{PN} and β_{PN} *today* that parameterize these theories in the weak field limit:

$$\gamma_{PN} = 1 - \frac{(dF/d\Phi)^2}{F + 2(dF/d\Phi)^2}, \tag{3.61}$$

$$\beta_{\text{PN}} = 1 + \frac{1}{4} \frac{F \, (\mathrm{d}F/\mathrm{d}\Phi)}{2F + 3(\mathrm{d}F/\mathrm{d}\Phi)^2} \frac{\mathrm{d}\gamma}{\mathrm{d}\Phi}, \tag{3.62}$$

as well as the quantity $\frac{\dot{G}_{\text{eff},0}}{G_{\text{eff},0}}$. The best present bounds are

$$\gamma_{\text{PN}} - 1 = (2.1 \pm 2.3) \cdot 10^{-5},$$

$$\beta_{\text{PN}} - 1 = (0 \pm 1) \cdot 10^{-4},$$

$$\frac{\dot{G}_{\text{eff},0}}{G_{\text{eff},0}} = (-0.2 \pm 0.5) \cdot 10^{-13} \, y^{-1}, \tag{3.63}$$

where the first bound was obtained from the Cassini mission (Bertotti, Iess and Tortora, 2003) and the other two from high precision ephemerides of planets (Pitjeva, 2005a,b). (The second bound has been recently confirmed by the Lunar Laser ranging by Williams, Turyshev and Boggs (2005) – their value is $\beta_{\text{PN}} - 1 = (1.2 \pm 1.1) \cdot 10^{-4}$.)

As a consequence of the smallness of $\gamma_{\text{PN}} - 1$, the Brans–Dicke parameter ω_{BD} satisfies *today* the inequality

$$\omega_{\text{BD},0} > 4 \times 10^4. \tag{3.64}$$

Hence we have from (3.53) $G_{\text{N},0} \approx G_{\text{eff},0}$, the gravitational constant measured in a Cavendish-type experiment; it is therefore very close to the usual Newton's constant G.

The basic microscopic functions $F(\Phi)$ and $U(\Phi)$ can be expressed as functions of z and systematically expanded into Taylor series in z (Gannouji *et al.*, 2006):

$$\frac{F(z)}{F_0} = 1 + F_1 \, z + F_2 \, z^2 + \cdots > 0, \tag{3.65}$$

$$\frac{U(z)}{3F_0 \, H_0^2} \equiv \Omega_{U,0} u = \Omega_{U,0} + u_1 \, z + u_2 \, z^2 + \cdots, \tag{3.66}$$

with $\Omega_{U,0} \equiv \frac{U_0}{3F_0 H_0^2}$ and $u = \frac{U}{U_0}$.

From (3.65) and (3.66), all other expansions can be derived, in particular:

$$w_{\text{de}}(z) = w_0 + w_1 \, z + w_2 \, z^2 + \cdots, \tag{3.67}$$

$$H_0^{-1} \frac{\dot{G}_{\text{eff}}}{G_{\text{eff}}} = g_0 + g_1 \, z + g_2 \, z^2 + \cdots. \tag{3.68}$$

It can be shown that the condition $|F_1| \ll 1$ is sufficient to ensure that all solar system constraints are satisfied (Gannouji *et al.*, 2006).

For $|F_1| \ll 1$, we can express all coefficients as functions of the post-Newtonian parameters γ, β and g_0 (we drop the subscript PN). The following results are obtained:

$$F_1 = g_0 \frac{\gamma - 1}{\gamma - 1 - 4(\beta - 1)}, \tag{3.69}$$

$$F_2 = -2\, g_0^2 \frac{\beta - 1}{[\gamma - 1 - 4(\beta - 1)]^2}, \tag{3.70}$$

$$\Omega_{de,0} - \Omega_{U,0} = -\frac{1}{6}\, g_0^2 \frac{\gamma - 1}{[\gamma - 1 - 4(\beta - 1)]^2}, \tag{3.71}$$

$$1 + w_{de,0} = -\frac{1}{3}\, g_0^2 \frac{4(\beta - 1) + \gamma - 1}{\Omega_{de,0}\,[\gamma - 1 - 4(\beta - 1)]^2}. \tag{3.72}$$

We see from the last equality that $w_{de,0}$ can be related to the post-Newtonian parameters; however, in view of (3.72) the prospects to measure its possible phantomness are certainly better with cosmological data.

Note that the relation between the Hubble parameter and the luminosity distance in scalar–tensor gravity is the same as in GR. However, the relation between the distance modulus and the luminosity distance is changed (Amendola, Corasaniti and Occhionero, 1999; Garcia-Berro *et al.*, 1999; Riazuelo and Uzan, 2002). Because the effective gravitational constant is varying, the connection between the quantities μ and d_L becomes more complicated than that given by Eq. (3.38). In scalar–tensor DE models, we should use

$$\mu_{th,i} = 5 \log \left((1 + z_i) \int_0^{z_i} \frac{dz}{h} \right) + \mu_0 + \frac{15}{4} \log \frac{G_{eff}(z_i)}{G_{eff,0}}. \tag{3.73}$$

The addition of the last term in Eq. (3.73) takes into account the variation of the effective gravitational constant. This term allows us to discriminate different scalar–tensor models using SN Ia data even for similar background expansion (see e.g. Gannouji and Polarski, 2008).

3.6.2 f(R) modified gravity DE models

We consider now the following Lagrangian density in the JF:

$$L = \frac{1}{2} f(R) + L_{rad} + L_m, \tag{3.74}$$

where $f(R)$ is some arbitrary function of the Ricci scalar R, and L_m and L_{rad} are the Lagrangian densities of dustlike matter and radiation respectively. Then the following equations are obtained:

$$3FH^2 = \rho_m + \rho_{\text{rad}} + \frac{1}{2}(FR - f) - 3H\dot{F}, \tag{3.75}$$

$$-2F\dot{H} = \rho_m + \frac{4}{3}\rho_{\text{rad}} + \ddot{F} - H\dot{F}, \tag{3.76}$$

where

$$F \equiv \frac{\mathrm{d}f}{\mathrm{d}R}. \tag{3.77}$$

In standard Einstein gravity one has $F = $ constant. We note that Eqs. (3.75) and (3.76) are similar to Eqs. (3.54) and (3.55) with a vanishing Brans–Dicke parameter $\omega_{\text{BD}} = 0$ and a specific potential $U = (FR - f)/2$ and we include here a radiation component. We can define DE, its energy density and pressure in a way similar to Eqs. (3.56) and (3.57):

$$3H^2 = \rho_{\text{de}} + \rho_m + \rho_{\text{rad}}, \tag{3.78}$$

$$-2\dot{H} = \rho_m + \frac{4}{3}\rho_{\text{rad}} + \rho_{\text{de}} + p_{\text{de}}, \tag{3.79}$$

leading to the equalities

$$\rho_{\text{de}} = \frac{1}{2}(FR - f) - 3H\dot{F} + 3H^2(1 - F), \tag{3.80}$$

$$p_{\text{de}} = \ddot{F} + 2H\dot{F} - \frac{1}{2}(FR - f) - (2\dot{H} + 3H^2)(1 - F), \tag{3.81}$$

in models with $G_* \approx G_{\text{eff},0} \approx G$, where G_* is the bare gravitational constant; we use here the convention $8\pi G_* = 1$ (see the discussion in Moraes *et al.*, 2009). Again, these theories admit a late-time accelerated expansion (Capozziello, 2002) and even a phantom phase. Actually, a model of this type was first introduced in the context of inflation by Starobinsky (1980). In that case, inflation is followed by standard cosmology, while in the context of DE, the order is inverted and as we will see this has surprising consequences. Here too one can define a meaningful DE equation of state with $w_{\text{de}} \equiv \frac{p_{\text{de}}}{\rho_{\text{de}}}$ for which the general formalism applies. However, these systems exhibit a very surprising property: many $f(R)$ models cannot produce a standard matter-dominated stage $a \propto t^{\frac{2}{3}}$ before the late-time accelerated expansion (Amendola, Polarski and Tsujikawa, 2007). In other words, they fail to have a viable cosmic expansion history at large redshifts. This came as a surprise because, for example, models of the kind $R + bR^{-1}$ (Carroll *et al.*, 2004) were naively thought to follow the usual expansion in the past when R is large, while the R^{-1} term would produce the accelerated expansion when it comes to dominate at late times. This unexpected property triggered renewed interest in the surprising properties of $f(R)$ models. This effect was discovered essentially with numerical calculations, but later a general scheme was suggested to study $f(R)$ models as a dynamical system for which one can find its phase portrait.

One can introduce the following (dimensionless) variables (Amendola *et al.*, 2007):

$$x_1 = -\frac{\dot{F}}{HF},\tag{3.82}$$

$$x_2 = -\frac{f}{6FH^2},\tag{3.83}$$

$$x_3 = \frac{R}{6H^2} = \frac{\dot{H}}{H^2} + 2,\tag{3.84}$$

$$x_4 = \frac{\rho_{\text{rad}}}{3FH^2}.\tag{3.85}$$

From Eq. (3.75) we have the algebraic identity

$$\frac{\rho_{\text{m}}}{3FH^2} = 1 - x_1 - x_2 - x_3 - x_4.\tag{3.86}$$

It is then straightforward to obtain the following autonomous system of equations

$$\frac{dx_1}{dN} = -1 - x_3 - 3x_2 + x_1^2 - x_1 x_3 + x_4,\tag{3.87}$$

$$\frac{dx_2}{dN} = \frac{x_1 x_3}{m} - x_2(2x_3 - 4 - x_1),\tag{3.88}$$

$$\frac{dx_3}{dN} = -\frac{x_1 x_3}{m} - 2x_3(x_3 - 2),\tag{3.89}$$

$$\frac{dx_4}{dN} = -2x_3 x_4 + x_1 x_4,\tag{3.90}$$

where N stands for $\ln a$ and

$$m \equiv \frac{d\log F}{d\log R} = \frac{RF_{,R}}{F},\tag{3.91}$$

$$r \equiv -\frac{d\log f}{d\log R} = -\frac{RF}{f} = \frac{x_3}{x_2},\tag{3.92}$$

where $F_{,R} \equiv dF/dR$. Inverting Eq. (3.92), we can express R as a function of x_3/x_2 that can be substituted in (3.91) to obtain m as a function of $x_3/x_2 = r$.

When $f(R) = \alpha R^{-n}$, $m = -n-1$ is a constant with $r = n = x_3/x_2$ and the system simplifies a lot and reduces to a three-dimensional one for the variables x_1, x_2 and x_4. However, for general $f(R)$ gravity models, m depends upon r. One can find conditions on $m(r)$ in the (m, r) plane in order to have a standard matter-dominated stage followed by a late-time accelerated expansion. In this way specific regions in the (m, r) plane are singled out. For viable models the $m(r)$ curve must be able to connect these regions and to satisfy the needed constraints. Many $f(R)$ are unable to satisfy these conditions. However, it is important to mention that some $f(R)$ models are still viable (Appleby and Battye, 2007; Hu and Sawicki, 2007; Starobinsky, 2007, and the discussion therein).

3.7 Reconstruction

The reconstruction of DE models is also a very instructive problem; it tells us how many 'ideal' observations we need in order to recover the DE model. We consider here quintessence and scalar–tensor models.

3.7.1 Quintessence

Let us recall the Friedmann equations for quintessence

$$3H^2 = 8\pi G\left(\rho_m + \frac{\dot{\Phi}^2}{2} + V\right),$$

$$\dot{H} = -4\pi G\left(\rho_m + \dot{\Phi}^2\right).$$

Combining both equations and transforming to the variable z, we can easily obtain

$$8\pi G V(z) = 3H^2(z) - \frac{1+z}{2}\frac{dH^2}{dz} - \frac{3}{2}\Omega_{m,0}H_0^2(1+z)^3. \qquad (3.93)$$

It is clear from (3.93) that we can reconstruct $V(z)$ if we have $H(z)$ (and the cosmological parameter $\Omega_{m,0}$) (Starobinsky, 1998). We can get $H(z)$ from the luminosity distances $H(z)^{-1} = \left(\frac{d_L(z)}{1+z}\right)'$, where a prime denotes a derivative with respect to z. We can then also find in principle $z = z(\Phi - \Phi_0)$ and therefore $V(\phi - \phi_0)$. Note the inequality

$$\frac{dH^2(z)}{dz} \geq 3\Omega_{m,0}H_0^2(1+z)^2, \qquad (3.94)$$

showing again that quintessence cannot be of the phantom type.

It should be stressed that one can also reconstruct the model using the matter perturbations (Starobinsky, 1998). In that case one reconstructs $H(z)$ from the matter

perturbations $\delta_m(z)$ (and the cosmological parameter H_0). This is a very important point because the reconstruction from the matter perturbations can give a way
to decide whether a DE model is inside or outside GR. Indeed, using the $\delta_m(z)$
data, we will reconstruct the correct $H(z)$ only if the dynamics of the matter perturbations corresponds to the one we have assumed. For the same Hubble diagram
$H(z)$ (same kinematics), two models inside and outside GR will give diffrent $\delta_m(z)$
because the evolution of matter perturbations depends on the prevailing gravitational dynamics (Ishak, Upadhye and Spergel, 2006; Polarski, 2006; Bludman,
2007). This opens a very interesting way to probe the nature of DE models: inside
or outside GR. We will return to this important point below.

3.7.2 Scalar–tensor DE models

Eliminating the quantity $\dot{\Phi}^2$ from (3.54) and (3.55), we obtain a master equation for the quantity F that takes the form, when all quantities are expressed as
functions of z:

$$
F'' + \left[(\ln h)' - \frac{4}{1+z} \right] F' + \left[\frac{6}{(1+z)^2} - \frac{2(\ln h)'}{1+z} \right] F
$$
$$
= \frac{6u}{(1+z)^2 \, h^2} F_0 \, \Omega_{U,0} + 3 \, (1+z) \, h^{-2} F_0 \, \Omega_{m,0}. \tag{3.95}
$$

The full reconstruction of the functions $F(\Phi)$, $U(\Phi)$ requires two independent
types of observations: $H(z)$ and the matter perturbations $\delta_m \equiv \delta\rho_m(z)/\rho_m(z)$ at
some comoving scale deep inside the Hubble radius.

As emphasized in Esposito-Farèse and Polarski (2001), powerful constraints can
be obtained even when one uses only the background expansion, and not the perturbations. This is possible when some additional condition is imposed on either
F or U; we can call this a partial reconstruction procedure. Maybe the most interesting case is when $U = 0$. When U vanishes, we can of course use Eq. (3.95) and
find $F(z)$ for given $H(z)$. Assuming the expansion obeys the law

$$
h^2 = 0.7 + 0.3(1+z)^3, \tag{3.96}
$$

we can use Eq. (3.95), substituting $U = 0$ in it, to recover the corresponding
function $F(z)$. Inside GR one would interpret (3.96) as a flat universe with a cosmological constant and $\Omega_{m,0} = 0.3$. We take (3.96) as a kinematical law while the
dynamics is given by scalar–tensor gravity with $U = 0$. As U vanishes, we intu-
itively expect that $F(z)$ goes 'crazy' if the model is to obey (3.96). The surprise,
however, is that it does so even at very low redshifts. Actually, as was shown in
Esposito-Farèse and Polarski (2001), $F(z)$ vanishes at a redshift $z \approx 0.66$ in a flat

universe, making the theory singular at redshifts that we can probe. Hence, cosmological observations rule out scalar–tensor gravity with a non-vanishing potential U. Remember that the solar system constraints are satisfied because of the smallness of the derivative of F today. The non-viability when $U = 0$ can be shown because we look at different times.

Let us now return to the general case where there are two non-vanishing functions F and U. In that case Eq. (3.95) must be supplemented by a second independent equation. This can be done if we use the equation for the matter perturbations and the $\delta_m(z)$ data.

We consider the perturbation equations in the longitudinal gauge

$$\mathrm{d}s^2 = -(1 + 2\phi)\,\mathrm{d}t^2 + a^2(1 - 2\psi)\mathrm{d}\mathbf{x}^2, \tag{3.97}$$

(see Boisseau *et al.*, 2000 for details). The idea is that, in the short wavelength limit, the leading terms are either those containing k^2, or those with δ_m. Then the following equation is obtained:

$$\delta\Phi \simeq (\phi - 2\psi)\,\frac{\mathrm{d}F}{\mathrm{d}\Phi} \simeq -\phi\,\frac{F\,\mathrm{d}F/\mathrm{d}\Phi}{F + 2(\mathrm{d}F/\mathrm{d}\Phi)^2}. \tag{3.98}$$

Hence, unlike in GR where the quintessence field is unclustered on small scales (hence it is in good approximation a genuine DE component), in ST gravity the dilaton remains partly clustered even for arbitrarily small scales. However, this clustering is small because ω_{BD} is large. Hence, the first thing we learn is that in some DE models, DE is no longer completely smooth and that it can cluster to some extent.

Even more important is that the perturbations will obey a modified equation. Indeed, in the same short wavelength limit, Poisson's equation has a form similar to that in GR, however, with the crucial difference that Newton's constant G_{N} is replaced by G_{eff}, defined in (3.53) above. Hence, the equation for the evolution of linear dustlike matter perturbations becomes

$$\ddot{\delta}_m + 2H\dot{\delta}_m - 4\pi G_{\mathrm{eff}}\rho_m\delta_m = 0, \tag{3.99}$$

which for our purposes is conveniently recast into

$$H^2\,\delta_m'' + \left(\frac{(H^2)'}{2} - \frac{H^2}{1+z}\right)\delta_m' \simeq \frac{3}{2}(1+z)H_0^2\,\frac{G_{\mathrm{eff}}(z)}{G_{\mathrm{N},0}}\,\Omega_{m,0}\,\delta_m. \tag{3.100}$$

So, we see here a second difference in the physical meaning of G_{eff} and G_{N}: while it is G_{N} that appears in the equation for the evolution of the perturbations in GR, in scalar–tensor models this role is played by G_{eff}.

Let us now sketch briefly the reconstruction itself. Extracting $H(z)$ and $\delta_m(z)$ from observations with sufficient accuracy, we first reconstruct $G_{\mathrm{eff}}(z)/G_{\mathrm{N},0}$ analytically. Since, as is seen from Eqs. (3.53) and (3.64), the quantities $G_{\mathrm{eff},0}$ and $G_{\mathrm{N},0}$ coincide with better than 0.001 25% accuracy, Eq. (3.100) taken at $z = 0$ also gives the value of $\Omega_{m,0}$ with the same accuracy. Thus, in principle, no independent measurement of $\Omega_{m,0}$ is required; we can extract it from (3.100) taken at $z = 0$.

We get an equation $G_{\mathrm{eff}}(z) = p(z)$, where $p(z)$ is a given function that can be determined solely using observational data, which can be transformed into a nonlinear second-order differential equation for $F(z)$ using the background equation

$$\Phi'^2 = - F'' - \left[(\ln H)' + \frac{2}{1+z} \right] F'$$

$$+ \frac{2(\ln H)'}{1+z} F - 3(1+z)\frac{H_0^2}{H^2}F_0\Omega_{m,0}. \tag{3.101}$$

Hence $F(z)$ can be determined by solving the equation $G_{\mathrm{eff}}(z) = p(z)$ after we supply the initial conditions $F_0 = \frac{1}{8\pi G_{\mathrm{N},0}}$ and F_0'. Actually F_0' must be very close to zero due to the solar system constraint (3.64). Once $F(z)$ is found, it can be substituted into Eq. (3.95) to yield the potential $U(z)$ as a function of redshift. Then, using Eq. (3.101), $\Phi(z)$ is found by simple integration, which, after inverting this relation, gives us $z = z(\Phi - \Phi_0)$. Finally, both unknown functions $F(\Phi)$ and $U(\Phi)$ are completely fixed as functions of $\Phi - \Phi_0$. Of course this reconstruction can only be implemented in the range probed by the data corresponding to $z \lesssim 2$. We note that we can reconstruct in particular $G_{\mathrm{eff}}(z)$.

3.8 The linear growth of matter perturbations

For a very large class of modified gravity DE models, the matter perturbations obey at the linear level a modified equation of the type

$$\ddot{\delta}_m + 2H\dot{\delta}_m - 4\pi G_{\mathrm{eff}}\rho_m\delta_m = 0, \tag{3.102}$$

where the gravitational constant G_{eff} depends on the specific model under consideration. As we have seen, such a modified equation holds for scalar–tensor models with G_{eff} given by (3.53) (Boisseau *et al.*, 2000). A similar equation is also found in many models (Neupane, 2007) including some higher dimensional DE models. The physics behind it is simple to understand, it is a modification of Poisson's equation (see e.g. Esposito-Farèse and Polarski, 2001) according to (we drop the subscript m)

$$\frac{k^2}{a^2}\phi = -4\pi G\rho\delta \rightarrow \frac{k^2}{a^2}\phi = -4\pi G_{\text{eff}}\rho\delta. \tag{3.103}$$

We note that this modification is scale independent and more drastic modifications can occur as well. One can further increase the degeneracy between models inside and outside GR (see e.g. Kunz and Sapone, 2007) in models that are even further away from ΛCDM, but we do not consider them here. When perturbations obey an equation like (3.102), their dynamics depends on the particular theory of gravity. In principle, it is then possible to check whether the measured quantities $\delta_m(z)$ and $H(z)$ are compatible with each other assuming GR. If this is not the case, we know we are dealing with a modified gravity model. This is why the growth of matter perturbations has attracted much interest recently and possible ways to measure it with great pecision have been proposed (see e.g. Heavens, Kitching and Verde, 2007; Acquaviva *et al.*, 2008).

It is convenient to introduce the quantity $f = \frac{d\ln\delta}{d\ln a}$. Then the linear perturbations obey the equation

$$\frac{df}{dx} + f^2 + \frac{1}{2}\left(1 - \frac{d\ln\Omega_m}{dx}\right)f = \frac{3}{2}\frac{G_{\text{eff}}}{G_{N,0}}\Omega_m, \tag{3.104}$$

and we have set here $x \equiv \ln a$. Equation (3.104) reduces to Eq. (B7) given in Wang and Steinhardt (1998) for $\frac{G_{\text{eff}}}{G_{N,0}} = 1$. The quantity δ is easily recovered using f as follows

$$\delta(a) = \delta_i \exp\left[\int_{x_i}^{x} f(x')\,dx'\right]. \tag{3.105}$$

We see that $f = p$ when $\delta \propto a^p$, in particular $f \rightarrow 1$ in ΛCDM for large z and $f = 1$ in an Einstein–de Sitter universe.

At large redshifts in the matter-dominated era, the solution to (3.102) can significantly depart from the behaviour $\delta \propto a$ valid in ΛCDM. When (3.104) boils down to

$$\frac{df}{dx} + f^2 + \frac{1}{2}f = \frac{3}{2}C, \tag{3.106}$$

with $C = $ constant, the general solutions are

$$\delta = D_1 a^{p_1} + D_2 a^{p_2}, \tag{3.107}$$

with

$$p_1 = \frac{1}{4}(-1 + \sqrt{1 + 24C}), \tag{3.108}$$

$$p_2 = \frac{1}{4}(-1 - \sqrt{1 + 24C}).\qquad(3.109)$$

Depending on the value of C, we can have a significant departure from the standard $p = 1$ behaviour. This can give a first important constraint.

Another important issue is to characterize departures on small redshifts for different models.

As is well known, for a ΛCDM universe it is possible to write $f \simeq \Omega_m^{\gamma}$, where γ is assumed to be constant, an approach pioneered in the literature some time ago (Peebles, 1984; Lahav *et al.*, 1991). There has been renewed interest lately for this approach in the hope of discriminating between models that are either inside or outside GR (Huterer and Linder, 2006; Acquaviva and Verde, 2007; Di Porto and Amendola, 2007; Hui and Parfree, 2007; Kiakotou, Elgaroy and Lahav, 2007; Linder and Cahn, 2007; Tsujikawa, 2007; Wang, 2007; Nesseris and Perivolaropoulos, 2008; Wei and Zhang, 2008). Clearly we can always write

$$f = \Omega_m^{\gamma(z)}.\qquad(3.110)$$

For ΛCDM we have $\gamma_0 \equiv \gamma(z=0) \approx 0.55$. In addition $\gamma_0' \equiv \frac{d\gamma}{dz}|_0 \approx -0.015$ (Polarski and Gannouji, 2008), for $\Omega_{m,0} = 0.3$, and $\gamma_0 = 0.555$, slightly higher than the constant $\frac{6}{11} = 0.5454$ derived in Wang and Steinhardt (1998) for a slowly varying DE equation of state and $\Omega_m \approx 1$.

Actually Eq. (3.106) yields the following equality (Polarski and Gannouji, 2008):

$$\gamma_0' = \left[\ln \Omega_{m,0}^{-1}\right]^{-1}\left[-\Omega_{m,0}^{\gamma_0} - 3(\gamma_0 - \frac{1}{2})w_{\text{eff},0} + \frac{3}{2}\Omega_{m,0}^{1-\gamma_0} - \frac{1}{2}\right],\qquad(3.111)$$

whenever $\frac{G_{\text{eff},0}}{G_{N,0}} = 1$ to very high accuracy. Hence we have a constraint

$$f(\gamma_0, \gamma_0', \Omega_{m,0}, w_{\text{de},0}) = 0,\qquad(3.112)$$

which can be shown to take the following form:

$$\gamma_0' \simeq -0.19 + d(\gamma_0 - 0.5),\qquad d \approx 3.\qquad(3.113)$$

The coefficient d depends on the background parameters $d = d(w_{\text{de},0}, \Omega_{m,0})$. For given parameters $\Omega_{m,0}$ and $w_{\text{de},0}$, γ_0' will take the corresponding value $\gamma_0'(\gamma_0)$.

The value of γ_0 realized depends on the particular model and can be obtained numerically. Typically we will have $\gamma_0' \neq 0$. For models inside GR satisfying (3.102) with $G_{\text{eff}} = G$, $|\gamma_0'| \lesssim 0.02$ seems to hold. Therefore if $|\gamma_0'|$ is substantially larger than this value, this could be a signature of a modified gravity model. The impossibility of a satisfactory fit with $\gamma = $ constant could hint at such a possibility.

3.9 Conclusions

As we have seen, we now have many classes of DE models able to explain the late-time accelerated expansion of the universe. Decisive progress will come with more observations that will probe both the background expansion and the evolution of matter perturbations with very high accuracy. Even after we have these data, it is not certain that only one model will emerge. Rather it is plausible to expect that several types of models will survive, with severely constrained parameters. This is also the situation prevailing for inflationary models. Not only have the latest data not ruled out a cosmological constant, on the contrary they seem to favour it. Therefore any variation of ρ_{de} or of w_{de}, even a very small one, would signal that a cosmological constant is not the solution. In that case, we will either learn something revolutionary about our microscopic world or about the laws of gravity. But a cosmological constant that is so tiny remains a mystery as well. One can conjecture that the way we view our universe and its expansion will be durably affected whatever the final solution to the dark energy puzzle.

Acknowledgments

It is a pleasure to thank my collaborators L. Amendola, B. Boisseau, M. Chevallier, G. Esposito-Farèse, R. Gannouji, A. Ranquet, A. Starobinsky and S. Tsujikawa.

References

Acquaviva, V., and Verde, L. (2007). [arXiv:0709.0082].
Acquaviva, V., Hajian, A., Spergel, D. N., and Das, S. (2008). [arXiv:0803.2236].
Amendola, L., Corasaniti, P. S., and Occhionero, F. (1999). [astro-ph/9907222].
Amendola, L., Polarski, D., and Tsujikawa, S. (2007). *Phys. Rev. Lett.* **98**, 131302.
Amendola, L., Gannouji, R., Polarski, D., and Tsujikawa, S. (2007). *Phys. Rev. D* **75**, 083504.
Appleby, S. A., and Battye, R. A. (2007). *Phys. Lett. B* **654**, 7.
Astier, P., *et al.* (2006). *Astron. Astrophys.* **447**, 31.
Barenboim, G., and Lykken, J. (2007). [arXiv:0711.3653].
Bartolo, N., and Pietroni, M. (2000). *Phys. Rev. D* **61**, 023518.
Bertotti, B., Iess, L., and Tortora, P. (2003). *Nature (London)* **425**, 374 (2003).
Bludman, S. (2007). [astro-ph/0702085].
Boisseau, B., Esposito-Farèse, G., Polarski, D., and Starobinsky, A. A. (2000). *Phys. Rev. Lett.* **85**, 2236.
Capozziello, S. (2002). *Int. J. Mod. Phys. D* **11**, 483.
Capozziello, S., Dunsby, P. K. S., Piedipalumbo, E., and Rubano, C. (2007). [arXiv:0706.2615].
Carroll, S. M., Duvvuri, V., Trodden, M., and Turner, M. S. (2004). *Phys. Rev. D* **70**, 043528.
Chevallier, M., and Polarski, D. (2001). *Int. J. Mod. Phys. D* **10**, 213.

Clarkson, C., Cortes, M., and Bassett, B. (2007). *JCAP* **0708**, 011.

Copeland, E. J., Sami, M., and Tsujikawa, S. (2006). [hep-th/0603057].

Dabrowski, M. P., Stachowiak, T., and Szydlowski, M. (2003). *Phys. Rev. D* **68**, 103519.

Davis, T. M., *et al.* (2007). *Astrophys. J.* **666**, 716.

Demianski, M., Piedipalumbo, E., Rubano, C., and Scudellaro, P. (2007). [arXiv:0711.1043].

Di Pietro, E., and Claeskens, J. F. (2003). *Mon. Not. R. Astron. Soc.* **341**, 1299.

Di Porto, C., and Amendola, L. (2007). [arXiv:0707.2686].

Durrer, R., and Maartens, R. (2008). *Gen. Relat. Grav.* **40**, 301.

Eisenstein, D. J., *et al.* (SDSS Collaboration) (2005). *Astrophys. J.* **633**, 560.

Esposito-Farèse, G., and Polarski, D. (2001). *Phys. Rev. D* **63**, 063504.

Faraoni, V. (2004). *Cosmology in Scalar-Tensor Gravity*, Dordrecht: Kluwer Academic Publishers.

Feng, B., Wang, X.-L., and Zhang, X.-M. (2005). *Phys. Lett. B* **607**, 35.

Fujii, Y., and Maeda, K. (2003). *The Scalar-tensor Theory of Gravitation*, Cambridge: Cambridge University Press.

Gannouji, R., Polarski, D., Ranquet, A., and Starobinsky, A. A. (2006). *JCAP* **0609**, 016.

Gannouji, R., and Polarski, D. (2008). *JCAP* **0805**, 018.

Garcia-Berro, E., Gaztanaga, E., Isern, J., Benvenuto, O., and Althaus, L. (1999). [astro-ph/9907440].

Gong, Y., and Zhang, Y.-Z. (2005). *Phys. Rev. D* **72**, 043518.

Heavens, A. F., Kitching, T. D., and Verde, L. (2007). [astro-ph/0703191].

Hu, W., and Sawicki, I. (2007). *Phys. Rev. D* **76**, 064004.

Hui, L., and Parfree, K. (2007). [arXiv:0712.1162].

Huterer, D., and Linder, E. (2006). [astro-ph/0608681].

Ichikawa, K., and Takahashi, T. (2007). *JCAP* **0702**, 001.

Ishak, M., Upadhye, A., and Spergel, D. N. (2006). *Phys. Rev. D* **74**, 043513.

Kiakotou, A., Elgaroy, O., and Lahav, O. (2007). [arXiv:0709.0253].

Kitching, T. D., Heavens, A. F., Verde, L., Serra, P., and Melchiorri, A. (2008). [arXiv:0801.4565].

Komatsu, E., *et al.* (2008). [arXiv:0803.0547].

Kunz, M., and Sapone, D. (2007). *Phys. Rev. Lett.* **98**, 121301.

Lahav, O., Lilje, P. B., Primack, J. R., and Rees, M. J. (1991). *Mon. Not. R. Astron. Soc.* **251**, 128.

Lazkoz, R., Nesseris, S., and Perivolaropoulos, L. (2005). *JCAP* **0511**, 010.

Liddle, A., and Scherrer, R. (1999). *Phys. Rev. D* **59**, 023509.

Linder, E. V. (2003). *Phys. Rev. Lett.* **90**, 091301.

Linder, E. V. (2005). *Astropart. Phys.* **24**, 391.

Linder, E., and Cahn, R. (2007). [astro-ph/0701317].

Martin, J., Schimd, C., and Uzan, J.-P. (2006). *Phys. Rev. Lett.* **96**, 061303.

Moraes, B., Gannouji, R., and Polarski, D. (2009). *JCAP* **0902**, 034.

Nesseris, S., and Perivolaropoulos, L. (2004). *Phys. Rev. D* **70**, 043531.

Nesseris, S., and Perivolaropoulos, L. (2008). *Phys. Rev. D* **77**, 023504.

Neupane, I. P. (2007). [arXiv:0711.3234].

Padmanabhan, T. (2003). *Phys. Rep.* **380**, 235.

Peebles, P. J. E. (1984). *Astrophys. J.* **284**, 439.

Percival, W. J, *et al.* (2007). [arXiv:0705.3323].

Perlmutter, S. J., *et al.* (1997). *Astrophys. J.* **483**, 565.

Perlmutter, S. J., *et al.* (1998). *Nature (London)* **391**, 51.

Perlmutter, S. J., *et al.* (1999). *Astrophys. J.* **517**, 565.

Perrotta, F., Baccigalupi, G., and Matarrese, S. (2000). *Phys. Rev. D* **61**, 023507.

Pitjeva, E. V. (2005a). *Astron. Lett.* **31**, 340.

Pitjeva, E. V. (2005b). *Sol. Sys. Res.* **39**, 176.

Polarski, D., and Gannouji, R. (2008). *Phys. Lett. B* **660**, 439.

Polarski, D., and Ranquet, A. (2005). *Phys. Lett. B* **627**, 1.

Polarski, D. (2006). *AIP Conf. Proc.* **861**, 1013 [astro-ph/0605532].

Ratra, B., and Peebles, P. J. E. (1988). *Phys. Rev. D* **37**, 3406.

Riazuelo, A., and Uzan, J.-P. (2002). *Phys. Rev. D* **66**, 023525.

Riess, A. G., *et al.* (1998). *Astron. J.* **116**, 1009.

Riess, A. G., *et al.* (2007). *Astrophys. J.* **659**, 98 [astro-ph/0611572].

Ruiz-Lapuente, P. (2007). *Class. Quant. Grav.* **24**, R91 [arXiv: 0704.1058].

Sahni, V., and Starobinsky, A. A. (2000). *Int. J. Mod. Phys. D* **9**, 373.

Sahni, V., and Starobinsky, A. A. (2006). *Int. J. Mod. Phys. D* **15**, 2105.

Starobinsky, A. A. (1980). *Phys. Lett. B* **91**, 99.

Starobinsky, A. A. (1998). *JETP Lett.* **68**, 757.

Starobinsky, A. A. (2007). *JETP Lett.* **86**, 157 [arXiv:0706.2041].

Torres, D. (2002). *Phys. Rev. D* **66**, 043522.

Tsujikawa, S. (2007). [arXiv:0705.1032].

Upadhye, A., Ishak, M., and Steinhardt, P. J. (2005). *Phys. Rev. D* **72**, 063501.

Verde, L. (2007). [arXiv:0712.3028].

Wang, L., and Steinhardt, P. J. (1998). *Astrophys. J.* **508**, 483.

Wang, Y., and Mukherjee, P. (2006). *Astrophys. J.* **650**, 1.

Wang, Y., and Mukherjee, P. (2007). *Phys. Rev. D* **76**, 103533.

Wang, Y. (2007). [arXiv:0710.3885].

Wei, H., Cai, R.-G., and Zeng, D.-F. (2005). *Class. Quant. Grav.* **22**, 3189.

Wei, H., and Zhang, S. N. (2008). [arXiv:0803.3292].

Wetterich, C. (1988). *Nucl. Phys. B* **302**, 668.

Williams, J. G., Turyshev S. G., and Boggs, D. K. (2005). [gr-qc/0507083].

Wood-Vasey, W. M., *et al.* (ESSENCE Collaboration) (2007). *Astrophys. J.* **666**, 694.

Zlatev, I., Wang, L., and Steinhardt, P. J. (1999). *Phys. Rev. Lett.* **82**, 896 (1999).

4

Emergent gravity and dark energy

THANU PADMANABHAN

4.1 The rise of the dark energy

Given the expansion rate of the universe in terms of the Hubble constant $H_0 = (\dot{a}/a)_0$, one can define a *critical energy density* $\rho_c = 3H_0^2/8\pi G$ that is required to make the spatial sections of the universe compact. It is convenient to measure the energy densities of the different species that drive the expansion of the universe in terms of this critical density using the dimensionless parameters $\Omega_i = \rho_i/\rho_c$ (with i denoting the different components: baryons, dark matter, radiation, etc.) The simplest possible universe one could imagine would have just baryons and radiation. However, a host of astronomical observations available since the mid-1970s indicated that the bulk of the matter in the universe is non-baryonic and dark. Around the same time, the theoretical prejudice for $\Omega_{tot} = 1$ gained momentum, largely led by the inflationary paradigm. During the eighties, this led many theoreticians to push (wrongly!) for a model of the universe with $\Omega_{tot} \approx \Omega_{DM} \approx 1$ in spite of the fact that a host of astronomical observations demanded that $\Omega_{DM} \simeq 0.2-0.3$.

The indications that the universe indeed has another component of energy density started accumulating in the late eighties and early nineties. Early analysis of several observations (Efstathiou *et al.*, 1990, 1992; Padmanabhan and Narasimha, 1992; Bagla, Padmanabhan and Narlikar, 1995; Ostriker and Steinhardt, 1995) indicated that this component is unclustered and has negative pressure. This is confirmed dramatically by the supernova observations since the late nineties (see Riess *et al.*, 1998, 2004; Perlmutter *et al.*, 1999; Tonry *et al.*, 2003; Barris, 2004; for a critical look at the current data, see Padmanabhan and Choudhury, 2003; Choudhury and Padmanabhan, 2005; Jassal *et al.*, 2005b; Nesseris and Perivolaropoulos, 2007; Wang and Mukherjee, 2007). The observations

Dark Energy: Observational and Theoretical Approaches, ed. Pilar Ruiz-Lapuente. Published by Cambridge University Press. © Cambridge University Press 2010.

suggest that the missing component has $w = p/\rho \lesssim -0.78$ and contributes $\Omega_{de} \cong 0.60-0.75$.

The simplest choice for such *dark energy* with negative pressure is the cosmological constant, which is a term that can be added to Einstein's equations. This term acts like a fluid with an equation of state $p_{de} = -\rho_{de}$. Combining this with all other observations (Balbi *et al.*, 2000; de Bernardis *et al.*, 2000; Hanany *et al.*, 2000; Mould *et al.*, 2000; Freedman *et al.*, 2001; Padmanabhan and Sethi, 2001; Percival *et al.*, 2001, 2002; Wang *et al.*, 2002; Bennett *et al.*, 2003; Pearson *et al.*, 2003; Mason *et al.*, 2003; Spergel *et al.*, 2003, 2006; for a review of Big Bang nucleosynthesis, see Sarkar, 1996), we end up with a weird composition for the universe with $0.98 \lesssim \Omega_{tot} \lesssim 1.08$ in which radiation (R), baryons (B), dark matter made of weakly interacting massive particles (DM) and dark energy (DE) contribute $\Omega_R \simeq 5 \times 10^{-5}$, $\Omega_B \simeq 0.04$, $\Omega_{DM} \simeq 0.26$, and $\Omega_{de} \simeq 0.7$, respectively. So the bulk of the energy density in the universe is contributed by dark energy, which is the theme of this article.

The remarkably successful paradigm of conventional cosmology is based on these numbers (for recent reviews of the cosmological paradigm, see, e.g., Padmanabhan, 2005a, 2006a) and works as follows. The key idea is that if there existed small fluctuations in the energy density in the early universe, then gravitational instability could amplify them leading to structures like galaxies etc. today. The popular procedure for generating these fluctuations is based on the idea that if the very early universe went through an inflationary phase (Starobinsky, 1979, 1981; Kazanas, 1980; Guth, 1981; Albrecht and Steinhardt, 1982; Linde, 1982; for a review, see, e.g., Narlikar and Padmanabhan, 1991; Alabidi and Lyth, 2005), then the quantum fluctuations of the field driving the inflation can lead to energy density fluctuations (Guth and Pi, 1982; Hawking, 1982; Starobinsky, 1982; Bardeen *et al.*, 1983; Abbott and Wise, 1984; Padmanabhan, 1988; Padmanabhan, Seshadri and Singh, 1989; for a recent discussion with detailed set of references, see Sriramkumar and Padmanabhan, 2005). While the inflationary models are far from unique and hence lack predictive power, it is certainly possible to construct models of inflation such that these fluctuations are described by a Gaussian random field and are characterized by a power spectrum of the form $P(k) = Ak^n$ with $n \simeq 1$. The inflationary models cannot predict the value of the amplitude A in an unambiguous manner. But it can be determined from CMBR observations and the inflationary model parameters can be fine tuned to reproduce the observed value. The CMBR observations are consistent with the inflationary model for the generation of perturbations and give $A \simeq (28.3h^{-1}\,\mathrm{Mpc})^4$ and $n \lesssim 1$. (The first results were from COBE (Smoot *et al.*, 1992) and WMAP has re-confirmed them with far greater accuracy). One can evolve the initial perturbations by linear perturbation theory when the perturbation is small. But when $\delta \approx (\delta\rho/\rho)$ is comparable to unity the perturbation theory breaks down and one has to resort to numerical simulations (for a pedagogical description, see Bagla and Padmanabhan, 1997a,b; Bagla, 2004) or theoretical models based on approximate ansatz (Zeldovich, 1970; Gurbatov *et al.*, 1989; Hamilton *et al.*, 1991; Brainerd *et al.*, 1993; Bagla and Padmanabhan, 1994, 1997a; Nityananda and Padmanabhan, 1994; Padmanabhan, 1996; Padmanabhan, 1996 Munshi *et al.*, 1997; Bagla *et al.*, 1998; Padmanabhan and Engineer, 1998; Engineer *et al.*, 2000; Kanenar *et al.*, 2001; Tatekawa, 2004; Padmanabhan and Ray, 2006) to understand their evolution – especially the baryonic part, which leads to observed structures in the universe. This rapid summary shows that modeling the universe and comparing the theory with observations is a rather involved affair; but the results obtained from all these attempts are broadly consistent with observations.

To the zeroth order, the universe is characterized by just seven numbers: $h \approx 0.7$ describing the current rate of expansion; $\Omega_{de} \simeq 0.7$, $\Omega_{DM} \simeq 0.26$, $\Omega_B \simeq 0.04$, $\Omega_R \simeq 5 \times 10^{-5}$ giving the composition of the universe; the amplitude $A \simeq (28.3h^{-1} \text{ Mpc})^4$ and the index $n \simeq 1$ of the initial perturbations.

4.2 A first look at the cosmological constant and its problems

The remaining challenge, of course, is to make some sense out of these numbers from a more fundamental point of view. Among all these components, the dark energy, which exerts negative pressure, is probably the weirdest one and has attracted most attention.

The key observational feature of dark energy is that, when treated as a fluid with a stress tensor $T_b^a = \text{dia}(\rho, -p, -p, -p)$, it has an equation state $p = w\rho$ with $w \lesssim -0.8$ at the present epoch. The spatial part **g** of the geodesic acceleration (which measures the relative acceleration of two geodesics in the spacetime) satisfies an *exact* equation in general relativity given by

$$\nabla \cdot \mathbf{g} = -4\pi G(\rho + 3p). \tag{4.1}$$

This shows that the source of geodesic acceleration is $(\rho + 3p)$ and not ρ. As long as $(\rho + 3p) > 0$, gravity remains attractive while $(\rho + 3p) > 0$ can lead to "repulsive" gravitational effects. In other words, dark energy with sufficiently negative pressure will accelerate the expansion of the universe, once it starts dominating over the normal matter. This is precisely what is established from the study of high redshift supernovae, which can be used to determine the expansion rate of the universe in the past (Perlmutter *et al.*, 1999; Riess *et al.*, 1998, 2004; Tonry *et al.*, 2002; Barris, 2004; Astier *et al.*, 2006).

The simplest model for a fluid with negative pressure is not a fluid at all but the cosmological constant with $w = -1$, $\rho = -p = $ constant (for a few of the recent reviews, see Padmanabhan, 2003a, 2005c; Nobbenhuis, 2006; Alcaniz, 2006; Hannestad, 2006; Perivolaropoulos, 2006; Copeland *et al.*, 2006; Dolgov, 2006). The cosmological constant introduces a fundamental length scale in the theory $L_\Lambda \equiv H_\Lambda^{-1}$, related to the constant dark energy density ρ_{de} by $H_\Lambda^2 \equiv (8\pi G\rho_{de}/3)$. Though, in classical general relativity, based on G, c and L_Λ, it is not possible to construct any dimensionless combination from these constants, when one introduces the Planck constant, $\hbar$, it is possible to form the dimensionless combination $\lambda = H_\Lambda^2 (G\hbar/c^3) \equiv (L_P^2/L_\Lambda^2)$. Observations then require $(L_P^2/L_\Lambda^2) \lesssim 10^{-123}$, requiring enormous fine tuning.

In the early days, this was considered puzzling but most people believed that this number λ is actually zero. The cosmological constant problem in those days was to understand why it is strictly zero. Usually, the vanishing of a constant (which could have appeared in the low energy sector of the theory) indicates an underlying symmetry of the theory. For example, the vanishing of the mass of the photon is closely related to the gauge invariance of electromagnetism. No such symmetry principle is known to operate at low energies, which made this problem very puzzling. There is a symmetry – called supersymmetry – which does ensure that $\lambda = 0$, but it is known that supersymmetry is broken at sufficiently high energies and hence cannot explain the observed value of λ.

Given the observational evidence for dark energy in the universe and the fact that the simplest candidate for dark energy, consistent with all observations today, is a cosmological constant with $\lambda \approx 10^{-123}$, the cosmological constant problem has become linked to the problem of dark energy in the universe. So, if we accept the simplest interpretation of the current observations, we need to explain why the cosmological constant is non-zero and has this small value. It should, however, be stressed that these are logically independent issues. *Even if all the observational evidence for dark energy goes away we still have a problem, i.e., explaining why λ is zero.*

There is another, related, aspect to the cosmological constant problem which needs to be stressed. In the conventional approach to gravity, one derives the equations of motion from a Lagrangian $\mathcal{L}_{tot} = \mathcal{L}_{grav}(g) + \mathcal{L}_{matt}(g, \phi)$ where $\mathcal{L}_{grav}$ is the gravitational Lagrangian dependent on the metric and its derivative and $\mathcal{L}_{matt}$ is the matter Lagrangian that depends on both the metric and the matter fields, symbolically denoted as ϕ. In such an approach, the cosmological constant can be introduced via two different routes that are conceptually different but operationally the same. First, one may decide to take the gravitational Lagrangian to be $\mathcal{L}_{grav} = (2\kappa)^{-1}(R - 2\Lambda_g)$ where Λ_g is a parameter in the (low energy effective) action just like the Newtonian gravitational constant κ. The second route is by shifting the matter Lagrangian by $\mathcal{L}_{\text{matt}} \rightarrow \mathcal{L}_{\text{matt}} - 2\lambda_m$. Such a shift is clearly equivalent to adding a cosmological constant $2\kappa\lambda_m$ to the $\mathcal{L}_{\text{grav}}$. In general, what can be observed through gravitational interaction is the combination $\Lambda_{\text{tot}} = \Lambda_g + 2\kappa\lambda_m$.

It is now clear that there are two distinct aspects to the cosmological constant problem. The first question is why Λ_{tot} is very small when expressed in natural units. Second, since Λ_{tot} could have had two separate contributions from the gravitational and matter sectors, why does the *sum* remain so fine tuned? This question is particularly relevant because it is believed that our universe went through several phase transitions in the course of its evolution, each of which shifts the energy momentum tensor of matter by $T_b^a \rightarrow T_b^a + L^{-4}\delta_b^a$ where L is the scale characterizing the transition. For example, the GUT and weak interaction scales are about $L_{\text{GUT}} \approx 10^{-29}$ cm and $L_{\text{SW}} \approx 10^{-16}$ cm, which are tiny compared to L_Λ. Even if we take a more pragmatic approach, the observation of the Casimir effect in the lab sets a bound that $L < \mathcal{O}(1)$ nanometer, leading to a ρ which is about 10^{12} times the observed value (Mahajan *et al.*, 2006).

Finally, I will comment on two other issues related to the cosmological constant that appear frequently in the literature. The first one is what could be called the "why now" problem of the cosmological constant. How come the energy density contributed by the cosmological constant (treated as the dark energy) is comparable to the energy density of the rest of the matter at the *current epoch* of the universe? I do not believe this is an *independent* problem; if we have a viable theory predicting a particular numerical value for λ, then the energy density due to this cosmological constant will be comparable to the rest of the energy density at *some* epoch. So the real problem is in understanding the numerical value of λ; once that problem is solved the "why now" issue will take care of itself. In fact, we do not have a viable theory to predict the current energy densities of any component that populates the universe, let alone the dark energy! For example, the energy density of radiation today is computed from its temperature, which is an observed parameter – there is no theory which tells us that this temperature has to be 2.73 K when, say, galaxy formation has taken place for a certain billion number of years.

One also notices in the literature a discussion of the contribution of the zero point energies of the quantum fields to the cosmological constant that is often misleading, if not incorrect. What is usually done is to attribute a zero point energy $(1/2)\hbar\omega$ to each mode of the field and add up all these energies with an ultraviolet cut-off. For an electromagnetic field, for example, this will lead to an integral proportional to

$$\rho_0 = \int_0^{k_{\text{max}}} dk \, k^2 \hbar k \propto k_{\text{max}}^4, \tag{4.2}$$

which will give $\rho_0 \propto L_P^{-4}$ if we invoke a Planck scale cut-off with $k_{max} = L_P^{-1}$. It is then claimed that this ρ_0 will contribute to the cosmological constant. There are several problems with such a naive analysis. First, the ρ_0 computed above can be easily eliminated by the normal ordering prescription in quantum field theory and what one really should compute is the fluctuations in the vacuum energy – not the vacuum energy itself. Second, even if we take the nonzero value of ρ_0 seriously, it is not clear this has anything to do with a cosmological constant. The energy–momentum tensor due to the cosmological constant has a very specific form $T_b^a \propto \delta_b^a$ and its trace is nonzero. The electromagnetic field, for example, has a stress tensor with zero trace, $T_a^a = 0$; hence, in the vacuum state the expectation value of the trace, $\langle vac|T_a^a|vac\rangle$, will vanish, showing that the equation of state of the bulk electromagnetic vacuum is still $\rho_0 = 3p_0$, which does not lead to a cosmological constant. (The trace anomaly will not work in the case of an electromagnetic field.) So the naive calculation of vacuum energy density with a cut-off and the claim that it contributes to the cosmological constant is not an accurate statement in many cases.

4.3 What if dark energy is not the cosmological constant?

A nice possibility would be to postulate that $\lambda = 0$ and come up with a symmetry principle that will explain why this is the case. One probably has a greater chance of success in such an attempt than in coming up with an explanation for $\lambda \approx 10^{-123}$. But then, one needs to provide an alternative explanation for the dark energy observations. We shall now discuss two classes of such explanations, one that uses conventional physics and another that is totally speculative, and conclude that both are not viable!

4.3.1 Conservative explanations of dark energy

One of the *least* esoteric ideas regarding the dark energy is that the cosmological constant term in the equations arises because we have not calculated the energy density driving the expansion of the universe correctly. This idea arises as follows. The energy–momentum tensor of the real universe, $T_{ab}(t, \mathbf{x})$ is inhomogeneous and anisotropic. If we could solve the exact Einstein's equations $G_{ab}[g] = \kappa T_{ab}$ with it as the source, we would be led to a complicated metric g_{ab}. The metric describing the large scale structure of the universe should be obtained by averaging this exact solution over a large enough scale, leading to $\langle g_{ab}\rangle$. But since we cannot solve exact Einstein's equations, what we actually do is to average the stress tensor *first* to get $\langle T_{ab}\rangle$ and *then* solve Einstein's equations. But since $G_{ab}[g]$ is a nonlinear function of the metric, $\langle G_{ab}[g]\rangle \neq G_{ab}[\langle g\rangle]$ and there is a discrepancy. This is most easily seen by writing

$$G_{ab}[\langle g \rangle] = \kappa[\langle T_{ab} \rangle + \kappa^{-1}(G_{ab}[\langle g \rangle] - \langle G_{ab}[g] \rangle)] \equiv \kappa[\langle T_{ab} \rangle + T_{ab}^{\text{corr}}]. \quad (4.3)$$

If – based on observations – we take the $\langle g_{ab} \rangle$ to be the standard Friedmann metric, this equation shows that it has, as its source, *two* terms: the first is the standard average stress tensor and the second is a purely geometrical correction term, $T_{ab}^{\text{corr}} = \kappa^{-1}(G_{ab}[\langle g \rangle] - \langle G_{ab}[g] \rangle)$, which arises because of nonlinearities in Einstein's theory that lead to $\langle G_{ab}[g] \rangle \neq G_{ab}[\langle g \rangle]$. If this term can mimic the cosmological constant at large scales there will be no need for dark energy and – as a bonus – one can solve the "why now" problem!

To make this idea concrete, we have to identify an effective expansion factor $a_{\text{eff}}(t)$ of an inhomogeneous universe (after suitable averaging), and determine the equation of motion satisfied by it. The hope is that it will be sourced by terms so as to have $\ddot{a}_{\text{eff}}(t) > 0$ while the standard matter (with $(\rho + 3p) > 0$) leads to deceleration of standard expansion factor $a(t)$. Since any correct averaging of positive quantities in $(\rho + 3p)$ will not lead to a negative quantity, the real hope is in defining $a_{\text{eff}}(t)$ and obtaining its dynamical equation such that $\ddot{a}_{\text{eff}}(t) > 0$. In spite of some recent attention this idea has received (Zalaletdinov, 1992; Rasanen, 2004; Ellis and Buchert, 2005; Hirata and Seljak, 2005; Kolb *et al.*, 2005a,b; Losic and Unruh, 2005; Ishibashi and Wald, 2006; Paranjape and Singh, 2008; for a review, see Buchert 2007) it is doubtful whether it will lead to the correct result when implemented properly. The reasons for my skepticism are the following:

- It is obvious that T_{ab}^{corr} is – mathematically speaking – nonzero (for an explicit computation, in the completely different context of an electromagnetic plane wave, see Padmanabhan, 1987a); the real question is how big is it compared to T_{ab}? It seems unlikely that, when properly done, we will get a large effect for the simple reason that the amount of mass contained in the nonlinear regimes in the universe today is subdominant.
- Any calculation in linear theory or any calculation in which special symmetries are invoked will be inconclusive in settling this issue. The key question, of identifying a suitable analogue of the expansion factor from an averaged geometry, is nontrivial and it is not clear that the answer will be unique. To illustrate this point by an extreme example, suppose we decide to call $a(t)^n$ with, say, $n > 2$ as the effective expansion factor i.e., $a_{\text{eff}}(t) = a(t)^n$; obviously $\ddot{a}_{\text{eff}}$ can be positive ("accelerating universe") even with $\ddot{a}$ being negative. So, unless one has a *unique* procedure to identify the expansion factor of the average universe, it is difficult to settle the issue.
- This approach is strongly linked to explaining the acceleration as observed by SN. Even if we decide to completely ignore all SN data, we still have reasonable evidence for dark energy and it is not clear how this approach can tackle such evidence.

Another equally conservative explanation for the cosmic acceleration could be that we are located in a large underdense region in the universe; so, locally, the underdensity acts like negative mass and produces a repulsive force. While there has been some discussion in the literature (Conley *et al.*, 2007; Jha *et al.*, 2007) as to whether observations indicate such a local "Hubble bubble", this does not seem to be a tenable explanation that one can take seriously at this stage. Again, CMBR observations indicating dark energy, for example, will not be directly affected by this feature though one does need to take into account the effect of the local void.

4.3.2 Dark energy from scalar fields

The most popular alternative to the cosmological constant uses a scalar field ϕ with a suitably chosen potential $V(\phi)$ so as to make the vacuum energy vary with time. The hope then is that one can find a model in which the current value can be explained naturally without any fine tuning. The scalar fields come in different shades and hues such as quintessence, K-essence and tachyonic fields amongst others.[1]

Since the quintessence field (or the tachyonic field) has an undetermined free function $V(\phi)$, it is possible to choose this function in order to produce a given expansion history of the universe characterized by the function $H(a) = \dot{a}/a$ expressed in terms of a. To see this explicitly, let us assume that the universe has two forms of energy density with $\rho(a) = \rho_{\text{known}}(a) + \rho_\phi(a)$, where $\rho_{\text{known}}(a)$ arises from all known forms of source (matter, radiation, etc.) and $\rho_\phi(a)$ is due to a scalar field. Let us first consider quintessence models with the Lagrangian

$$\mathcal{L}_{\text{quin}} = \frac{1}{2}\partial_a\phi\,\partial^a\phi - V(\phi). \tag{4.4}$$

Here, the potential is given implicitly by the form (Padmanabhan, 2002a; Padmanabhan and Choudhury, 2002; Bagla *et al.*, 2003; Ellis and Madsen, 1991; Schunck and Mielke, 1994)

$$V(a) = \frac{1}{16\pi G}H(1-Q)\left[6H + 2aH' - \frac{aHQ'}{1-Q}\right], \tag{4.5}$$

$$\phi(a) = \left[\frac{1}{8\pi G}\right]^{1/2}\int\frac{da}{a}\left[aQ' - (1-Q)\frac{d\ln H^2}{d\ln a}\right]^{1/2}, \tag{4.6}$$

---

[1] For a small sample of recent papers, see: Calcagni and Liddle (2006); Carneiro *et al.* (2006); Das *et al.* (2006); Fabris *et al.* (2006); Fang *et al.* (2006); Fuzfa and Alimi (2006); Gannouji *et al.* (2006); Gao *et al.* (2006); Grande *et al.* (2006); Johri and Rath (2006); Lima *et al.* (2006); Nojiri and Odintsov (2006); Panotopoulos (2006, 2007); Papantonopoulos *et al.* (2006); Ren *et al.* (2006); Sen (2006); Setare (2006, 2007); Sola and Stefancic (2006); Wang (2006); Andrianov *et al.* (2007); Capozziello (2007); Chattopadhyay *et al.* (2007); Costa *et al.* (2007); de Putter and Linder (2007); Jain *et al.* (2007); Movahed *et al.* (2007); Panigrahi *et al.* (2007); Shao *et al.* (2007); Sola (2007); Stojkovic *et al.* (2007); Cannata and Kamenshchik (2008); Gu (2008); Setare and Vagenas (2008); Shapiro (2008); Zhang *et al.* (2008).

where $Q(a) \equiv [8\pi G\rho_{known}(a)/3H^2(a)]$ and prime denotes differentiation with respect to a. Given any $H(a)$, $Q(a)$, these equations determine $V(a)$ and $\phi(a)$ and thus the potential $V(\phi)$. *Every quintessence model studied in the literature can be obtained from these equations.*

Similar results exist for the tachyonic scalar field (Padmanabhan, 2002a; Padmanabhan and Choudhury, 2002; Bagla *et al.*, 2003), which has the Lagrangian

$$\mathcal{L}_{tach} = -V(\phi)[1 - \partial_a\phi\, \partial^a\phi]^{1/2}. \tag{4.7}$$

Given any $H(a)$, one can construct a tachyonic potential $V(\phi)$ so that the scalar field is the source for the cosmology. The equations determining $V(\phi)$ are now given by

$$\phi(a) = \int \frac{da}{aH} \left(\frac{aQ'}{3(1-Q)} - \frac{2}{3}\frac{aH'}{H} \right)^{1/2}, \tag{4.8}$$

$$V(a) = \frac{3H^2}{8\pi G}(1-Q)\left(1 + \frac{2}{3}\frac{aH'}{H} - \frac{aQ'}{3(1-Q)}\right)^{1/2}. \tag{4.9}$$

Equations (4.8) and (4.9) completely solve the problem. Given any $H(a)$, these equations determine $V(a)$ and $\phi(a)$ and thus the potential $V(\phi)$. A wide variety of phenomenological models with time-dependent cosmological constant have been considered in the literature; all of these can be mapped to a scalar field model with a suitable $V(\phi)$.

It is very doubtful whether this – rather popular – approach, based on scalar fields, has helped us to understand the nature of the dark energy at any deeper level. These models, viewed objectively, suffer from several shortcomings:

- The most serious problem with them is that they have no predictive power. As explicitly demonstrated above, virtually every form of $a(t)$ can be modeled by a suitable "designer" $V(\phi)$.
- We see from the above discussion that even when $w(a)$ is determined by observations, it is not possible to proceed further and determine the nature of the scalar field Lagrangian. The explicit examples given above show that there are *at least* two different forms of scalar field Lagrangians – corresponding to the quintessence or the tachyonic field – that could lead to the same $w(a)$. (See Padmanabhan and Choudhury (2003) for an explicit example of such a construction.)
- By and large, the potentials used in the literature have no natural field theoretical justification. All of them are nonrenormalizable in the conventional sense and have to be interpreted as a low energy effective potential in an ad hoc manner.
- One key difference between cosmological constant and scalar field models is that the latter lead to a $(p/\rho) \equiv w(a)$ which varies with time. So they are worth considering if

the observations have suggested a varying w, or if observations have ruled out $w = -1$ at the present epoch. However, all available observations are consistent with a cosmological constant ($w = -1$) and, in fact, the possible variation of w is strongly constrained (Jassal *et al.*, 2006).

As an aside, let us note that in drawing conclusions from the observational data, one should be careful about the hidden assumptions in the statistical analysis. Claims regarding w depend crucially on the data sets used, priors that are assumed and possible parameterizations that are adopted. (For more details related to these issues, see Jassal *et al.* (2005a).) It is fair to say that all currently available data are consistent with $w = -1$. Further, there is some amount of tension between WMAP and SN-Gold data with the recent SNLS (Astier *et al.*, 2006) being more concordant with WMAP than the SN Gold data.

One also needs to remember that, for the scalar field models to work, we first need to find a mechanism that will make the cosmological constant vanish. In other words, all the scalar field potentials require fine tuning of the parameters in order to be viable. This is obvious in the quintessence models in which adding a constant to the potential is the same as invoking a cosmological constant. If we shift $\mathcal{L} \to \mathcal{L}_{\text{matt}} - 2\lambda_m$ in an otherwise successful scalar field model for dark energy, we end up "switching on" the cosmological constant and raising the problems again.

4.4 Cosmological constant as dark energy

Even if all the evidence for dark energy disappears within a decade, we still need to understand why the cosmological constant is zero, and much of what I have to say in the following will remain relevant. I stress this because there is a recent tendency to forget the fact that the problem of the cosmological constant existed (and was recognized as a problem) long before the observational evidence for dark energy, accelerating universe, etc. cropped up. In this sense, the cosmological constant problem has an important theoretical dimension that is distinct from what has been introduced by the observational evidence for dark energy.

Though invoking the cosmological constant as the candidate for dark energy leads to well-known problems mentioned earlier, it is also the most economical (just one number) explanation for all the observations. Therefore it is worth examining this idea in detail and asking how these problems can be tackled.

If the cosmological constant is nonzero, then classical gravity will be described by the three constants G, c and $\Lambda \equiv L_\Lambda^{-2}$. Since $\Lambda (G\hbar/c^3) \equiv (L_P/L_\Lambda)^2 \approx 10^{-123}$, it is obvious that the cosmological constant is actually telling us something regarding *quantum gravity*, indicated by the combination $G\hbar$. *An acid test for any quantum gravity model will be its ability to explain this value*; needless to say, all the currently available models – strings, loops, etc. – flunk this test.

While the occurrence of $\hbar$ in $\Lambda(G\hbar/c^3)$ shows that it is a relic of a quantum gravitational effect (or principle) of unknown nature, the cosmological constant problem is an infrared problem *par excellence* in terms of the energy scales involved. This is a somewhat unusual possibility of a high energy phenomenon leaving a low energy relic and an analogy will be helpful to illustrate this idea (Padmanabhan and Choudhury, 2000). Suppose we solve the Schrodinger equation for the helium atom for the quantum states of the two electrons $\psi(x_1, x_2)$. When the result is compared with observations, we will find that only half the states – those in which $\psi(x_1, x_2)$ is antisymmetric under $x_1 \longleftrightarrow x_2$ interchange – are realized in nature. But the low energy Hamiltonian for electrons in the helium atom has no information about this effect! Here is a low energy (IR) effect which is a relic of relativistic quantum field theory (spin-statistics theorem) that is totally nonperturbative, in the sense that writing corrections to the Hamiltonian of the helium atom in some $(1/c)$ expansion will *not* reproduce this result. I suspect the current value of the cosmological constant is related to quantum gravity in a similar spirit. There must exist a deep principle in quantum gravity which leaves its nonperturbative trace even in the low energy limit that appears as the cosmological constant. I shall now attempt a more quantitative discussion of these possibilities.

4.4.1 Area scaling law for energy fluctuations

Given the theory with two length scales L_P and L_Λ, one can construct two energy scales $\rho_{UV} = 1/L_P^4$ and $\rho_{IR} = 1/L_\Lambda^4$ in natural units ($c = \hbar = 1$). There is sufficient justification from different theoretical perspectives to treat L_P as the zero point length of spacetime (Snyder, 1947; DeWitt, 1964; Yoneya, 1976; Padmanabhan, 1985a, 1987b, 1997, 1998a; Ashtekar *et al.*, 1992; Srinivasan *et al.*, 1998; Calmet *et al.*, 2004; Fontanini *et al.*, 2006; for a review, see Garay, 1995), giving a natural interpretation to ρ_{UV}. The second scale, ρ_{IR}, also has a natural interpretation. Since the universe dominated by a cosmological constant at late times will be asymptotically de Sitter with $a(t) \propto \exp(t/L_\Lambda)$ at late times, it will have a horizon and associated thermodynamics (Gibbons and Hawking, 1977; Padmanabhan, 2004a) with a temperature $T = H_\Lambda/2\pi$. The corresponding thermal energy density is $\rho_{thermal} \propto T^4 \propto 1/L_\Lambda^4 = \rho_{IR}$. Thus L_P determines the *highest* possible energy density in the universe while L_Λ determines the *lowest* possible energy density in this universe. As the energy density of normal matter drops below this value, ρ_{IR}, the thermal ambiance of the de Sitter phase will remain constant and provide the irreducible "vacuum noise". The observed dark energy density is the geometric mean

$$\rho_{de} = \sqrt{\rho_{IR}\rho_{UV}} = \frac{1}{L_P^2 L_\Lambda^2} \qquad (4.10)$$

of these two energy densities. If we define a dark energy length scale L_{de} such that $\rho_{de} = 1/L_{de}^4$ then $L_{de} = \sqrt{L_P L_\Lambda}$ is the geometric mean of the two length scales in the universe.

It is possible to interpret this relation along the following lines. Consider a three-dimensional region of size L with a bounding area that scales as L^2. Let us assume that we associate with this region N microscopic cells of size L_P each having a Poissonian fluctuation in energy of amount $E_P \approx 1/L_P$. Then the mean square fluctuation of energy in this region will be $(\Delta E)^2 \approx N L_P^{-2}$ corresponding to the energy density $\rho = \Delta E/L^3 = \sqrt{N}/L_P L^3$. If we make the usual assumption that $N = N_{vol} \approx (L/L_P)^3$, this will give

$$\rho = \frac{\sqrt{N_{vol}}}{L_P L^3} = \frac{1}{L_P^4}\left(\frac{L_P}{L}\right)^{3/2} \qquad \text{(bulk fluctuations).} \qquad (4.11)$$

On the other hand, if we assume that (for reasons that are unknown) the relevant degrees of freedom scale as the surface area of the region, then $N = N_{sur} \approx (L/L_P)^2$ and the relevant energy density is

$$\rho = \frac{\sqrt{N_{sur}}}{L_P L^3} = \frac{1}{L_P^4}\left(\frac{L_P}{L}\right)^2 = \frac{1}{L_P^2 L^2} \qquad \text{(surface fluctuations).} \qquad (4.12)$$

If we take $L \approx L_\Lambda$, the surface fluctuations in Eq. (4.12) give precisely the geometric mean in Eq. (4.10) that is observed. On the other hand, the *bulk fluctuations* lead to an energy density that is larger by a factor $(L/L_P)^{1/2}$. Of course, if we do not take fluctuations in energy but coherently add them, we will get $N/L_P L^3$ which is $1/L_P^4$ for the bulk and $(1/L_P)^4(L_P/L)$ for the surface. In summary, we have the following hierarchy:

$$\rho = \frac{1}{L_P^4} \times \left[1, \left(\frac{L_P}{L}\right), \left(\frac{L_P}{L}\right)^{3/2}, \left(\frac{L_P}{L}\right)^2, \left(\frac{L_P}{L}\right)^4, \dots\right], \qquad (4.13)$$

in which the first one arises by coherently adding energies $(1/L_P)$ per cell with $N_{vol} = (L/L_P)^3$ cells; the second arises from coherently adding energies $(1/L_P)$ per cell with $N_{sur} = (L/L_P)^2$ cells; the third one is obtained by taking fluctuations in energy and using N_{vol} cells; the fourth from energy fluctuations with N_{sur} cells; and finally the last one is the thermal energy of the de Sitter space if we take $L \approx L_\Lambda$; clearly the further terms are irrelevant due to this vacuum noise.

Of all these, the only viable possibility is what arises if we assume that: (a) the number of active degrees of freedom in a region of size L scales as

$N_{sur} = (L/L_P)^2$; (b) it is the *fluctuations* in the energy that contribute to the cosmological constant (Padmanabhan, 2005d,e; for earlier attempts in similar spirit, see
Sorkin, 1997; Padmanabhan, 2002b; for related work, see Lindesay *et al.*, 2004;
Myung, 2004; Volovik, 2004; Elizalde *et al.*, 2005; Barrow, 2006) and the bulk
energy does not gravitate.

It has been demonstrated recently (Padmanabhan, 2002e, 2003b, 2004b, 2005f;
for related ideas, see, e.g., Mottola and Vaulin, 2006; Cremades *et al.*, 2006) that
it is possible to obtain classical relativity from purely thermodynamic considerations in which the surface term of the gravitational action plays a crucial role.
The area scaling is familiar from the usual result that entropy of horizons scales as
area. (Further, in cases like Schwarzschild black hole, one cannot even properly
define the volume inside a horizon.) In fact, one can argue from general considerations that the entropy associated with *any* null surface should be $(1/4)$ per
unit area and will be observer dependent. A null surface, obtained as a limit of a
sequence of timelike surfaces (like the $r = 2M$ obtained from $r = 2M + k$ surfaces
with $k \to 0^+$ in the case of the Schwarzschild black hole), "loses" one dimension in the process (e.g., $r = 2M + k$ is three-dimensional and timelike for $k > 0$
but is two-dimensional and null for $k = 0$) suggesting that the scaling of degrees
of freedom has to change appropriately. It is difficult to imagine that these features are unconnected and accidental and I shall discuss these ideas further in the
next section.

4.5 An alternative perspective: emergent gravity

I shall now describe an alternative perspective in which gravity is treated as an
emergent phenomenon – like elasticity – and argue that such a perspective is indeed
necessary to succeed in solving the cosmological constant problem. To do this, I
first identify the key ingredient of the cosmological constant problem and try to
address it head on.

4.5.1 Why do we need a new perspective on gravity?

The equations of motion of gravity are obtained in the conventional approach to
gravity from a Lagrangian $\mathcal{L}_{tot} = \mathcal{L}_{grav}(g) + \mathcal{L}_{matt}(g, \phi)$ where $\mathcal{L}_{grav}$ is the gravitational Lagrangian dependent on the metric and its derivative and $\mathcal{L}_{matt}$ is the matter
Lagrangian, which depends on both the metric and the matter fields, symbolically
denoted as ϕ. This total Lagrangian is integrated over the spacetime volume with
the covariant measure $\sqrt{-g}\mathrm{d}^4 x$ to obtain the action.

Suppose we now add a constant $(-2\lambda_m)$ to the matter Lagrangian thereby
inducing the change $\mathcal{L}_{matt} \to \mathcal{L}_{matt} - 2\lambda_m$. The equations of motion for matter are

invariant under such a transformation, which implies that, in the absence of gravity, we cannot determine the value of λ_m. The transformation $\mathcal{L} \to \mathcal{L}_{\text{matt}} - 2\lambda_m$ is a symmetry of the matter sector (at least at scales below the scale of supersymmetry breaking; we shall ignore supersymmetry in what follows). But, in the conventional approach, gravity breaks this symmetry. *This is the root cause of the cosmological constant problem.* As long as gravitational field equations are of the form $E_{ab} = \kappa T_{ab}$ where E_{ab} is some geometrical quantity (which is G_{ab} in Einstein's theory) the theory cannot be invariant under the shifts of the form $T_b^a \to T_b^a + \rho\delta_b^a$. Since such shifts are allowed by the matter sector, it is very difficult to imagine a definitive solution to the cosmological constant problem within the conventional approach to gravity.

If the metric represents the gravitational degree of freedom that is varied in the action and we demand full general covariance, we cannot avoid $\mathcal{L}_{\text{matt}}\sqrt{-g}$ coupling and cannot obtain the equations of motion that are invariant under the shift $T_{ab} \to T_{ab} + \Lambda g_{ab}$. Clearly a new, drastically different, approach to gravity is required. We need to look for an approach that has the following ingredients (Padmanabhan, 2006a,b,c).

To begin with, the field equations must remain invariant under the shift $\mathcal{L}_{\text{matt}} \to \mathcal{L}_{\text{matt}} + \lambda_m$ of the matter Lagrangian $\mathcal{L}_{\text{matt}}$ by a constant λ_m. That is, we need to have some kind of "gauge freedom" to absorb any λ_m. General covariance requires using the integration measure $\sqrt{-g}d^D x$ in actions. Since we do not want to restrict general covariance but at the same time do not want this coupling to the metric tensor via $\sqrt{-g}$, it follows that the *metric cannot be the dynamical variable in our theory.* Second, even if we manage to obtain a theory in which gravitational action is invariant under the shift $T_{ab} \to T_{ab} + \Lambda g_{ab}$, we would have only succeeded in making gravity decouple from the bulk vacuum energy. While this is considerable progress, there still remains the second issue of explaining the observed value of the cosmological constant. Once the bulk value of the cosmological constant (or vacuum energy) decouples from gravity, *classical* gravity becomes immune to the cosmological constant; that is, the bulk classical cosmological constant can be gauged away. Any observed value of the cosmological constant has to be necessarily a *quantum* phenomenon arising as a relic of microscopic spacetime fluctuations. The discussion in Section 4.4.1, especially Eq. (4.12), shows that the relevant degrees of freedom should be linked to surfaces in spacetime rather than bulk regions. The observed cosmological constant is a relic of quantum gravitational physics and should arise from degrees of freedom that scale as the surface area.

Thus, in an approach in which the surface degrees of freedom play the dominant role, rather than bulk degrees of freedom, we have some hope of obtaining the correct value for the cosmological constant. One should then obtain a theory of gravity that is more general than Einstein's theory with the latter emerging as a low energy approximation.

4.5.2 Microstructure of the spacetime

For reasons described above, we abandon the usual picture of treating the metric as the fundamental dynamical degrees of freedom of the theory and treat it as providing a coarse grained description of the spacetime at macroscopic scales, somewhat like the density of a solid – which has no meaning at atomic scales (Sakharov, 1968; Jacobson, 1995; Barcelo *et al.*, 2001; Volovik, 2001, 2003, 2006; Padmanabhan, 2002d, 2003c, 2004c,d; Visser, 2002; Huang and Sun, 2007; Makela, 2007). The unknown, microscopic degrees of freedom of spacetime (which should be analogous to the atoms in the case of solids) will play a role only when spacetime is probed at Planck scales (which would be analogous to the lattice spacing of a solid: Snyder, 1947; DeWitt, 1964; Yoneya, 1976; Padmanabhan, 1985a, 1987b, 1997, 1998a; Ashtekar *et al.*, 1992; Srinivasan *et al.*, 1998; Calmet *et al.*, 2004; Fontanini *et al.*, 2006).

Some further key insight can be obtained by noticing that in the study of ordinary solids, one can distinguish between three levels of description. At the macroscopic level, we have the theory of elasticity, which has a life of its own and can be developed purely phenomenologically. At the other extreme, the microscopic description of a solid will be in terms of the statistical mechanics of a lattice of atoms and their interaction.

Both of these are well known; but interpolating between these two limits is the thermodynamic description of a solid at finite temperature, *which provides a crucial window into the existence of the corpuscular substructure of solids.* As Boltzmann told us, heat is a form of motion and we would not have the thermodynamic layer of description if matter were a continuum all the way to the finest scales and atoms did not exist! *The mere existence of a thermodynamic layer in the description is proof enough that there are microscopic degrees of freedom.*

The situation is similar in the case of the spacetime (Padmanabhan, 2007). Again we should have three levels of description. The macroscopic level is the smooth spacetime continuum with a metric tensor $g_{ab}(x^i)$ and the equations governing the metric have the same status as the phenomenological equations of elasticity. At the microscopic level, we expect a quantum description in terms of the "atoms of spacetime" and some associated degrees of freedom q_A, which are still elusive. But what is crucial is the existence of an interpolating layer of thermal phenomena associated with null surfaces in the spacetime. Just as a solid cannot exhibit thermal phenomena if it does not have microstructure, *the thermal nature of horizon, for example, cannot arise without the spacetime having a microstructure.*

In such a picture, we normally expect the microscopic structure of spacetime to manifest itself only at Planck scales or near singularities of the classical theory. However, in a manner that is not fully understood, the horizons, which block information from certain classes of observers, link certain aspects of microscopic physics with the bulk dynamics (see, e.g., Padmanabhan, 1998b, 1999 and references therein), just as thermodynamics can provide a link between statistical mechanics and (zero temperature) dynamics of a solid. The reason is probably related to the fact that horizons lead to infinite redshift, which probes *virtual* high energy processes.

The following three results, showing a fundamental relationship between the dynamics of gravity and thermodynamics of horizons (for a recent review, see Padmanabhan, 2002d, 2005g), strongly support the above point of view:

- The dynamical equations governing the metric can be interpreted as a thermodynamic relation closely related to the thermodynamics of horizons. An explicit example was provided in Padmanabhan (2002b), in the case of spherically symmetric horizons in four dimensions, in which it was shown that Einstein's equations can be interpreted as a thermodynamic relation $T\,dS = dE + P\,dV$ arising out of virtual radial displacements of the horizon. Further work showed that this result is valid in *all* the cases for which explicit computation can be carried out – as in the Friedmann models (Akbar and Cai, 2006a; Cai and Cao, 2006; Akbar, 2007; Zhou *et al.*, 2007) as well as for rotating and time-dependent horizons in Einstein's theory (Kothawala, Sarkar and Padmanabhan, 2007).
- The standard Lagrangian in Einstein's theory has the structure $\mathcal{L}_{\mathrm{EH}} \propto R \sim (\partial g)^2 + \partial^2 g$. In the usual approach the surface term arising from $\mathcal{L}_{\mathrm{sur}} \propto \partial^2 g$ has to be ignored or canceled to get Einstein's equations from $\mathcal{L}_{\mathrm{bulk}} \propto (\partial g)^2$. But there is a peculiar (unexplained) relationship (Padmanabhan, 2002c, 2003b, 2004b, 2005f) between $\mathcal{L}_{\mathrm{bulk}}$ and $\mathcal{L}_{\mathrm{sur}}$:

$$\sqrt{-g}\,\mathcal{L}_{\mathrm{sur}} = -\partial_a \left(g_{ij} \frac{\partial \sqrt{-g}\,\mathcal{L}_{\mathrm{bulk}}}{\partial(\partial_a g_{ij})} \right). \tag{4.14}$$

This shows that the gravitational action is "holographic" with the same information being coded in both the bulk and surface terms and one of them should be sufficient.
- One can indeed obtain Einstein's equations from an action principle that uses *only* the surface term and the virtual displacements of horizons (Padmanabhan, 2005h, 2006a,b,d,e). It is possible to determine the form of this surface term from general considerations. If we now demand that the action should not receive contributions for radial displacements of the horizons, defined in a particular manner using local Rindler horizons, one can obtain, at the lowest order, the equations

$$(G_{ab} - \kappa T_{ab})\xi^a \xi^b = 0, \tag{4.15}$$

where ξ^a is a null vector. Demanding the validity of Eq. (4.15) in all local Rindler frames then leads to Einstein's theory with the cosmological constant emerging as an integration constant. Note that Eq. (4.15) is invariant under the constant shift of the matter Lagrangian, making gravity immune to the bulk cosmological constant. Since the surface term has a thermodynamic interpretation as the entropy of horizons, this establishes a direct connection between spacetime dynamics and horizon thermodynamics.

- Further work has shown that *all the above results extend beyond Einstein's theory.* The connection between field equations and the thermodynamic relation $T\, dS = dE + P\, dV$ is not restricted to Einstein's theory alone, but is in fact true for the case of the generalized, higher derivative Lanczos–Lovelock gravitational theory in D dimensions as well (Allemandi *et al.*, 2003; Akbar and Cai, 2006b; Paranjape, Sarkar and Padmanabhan, 2006; Ge, 2007; Sheykhi *et al.*, 2007). The same is true (Mukhopadhyay and Padmanabhan, 2006) for the holographic structure of the action functional: the Lanczos–Lovelock action has the same structure and, again, the entropy of the horizons is related to the surface term of the action. *These results show that the thermodynamic description is far more general than just Einstein's theory* and occurs in a wide class of theories in which the metric determines the structure of the light cones and null surfaces exist blocking the information.

The conventional approach to gravity fails to provide any clue regarding the thermodynamic aspects of gravity, just as Newtonian continuum mechanics – without corpuscular, discrete, substructure for matter – cannot explain thermodynamic phenomena. A natural explanation for these results requires a different approach to spacetime dynamics that I shall now outline. (More details can be found in Padmanabhan and Paranjape, 2007; Padmanabhan, 2008).

4.5.3 Surfaces in spacetime: key to the new paradigm

In obtaining the relation between gravitational dynamics and horizon thermodynamics, one treats the null surfaces (which act as horizons) as the limit of a sequence of, say, timelike surfaces. The virtual displacements of the horizon in the direction normal to the surfaces will be used in the action principle. All these suggest that one may be able to obtain a more formal description of the theory in terms of deformation of surfaces in spacetime. I now describe one such model that is unreasonably successful.

To set the stage, let us suppose there are certain microscopic, as yet unknown, degrees of freedom q_A, analogous to the atoms in the case of solids, described by some microscopic action functional $A_{\text{micro}}[q_A]$. In the case of a solid, the relevant long-wavelength elastic dynamics is captured by the *displacement vector field* that occurs in the equation $x^a \to x^a + \xi^a(x)$, which is only very indirectly connected with the microscopic degrees of freedom. Similarly, in the case of spacetime, we need to introduce some other degrees of freedom, analogous to ξ^a in the case of elasticity, and an effective action functional based on it. (As explained above, we do not want to use the metric as a dynamical variable.) Normally, varying an action functional with respect to certain degrees of freedom will lead to equations of motion determining *those* degrees of freedom. But we now make an unusual demand that varying our action principle with respect to some (nonmetric) degrees of freedom should lead to an equation of motion *determining the background metric* which remains nondynamical.

Based on the role expected to be played by surfaces in spacetime, we take the relevant degrees of freedom to be the normalized vector fields $n_i(x)$ in the spacetime (Padmanabhan and Paranjape, 2007; Padmanabhan, 2008) with a norm that is fixed at every event but might vary from event to event; i.e., $n_i n^i \equiv \epsilon(x)$ with $\epsilon(x)$ being a fixed function that takes the values 0, ± 1 at each event. Just as the displacement vector ξ^a captures the macro-description in the case of solids, the normalized vectors (e.g., local normals to surfaces) capture the essential macro-description in the case of gravity in terms of an effective action $S[n^a]$. More formally, we expect the coarse graining of microscopic degrees of freedom to lead to an effective action in the long-wavelength limit:

$$\sum_{q_A} \exp(-A_{\text{micro}}[q_A]) \longrightarrow \exp(-S[n^a]). \tag{4.16}$$

To proceed further we need to determine the nature of $S[n^a]$. The general form of $S[n^a]$ in such an effective description, at the quadratic order, will be

$$S[n^a] = \int_{\mathcal{V}} d^D x \sqrt{-g} \left(4 P_{ab}{}^{cd} \nabla_c n^a \nabla_d n^b - T_{ab} n^a n^b \right), \tag{4.17}$$

where P_{ab}^{cd} and T_{ab} are two tensors and the signs, notation, etc. are chosen with hindsight. (We will see later that T_{ab} can be identified with the matter stress-tensor.) The full action for gravity plus matter will be taken to be $S_{\text{tot}} = S[n^a] + S_{\text{matt}}$ with

$$S_{\text{tot}} = \int_{\mathcal{V}} d^D x \sqrt{-g} \left(4 P_{ab}{}^{cd} \nabla_c n^a \nabla_d n^b - T_{ab} n^a n^b \right) + \int_{\mathcal{V}} d^D x \sqrt{-g} \mathcal{L}_{\text{matt}}, \tag{4.18}$$

with an important extra prescription: since the gravitational sector is related to spacetime microstructure, we must *first* vary the n^a and *then* vary the matter degrees of freedom. In the language of path integrals, we should integrate out the gravitational degrees of freedom n^a first and use the resulting action for the matter sector.

We next address one crucial difference between the dynamics in gravity and say, elasticity, which we mentioned earlier. In the case of solids, one could write a similar functional for thermodynamic potentials in terms of the displacement vector ξ^a and extremizing it would lead to an equation *that determines* ξ^a. In the case of spacetime, we expect the variational principle to hold for all vectors n^a with a fixed norm and lead to a condition on the *background metric*. Obviously, the action functional in Eq. (4.17) must be rather special to accomplish this and one needs to impose two restrictions on the coefficients P^{cd}_{ab} and T_{ab} to achieve this. First, the tensor P_{abcd} should have algebraic symmetries similar to the Riemann tensor R_{abcd} of the D-dimensional spacetime. Second, we need

$$\nabla_a P^{abcd} = 0 = \nabla_a T^{ab}. \tag{4.19}$$

In a complete theory, the explicit form of P^{abcd} will be determined by the long-wavelength limit of the microscopic theory just as the elastic constants can, in principle, be determined from the microscopic theory of the lattice. In the absence of such a theory, we can take a cue from the renormalization group theory and expand P^{abcd} in powers of derivatives of the metric (Padmanabhan, 2005h, 2006a, 2008; Padmanabhan and Paranjape, 2007). That is, we expect

$$P^{abcd}(g_{ij}, R_{ijkl}) = c_1 \overset{(1)}{P}{}^{abcd}(g_{ij}) + c_2 \overset{(2)}{P}{}^{abcd}(g_{ij}, R_{ijkl}) + \cdots, \tag{4.20}$$

where $c_1, c_2, \ldots$ are coupling constants and the successive terms progressively probe smaller and smaller scales. The lowest order term must clearly depend only on the metric with no derivatives. The next term depends (in addition to the metric) linearly on the curvature tensor and the next one will be quadratic in curvature, etc. It can be shown that the mth order term that satisfies our constraints is *unique* and is given by

$$\overset{(m)}{P}{}^{cd}_{ab} \propto \delta^{cda_3\ldots a_{2m}}_{abb_3\ldots b_{2m}} R^{b_3 b_4}_{a_3 a_3} \ldots R^{b_{2m-1} b_{2m}}_{a_{2m-1} a_{2m}} = \frac{\partial \mathcal{L}^{(D)}_m}{\partial R^{ab}_{cd}}. \tag{4.21}$$

where $\delta^{cda_3\ldots a_{2m}}_{abb_3\ldots b_{2m}}$ is the alternating tensor and the last equality shows that it can be expressed as a derivative of the mth order Lanczos–Lovelock Lagrangian (Padmanabhan, 2006a, 2005h, 2008; Padmanabhan and Paranjape, 2007; Lanczos, 1932, 1938; Lovelock, 1971), given by

$$\mathcal{L}^{(D)} = \sum_{m=1}^{K} c_m \mathcal{L}^{(D)}_m; \qquad \mathcal{L}^{(D)}_m = \frac{1}{16\pi} 2^{-m} \delta^{a_1 a_2 \ldots a_{2m}}_{b_1 b_2 \ldots b_{2m}} R^{b_1 b_2}_{a_1 a_2} R^{b_{2m-1} b_{2m}}_{a_{2m-1} a_{2m}}, \tag{4.22}$$

where the c_m are arbitrary constants and $\mathcal{L}^{(D)}_m$ is the mth order Lanczos–Lovelock term and we assume $D \geq 2K + 1$. The lowest order term (which leads to Einstein's theory) is

$$\overset{(1)}{P}{}^{ab}_{cd} = \frac{1}{16\pi}\frac{1}{2}\delta^{a_1 a_2}_{b_1 b_2} = \frac{1}{32\pi}(\delta^a_c \delta^b_d - \delta^a_d \delta^b_c). \tag{4.23}$$

while the first order term gives the Gauss–Bonnet correction. All higher order terms are obtained in a similar manner.

In our paradigm based on Eq. (4.16), the field equations for gravity arise from extremizing S with respect to variations of the vector field n^a, with the constraint $\delta(n_a n^a) = 0$, and demanding that the resulting condition holds for *all normalized vector fields*. One can show (Padmanabhan, 2006a, 2005h, 2008; Padmanabhan and Paranjape, 2007) that this leads to the field equations

$$16\pi \left[P^{ijk}_b R^a_{ijk} - \frac{1}{2}\delta^a_b \mathcal{L}^{(D)}_m \right] = 8\pi T^a_b + \Lambda \delta^a_b, \tag{4.24}$$

where Λ is an undetermined integration constant. These are identical to the field equations for the Lanczos–Lovelock gravity with a cosmological constant arising as an undetermined integration constant. To the lowest order, when we use Eq. (4.23) for P^{ijk}_b, Eq. (4.24) reproduces Einstein's theory. More generally, we get Einstein's equations with higher order corrections, which are to be interpreted as emerging from the derivative expansion of the action functional as we probe smaller and smaller scales. Remarkably enough, we can derive not only Einstein's theory but even Lanczos–Lovelock theory from a dual description in terms on the normalized vectors in spacetime, *without varying g_{ab} in an action functional!*

The crucial feature of the coupling between matter and gravity through $T_{ab}n^a n^b$ in Eq. (4.18) is that, under the shift $T_{ab} \to T_{ab} + \rho_0 g_{ab}$, the ρ_0 term in the action in Eq. (4.17) decouples from n^a and becomes irrelevant:

$$\int_V d^D x \sqrt{-g}\, T_{ab}n^a n^b \to \int_V d^D x \sqrt{-g}\, T_{ab}n^a n^b + \int_V d^D x \sqrt{-g}\,\epsilon\rho_0. \tag{4.25}$$

Since $\epsilon = n_a n^a$ is not varied when n_a is varied there is no coupling between ρ_0 and the dynamical variables n_a and the theory is invariant under the shift $T_{ab} \to T_{ab} + \rho_0 g_{ab}$. We see that the condition $n_a n^a = $ constant on the dynamical variables has led to a "gauge freedom", which allows an arbitrary integration constant to appear in the theory that can absorb the bulk cosmological constant.

To gain a bit more insight into what is going on, let us consider the *on-shell* value of the action functional in Eq. (4.18). It can be shown that the on-shell value is given by a surface term that will lead to the entropy of the horizons (which will be 1/4 per unit transverse area in the case of general relativity). Even in the case of a theory with a general P^{ab}_{cd} it can be shown that the on-shell value of the action reduces to the entropy of the horizons (Padmanabhan and Paranjape, 2007; Padmanabhan, 2008). The general expression is

$$S|_{\mathcal{H}} = \sum_{m=1}^{K} 4\pi m c_m \int_{\mathcal{H}} d^{D-2} x_{\perp} \sqrt{\sigma} \mathcal{L}_{(m-1)}^{(D-2)} = \frac{1}{4}[\text{Area}]_{\perp} + \text{corrections}, \quad (4.26)$$

where $x_{\perp}$ denotes the transverse coordinates on the horizon $\mathcal{H}$, σ is the determinant of the intrinsic metric on $\mathcal{H}$ and we have restored a summation over m, thereby giving the result for the most general Lanczos–Lovelock case obtained as a sum of individual Lanczos–Lovelock Lagrangians. The expression in Eq. (4.26) *is precisely the entropy of a general Killing horizon in Lanczos–Lovelock gravity* based on the general prescription given by Wald and computed by several authors.

This result shows that in the semiclassical limit, in which the action can possibly be related to entropy, we reproduce the conventional entropy, which scales as the area in Einstein's theory. Since the entropy counts the relevant degrees of freedom, this shows that the degrees of freedom that survive and contribute in the long-wavelength limit scale as the area. The quantum fluctuations in these degrees of freedom can then lead to the correct, observed, value of the cosmological constant. I shall discuss this aspect briefly in the next section.

Our action principle is somewhat peculiar compared to the usual action principles in the sense that we have varied n_a and demanded that the resulting equations hold for *all* vector fields of constant norm. Our action principle actually stands for an infinite number of action principles, one for each vector field of constant norm! This class of *all* n^i allows an effective, coarse grained description of some (unknown) aspects of spacetime microphysics. This is why we need to first vary n_a, obtain the equations constraining the background metric and then use the reduced action to obtain the equations of motion for matter. Of course, in most contexts, $\nabla_a T_b^a = 0$ will take care of the dynamical equations for matter and these issues are irrelevant.

At this stage, it is not possible to proceed further and relate n^i to some microscopic degrees of freedom q^A. This issue is conceptually similar to asking one to identify the atomic degrees of freedom given the description of an elastic solid in terms of a displacement field ξ^a – which we know is virtually impossible. However, the same analogy tells us that the relevant degrees of freedom in the long-wavelength limit (i.e., ξ^a or n^i) can be completely different from the microscopic degrees of freedom and it is best to proceed phenomenologically.

4.5.4 Gravity as a detector of the vacuum fluctuations

The description of gravity given above provides a natural backdrop for gauging away the bulk value of the cosmological constant since it decouples from the dynamical degrees of freedom in the theory. Once the bulk term is eliminated, what is observable through gravitational effects, in the correct theory of quantum gravity, should be the *fluctuations* in the vacuum energy. These fluctuations will be nonzero if the universe has a de Sitter horizon that provides a confining volume. In this paradigm the vacuum structure can readjust to gauge away the bulk energy density $\rho_{UV} \simeq L_P^{-4}$, while *quantum fluctuations* can generate the observed value ρ_{de}.

The role of energy fluctuations contributing to gravity also arises, more formally, when we study the question of *detecting* the energy density using the gravitational field as a probe. Recall that a detector with a linear coupling to the *field* ϕ actually responds to $\langle 0|\phi(x)\phi(y)|0\rangle$ rather than to the field itself (Fulling, 1973; Unruh, 1976; DeWitt, 1979; Padmanabhan, 1985b; Srinivasan *et al.*, 1999; Sriramkumar *et al.*, 2002). Similarly, one can use the gravitational field as a natural "detector" of energy–momentum tensor T_{ab} with the standard coupling $L = \kappa h_{ab} T^{ab}$. Such a model was analyzed in detail in Padmanabhan and Singh (1987) and it was shown that the gravitational field responds to the two-point function $\langle 0|T_{ab}(x)T_{cd}(y)|0\rangle$. In fact, it is essentially these fluctuations in the energy density that are computed in the inflationary models (Starobinsky, 1979, 1980; Kazanas, 1980; Guth, 1981; Albrecht and Steinhardt, 1982; Linde, 1982) as the *source* for the gravitational field, as stressed in Padmanabhan (1988) and Padmanabhan *et al.* (1989). All these suggest treating the energy fluctuations as the physical quantity "detected" by gravity, when one incorporates quantum effects.

Quantum theory, especially the paradigm of the renormalization group has taught us that the concept of the vacuum state depends on the scale at which it is probed. The vacuum state we use to study the lattice vibrations in a solid, say, is not the same as the vacuum state of QED and it is not appropriate to ask questions about the vacuum without specifying the scale. If the cosmological constant arises due to fluctuations in the energy density of the vacuum, then one needs to understand the structure of the quantum gravitational vacuum at cosmological scales. If the spacetime has a cosmological horizon that blocks information, the natural scale is provided by the size of the horizon, L_Λ, and we should use observables defined within the accessible region. The operator $H(<L_\Lambda)$, corresponding to the total energy inside a region bounded by a cosmological horizon, will exhibit fluctuations ΔE since the vacuum state is not an eigenstate of *this* operator. A rigorous calculation (see Padmanabhan, 2005b) shows that the fluctuations in the energy density of the vacuum in a sphere of radius L_Λ are given by

$$\Delta\rho_{vac} = \frac{\Delta E}{L_\Lambda^3} \propto L_P^{-2} L_\Lambda^{-2}. \tag{4.27}$$

The numerical coefficient will depend on c_1 as well as the precise nature of the infrared cut-off radius; but it is a fact of life that a fluctuation of magnitude $\Delta\rho_{vac} \simeq H_\Lambda^2/G$ will exist in the energy density inside a sphere of radius H_Λ^{-1} if the Planck length is the UV cut-off. On the other hand, since observations suggest that there is a ρ_{vac} of similar magnitude in the universe it seems natural to identify the two. Our approach explains why there is a *surviving* cosmological constant that satisfies $\rho_{de} = \sqrt{\rho_{IR}\rho_{UV}}$.

Such a computation of energy fluctuations is completely meaningless in the models of gravity in which the metric couples to the bulk energy density. Once a UV cut-off at Planck scale is imposed, one will always get a bulk contribution $\rho_{UV} \approx L_P^{-4}$ with the usual problems. It is only because we have a way of decoupling the bulk term from contributing to the dynamical equations that we have a right to look at the subdominant term $L_P^{-4}(L_P/L_\Lambda)^2$. Approaches in which the subdominant term is introduced in an ad hoc manner are technically flawed since the bulk term cannot be ignored in these usual approaches to gravity. Getting the correct value of the cosmological constant from the energy fluctuations is not as difficult as understanding why the bulk value (which is larger by 10^{120}!) can be ignored. Our approach provides a natural backdrop for ignoring the bulk term, and as a bonus we get the right value for the cosmological constant from the fluctuations. The cosmological constant is small because it is a quantum relic.

4.6 Conclusions

The simplest choice for the negative pressure component in the universe is the cosmological constant; other models based on scalar fields (as well as those based on branes etc., which I have not discussed) do not alleviate the difficulties faced by the cosmological constant and, in fact, make them worse. The cosmological constant is most likely to be a low energy relic of a quantum gravitational effect or principle and its explanation will require a radical shift in our current paradigm.

The new approach to gravity described here could provide a possible broad paradigm to understand the cosmological constant. The conceptual basis for this approach rests on the following logical ingredients. I have shown that it is impossible to solve the cosmological constant problem unless the gravitational sector of the theory is invariant under the shift $T_{ab} \to T_{ab} + \lambda_m g_{ab}$. Any approach that does not address this issue cannot provide a comprehensive solution to the cosmological constant problem. But general covariance requires us to use the measure $\sqrt{-g}d^D x$ in D dimensions in the action. This will couple the metric (through its determinant) to the matter sector. Hence, as long as we insist on the metric as the fundamental variable describing gravity, one cannot address this issue. So we need to introduce some other degrees of freedom and an effective action which, however, is capable of constraining the background metric.

An action principle, based on the normalized vector fields in spacetime, satisfies all the criteria mentioned above. The new action does not couple to the bulk energy density and maintains invariance under the shift $T_{ab} \rightarrow T_{ab} + \lambda_m g_{ab}$. What is more, the on-shell value of the action is related to the entropy of horizons showing the relevant degrees of freedom scale as the area of the bounding surface. Since our formalism ensures that the bulk energy density does not contribute to gravity – and only because of that – it makes sense to compute the next order correction due to fluctuations in the energy density. This is impossible to do rigorously with the machinery available, but a plausible case can be made as to how this will lead to the correct, observed, value of the cosmological constant.

An effective theory can capture the relevant physics at the long-wavelength limit using the degrees of freedom contained in the fluctuations of the normal-ized vectors. The resulting theory is more general than Einstein gravity since the thermodynamic interpretations should transcend classical considerations and incorporate some of the microscopic corrections. Einstein's equations provide the lowest order description of the dynamics and *calculable*, higher order corrections arise as we probe smaller scales. The mechanism for ignoring the bulk cosmologi-cal constant is likely to survive quantum gravitational corrections, which are likely to bring in additional, higher derivative terms to the action.

References

Abbott, L. F., and Wise, M. B. (1984). *Nucl. Phys. B* **244**, 541.

Akbar, M. (2007). [hep-th/0702029].

Akbar, M., and Cai, R.-G. (2006a). [hep-th/0609128].

Akbar, M., and Cai, R.-G. (2006b). [gr-qc/0612089].

Alabidi, L., and Lyth, D. H. (2005). [astro-ph/0510441].

Albrecht, A., and Steinhardt, P. J. (1982). *Phys. Rev. Lett.* **48**, 1220.

Alcaniz, J. S. (2006). [astro-ph/0608631].

Allemandi, G., *et al.* (2003). [gr-qc/0308019].

Andrianov, A. A., *et al.* (2007). [arXiv:0711.4300].

Ashtekar, A., *et al.* (1992). *Phys. Rev. Lett.* **69**, 237.

Astier, P., *et al.* (2006). *Astron. Astrophys.* **447**, 31 [astro-ph/0510447].

Bagla, J. S., and Padmanabhan, T. (1994). *Mon. Not. R. Astron. Soc.* **266**, 227 [gr-qc/9304021].

Bagla, J. S., and Padmanabhan, T. (1997a). *Mon. Not. R. Astron. Soc.* **286**, 1023 [astro-ph/9605202].

Bagla, J. S., and Padmanabhan, T. (1997b). *Pramana* **49**, 161 [astro-ph/0411730].

Bagla, J. S., Padmanabhan, T., and Narlikar, J. V. (1996). *Comments Astrophys.* **18**, 275 [astro-ph/9511102].

Bagla, J. S., *et al.* (1998). *Astrophys. J.* **495**, 25 [astro-ph/9707330].

Bagla, J. S., *et al.* (2003). *Phys. Rev. D* **67**, 063504 [astro-ph/0212198].

Bagla, J. S. (2004). [astro-ph/0411043].

Balbi, A., *et al.* (2000). *Astrophys. J.* **545**, L1.

Barcelo, C., *et al.* (2001). *Int. J. Mod. Phys. D* **10**, 799 [gr-qc/0106002].

Bardeen, J. M., *et al.* (1983). *Phys. Rev D* **28**, 679.

Barris, B. J. (2004). *Astrophys. J* **602**, 571.

Barrow, J. D. (2006). [gr-qc/0612128].

Bennett, C., *et al.* (2003). *Astrophys. J. Suppl.* **148**, 1.

Brainerd, T. G., *et al.* (1993). *Astrophys. J.* **418**, 570.

Buchert, T. (2007). [arXiv:0707.2153].

Cai, R.-G., and Cao, L.-M. (2006). [gr-qc/0611071].

Calcagni, G., and Liddle, A. R. (2006). [astro-ph/0606003].

Calmet, X., *et al.* (2004). *Phys. Rev. Lett.* **93**, 211101 [hep-th/0505144].

Cannata, F., and Kamenshchik, A.Y. (2006). [gr-qc/0603129].

Cannata, F., and Kamenshchik, A.Y. (2008). [arXiv:0801.2348].

Capozziello, S. (2007). [arXiv:0706.3587].

Carneiro, S., *et al.* (2006). [astro-ph/0605607].

Chattopadhyay, S., *et al.* (2007). [arXiv:0712.3107].

Choudhury, T. R., and Padmanabhan, T. (2005). *Astron. Astrophys.* **429**, 807 [astro-ph/0311622].

Conley, A., *et al.* (2007). [arXiv:0705.0367v1].

Copeland, E. J., *et al.* (2006). *Int. J. Mod. Phys. D* **15**, 1753 (2006).

Costa, F. E. M., *et al.* (2007). [arXiv:0708.3800].

Cremades, D., *et al.* (2006). [hep-th/0608174].

Das, A., *et al.* (2006). [gr-qc/0610097].

de Bernardis, P., *et al.* (2000). *Nature (London)* **404**, 955.

de Putter, R, and Linder, E. V. (2007). [arXiv:0705.0400].

DeWitt, B. S. (1964). *Phys. Rev. Lett.* **13**, 114.

DeWitt, B. S. (1979). In *General Relativity: An Einstein Centenary Survey*, S. W. Hawking and W. Israel, Eds., Cambridge: Cambridge University Press, p. 680.

Dolgov, A. D. (2006). [hep-ph/0606230].

Efstathiou, G., *et al.* (1990). *Nature (London)* **348**, 705.

Efstathiou, G., *et al.* (1992). *Mon. Not. R. Astron. Soc.* **258**, 1.

Elizalde, E., *et al.* (2005). [hep-th/0502082].

Ellis, G. F. R., and Buchert, T. (2005). *Phys. Lett. A* **347**, 38.

Ellis, G. F. R., and Madsen, M. S. (1991). *Class. Quant. Grav.* **8**, 667 (1991).

Engineer, S., *et al.* (2000). *Mon. Not. R. Astron. Soc.* **314**, 279 [astro-ph/9812452].

Fabris, J. C., *et al.* (2006). [gr-qc/0609017].

Fang, W., *et al.* (2006). [hep-th/0606033].

Fontanini, M., *et al.* (2006). *Phys. Lett. B* **633**, 627 [hep-th/0509090].

Freedman, W., *et al.* (2001). *Astrophys. J.* **553**, 47.

Fulling, S. A. (1973). *Phys. Rev. D* **7**, 2850

Fuzfa, A., and Alimi, J.-M. (2006). [astro-ph/0611284].

Gannouji, R., *et al.* (2006). [astro-ph/0606287].

Gao, C. J., *et al.* (2006). [astro-ph/0605682].

Garay, L. J. (1995). *Int. J. Mod. Phys. A* **10**, 145.

Ge, X.-H. (2007). [hep-th/0703253].

Gibbons, G. W., and Hawking, S. W. (1977). *Phys. Rev. D* **15**, 2738.

Grande, J., *et al.* (2006). [gr-qc/0604057].

Gu, J. (2008). [arXiv:0801.4737].

Gurbatov, S. N., *et al.* (1989). *Mon. Not. R. Astron. Soc.* **236**, 385.

Guth, A. H. (1981). *Phys. Rev. D* **23**, 347.

Guth, A. H,, and Pi, S.-Y. (1982). *Phys. Rev. Lett.* **49**, 1110.

Hamilton, A. J. S., *et al.* (1991). *Astrophys. J.* **374**, L1.

Hanany, S., *et al.* (2000). *Astrophys. J.* **545**, L5.

Hannestad, S. (2006). *Int. J. Mod. Phys. A* **21**, 1938.

Hawking, S. W. (1982). *Phys. Lett. B* **115**, 295.

Hirata, C. M., and Seljak, U. (2005). *Phys. Rev. D* **72**, 083501.

Huang, C.-G., and Sun, J.-R. (2007). [gr-qc/0701078].

Ishibashi, A., and Wald, R. M. (2006). *Class. Quant. Grav.* **23**, 235.

Jacobson, T. (1995). *Phys. Rev. Lett.* **75**, 1260.

Jain, D., *et al.* (2007). *Phys. Lett. B* **656**, 15 [arXiv:0709.4234].

Jassal, H. K., *et al.* (2005a). *Mon. Not. R. Astron. Soc.* **356**, L11 [astro-ph/0404378].

Jassal, H. K., *et al.* (2005b). *Phys. Rev. D* **72**, 103503 [astro-ph/0506748].

Jassal, H. K., *et al.* (2006). [astro-ph/0601389].

Jha, S., *et al.* (2007). *Astrophys. J.* **659**, 122.

Johri, V. V., and Rath, P. K. (2006). *Phys. Rev. D* **74**, 123516 [astro-ph/0603786].

Kanekar, N., *et al.* (2001). *Mon. Not. R. Astron. Soc.* **324**, 988 [astro-ph/0101562].

Kazanas, D. (1980). *Astrophys. J. Lett.* **241**, 59.

Kolb, E. W., *et al.* (2005a). [astro-ph/0506534]

Kolb, E. W., *et al.* (2005b). [hep-th/0503117].

Kothawala, D., Sarkar, S., and Padmanabhan, T. (2007). *Phys. Lett. B* **652**, 338 [gr-qc/0701002].

Lanczos, C. (1932). *Z. Phys.* **73**, 147.

Lanczos, C. (1938). *Ann. Math.* **39**, 842.

Lima, J. A. S., *et al.* (2006). [astro-ph/0611007].

Linde, A. D. (1982). *Phys. Lett. B* **108**, 389.

Lindesay, J. V., *et al.* (2004). [astro-ph/0412477].

Losic, B., and Unruh, W. G. (2005). *Phys. Rev. D* **72**, 123510.

Lovelock, D. (1971). *J. Math. Phys.* **12**, 498.

Mahajan, G., *et al.* (2006). *Phys. Lett. B* **641**, 6 [astro-ph/0604265].

Makela, J. (2007). [gr-qc/0701128].

Mason, B. S., *et al.* (2003). *Astrophys. J.* **591**, 540.

Mottola, E., and Vaulin, R. (2006). [gr-qc/0604051].

Mould, J. R., *et al.* (2000). *Astrophys. J.* **529**, 786.

Movahed, M. S., *et al.* (2007). [astro-ph/0701339].

Mukhopadhyay, A., and Padmanabhan, T. (2006). *Phys. Rev. D* **74**, 124023 [hep-th/0608120].

Munshi, D., *et al.* (1997). *Mon. Not. R. Astron. Soc.* **290**, 193 [astro-ph/9606170].

Myung, Y. S. (2004). [hep-th/0412224].

Narlikar, J. V., and Padmanabhan, T. (1991). *Ann. Rev. Astron. Astrophys.* **29**, 325.

Nesseris, S., and Perivolaropoulos, L. (2007). *JCAP* **0702**, 025.

Nityananda, R., and Padmanabhan, T. (1994). *Mon. Not. R. Astron. Soc.* **271**, 976 [gr-qc/9304022].

Nobbenhuis, S. (2006). [gr-qc/0609011].

Nojiri, S., and Odintsov, S. D. (2006a). [hep-th/0601213].

Nojiri, S., and Odintsov, S. D. (2006b). [hep-th/0606025].

Ostriker, J. P., and Steinhardt, P. J. (1995). *Nature (London)* **377**, 600.

Padmanabhan, T. (1985a). *Ann. Phys. (N.Y.)* **165**, 38.

Padmanabhan, T. (1985b). *Class. Quant. Grav.* **2**, 117.

Padmanabhan, T. (1987a). *Gen. Relat. Grav.* **19**, 927.

Padmanabhan, T. (1987b). *Class. Quant. Grav.* **4**, L107.

Padmanabhan, T. (1988). *Phys. Rev. Lett.* **60**, 2229.

Padmanabhan, T. (1996). *Mon. Not. R. Astron. Soc.* **278**, L29 [astro-ph/9508124].

Padmanabhan, T. (1997). *Phys. Rev. Lett.* **78**, 1854 [hep-th/9608182].

Padmanabhan, T. (1998a). *Phys. Rev. D* **57**, 6206.

Padmanabhan, T. (1998b). *Phys. Rev. Lett.* **81**, 4297 [hep-th/9801015].

Padmanabhan, T. (1999). *Phys. Rev. D* **59**, 124012 [hep-th/9801138].

Padmanabhan, T. (2002a). *Phys. Rev. D* **66**, 021301 [hep-th/0204150].

Padmanabhan, T. (2002b). *Class. Quant. Grav.* **19**, L167 [gr-qc/0204020].

Padmanabhan, T. (2002c). *Gen. Relat. Grav.* **34**, 2029 (2002) [gr-qc/0205090].

Padmanabhan, T. (2002d). *Mod. Phys. Lett. A* **17**, 1147 [hep-th/0205278].

Padmanabhan, T. (2002e). *Class. Quant. Grav.* **19**, 5387 [gr-qc/0204019].

Padmanabhan, T. (2003a). *Phys. Rep.* **380**, 235 [hep-th/0212290].

Padmanabhan, T. (2003b). *Gen. Relat. Grav.* **35**, 2097.

Padmanabhan, T. (2003c). *Mod. Phys. Lett. A* **18**, 2903 (2003) [hep-th/0302068].

Padmanabhan, T. (2004a). *Mod. Phys. Lett. A* **19**, 2637 [gr-qc/0405072].

Padmanabhan, T. (2004b). *Class. Quant. Grav.* **21**, L1 [gr-qc/0310027].

Padmanabhan, T. (2004c). *Int. J. Mod. Phys. D* **13**, 2293 [gr-qc/0408051].

Padmanabhan, T. (2004d). *Class. Quant. Grav.* **21**, 4485 (2004) [gr-qc/0308070].

Padmanabhan, T. (2005a). In *100 Years of Relativity – Space-time Structure: Einstein and Beyond*, A. Ashtekar, Ed. Singapore: World Scientific, p. 175 [gr-qc/0503107].

Padmanabhan, T. (2005b). [astro-ph/0512077].

Padmanabhan, T. (2005c). *Curr. Sci.* **88**, 1057 [astro-ph/0411044].

Padmanabhan, T. (2005d). Gravity: a new holographic perspective (Lecture at the International Conference on Einstein's Legacy in the New Millennium, Puri, India, Dec. 2005) *Int. J. Mod. Phys. D* **15**, 1659 [gr-qc/0606061].

Padmanabhan, T. (2005e). *Class. Quant. Grav.* **22**, L107 [hep-th/0406060].

Padmanabhan, T. (2005f). *Brazilian J. Phys.* (Special Issue) **35**, 362 [gr-qc/0412068].

Padmanabhan, T. (2005g). *Phys. Rep.* **406**, 49 [gr-qc/0311036].

Padmanabhan, T. (2005h). *Int. J. Mod. Phys D* **14**, 2263-2270 [gr-qc/0510015].

Padmanabhan, T. (2006a). *AIP Conf. Proc.* **843**, 111 [astro-ph/0602117].

Padmanabhan, T. (2006b). *Gravity's Immunity from Vacuum: The Holographic Structure of Semiclassical Action*, Third prize essay; Gravity Essay Contest.

Padmanabhan, T. (2006c). *Gen. Relat. Grav.* **38**, 1547.

Padmanabhan, T. (2006d). *Int. J. Mod. Phys. D* **15**, 2029 [gr-qc/0609012].

Padmanabhan, T. (2006e). *AIP Conf. Proc.* **861**, 858 [astro-ph/0603114].

Padmanabhan, T. (2007). In *Gravity as an Emergent Phenomenon: A Conceptual Description*, *AIP Conf. Proc.* **989**, 114 [arXiv:0706.1654].

Padmanabhan, T. (2008). *Gen. Relat. Grav.* **40**, 529 [arXiv:0705.2533].

Padmanabhan, T., and Choudhury, T. R. (2000). *Mod. Phys. Lett. A* **15**, 1813 [gr-qc/0006018].

Padmanabhan, T., and Choudhury, T. R. (2002). *Phys. Rev. D* **66**, 081301 [hep-th/0205055].

Padmanabhan, T., and Choudhury, T. R. (2003). *Mon. Not. R. Astron. Soc.* **344**, 823 [astro-ph/0212573].

Padmanabhan, T., and Engineer, S. (1998). *Astrophys. J.* **493**, 509 [astro-ph/9704224].

Padmanabhan, T., and Narasimha, D. (1992). *Mon. Not. R. Astron. Soc.* **259**, 41P.

Padmanabhan, T., and Paranjape, A. (2007). *Phys. Rev. D* **75**, 064004 [gr-qc/0701003].

Padmanabhan, T., and Ray, S. (2006). *Mon. Not. R. Astron. Soc. Lett.* **372**, L53 [astro-ph/0511596].

Padmanabhan, T., and Sethi, S. (2001). *Astrophys. J.* **555**, 125 [astro-ph/0010309].

Padmanabhan, T., and Singh, T.P. (1987). *Class. Quant. Grav* **4**, 1397.

Padmanabhan, T., Seshadri, T. R., and Singh, T. P. (1989). *Phys. Rev. D* **39**, 2100.

Padmanabhan, T., *et al.* (1996). *Astrophys. J.* **466**, 604 [astro-ph/9506051].

Panigrahi, K. L., *et al.* (2007). [arXiv:0708.1679].

Panotopoulos, G. (2006). [astro-ph/0606249].

Panotopoulos, G. (2007). [arXiv:0712.1177].

Papantonopoulos, E., *et al.* (2006). [hep-th/0601152].

Paranjape, A., and Singh, T. P. (2008). [arXiv:0801.1546].

Paranjape, A., Sarkar, S., and Padmanabhan, T. (2006). *Phys. Rev. D* **74**, 104015 [hep-th/0607240].

Pearson, T. J., *et al.* (2003). *Astrophys. J.* **591**, 556.

Percival, W. J., *et al.* (2001). *Mon. Not. R. Astron. Soc.* **327**, 1297.

Percival, W. J., *et al.* (2002). *Mon. Not. R. Astron. Soc.* **337**, 1068.

Perivolaropoulos, L. (2006). *AIP Conf. Proc.* **848**, 698.

Perlmutter, S. J., *et al.* (1999). *Astrophys. J.* **517**, 565.

Rasanen, S. (2004). *JCAP* **0402**, 003.

Ren, J., *et al.* (2006). [astro-ph/0610266].

Riess, A. G., *et al.* (1998). *Astron. J.* **116**, 1009.

Riess, A. G., *et al.* (2004). *Astrophys. J.* **607**, 665.

Sakharov, A. D. (1968). *Sov. Phys. Dokl.* **12**, 1040.

Sarkar, S. (1996). *Rep. Prog. Phys.* **59**, 1493.

Schunck, F. E., and Mielke, E. W. (1994). *Phys. Rev. D* **50**, 4794.

Sen, A. A. (2006). [gr-qc/0604050].

Setare, M. R. (2006). *Phys. Lett. B* **642**, 1.

Setare, M. R. (2007). *Phys. Lett. B* **644**, 99.

Setare, M. R., and Vagenas, E. C. (2008). [arXiv:0801.4478].

Shao, Y., *et al.* (2007). [gr-qc/0703112].

Shapiro, I. L. (2008). [arXiv:0801.0216].

Sheykhi, A., *et al.* (2007). [hep-th/0701198].

Smoot, G. F., *et al.* (1992). *Astrophys. J.* **396**, L1.

Snyder, H. S. (1947). *Phys. Rev.* **71**, 38.

Sola, S., and Stefancic, H. (2006). *Mod. Phys. Lett. A* **A21**, 479.

Sola, J. (2007). [arXiv:0710.4151].

Sorkin, D. (1997). *Inl. J. Theor. Phys.* **36**, 2759.

Spergel, D. N., *et al.* (2003). *Astrophys. J. Suppl.* **148**, 175.

Spergel, D. N, *et al.* (2006). [astro-ph/0603449].

Srinivasan, K., *et al.* (1998). *Phys. Rev. D* **58**, 044009 [gr-qc/9710104].

Srinivasan, K., *et al.* (1999). *Phys. Rev. D* **60**, 24007 [gr-qc/9812028].

Sriramkumar, L., and Padmanabhan, T. (2005). *Phys. Rev. D* **71**, 103512 [gr-qc/0408034].

Sriramkumar, L., *et al.* (2002). *Int. J. Mod. Phys. D* **11**, 1 [gr-qc/9903054].

Starobinsky, A. A. (1979). *JETP Lett.* **30**, 682.

Starobinsky, A. A. (1981). *Phys. Lett. B* **91**, 99.

Starobinsky, A. A. (1982). *Phys. Lett. B* **117**, 175.

Stojkovic, D., *et al.* (2007). [hep-ph/0703246].

Tatekawa, T. (2004). [astro-ph/0412025].

Tonry, J. L., *et al.* (2002). *Astrophys. J.* **594**, 1.

Unruh, W. G. (1976). *Phys. Rev. D* **14**, 870.

Visser, M. (2002). *Mod. Phys. Lett. A* **17**, 977 [gr-qc/0204062].

Volovik, G. E. (2001). *Phys. Rep.* **351**, 195.

Volovik, G. E. (2003). *The Universe in a Helium Droplet*, Oxford: Oxford University Press.

Volovik, G. E. (2004). [gr-qc/0405012].

Volovik, G. E. (2006). [gr-qc/0604062].

Wang, M. (2006). [hep-th/0601189].

Wang, Y., and Mukherjee, P. (2007). *Phys. Rev. D* **76**, 103533.

Wang, X., *et al.* (2002). *Phys. Rev. D* **65**, 123001.

Yoneya, T. (1976). *Prog. Theor. Phys.* **56**, 1310.

Zalaletdinov, R. M. (1992). *Gen. Relat. Grav.* **24**, 1015.

Zeldovich, Ya. B. (1970). *Astron. Astrophys.* **5**, 84.

Zhang, J., *et al.* (2008). [arXiv:0801.2809].

Zhou, J., *et al.* (2007). [arXiv:0705.1264].

Part II

Observations

5

Foundations of supernova cosmology

ROBERT P. KIRSHNER

5.1 Supernovae and the discovery of the expanding universe

Supernovae have been firmly woven into the fabric of cosmology from the very beginning of modern understanding of the expanding, and now accelerating universe. Today's evidence for cosmic acceleration is just the perfection of a long quest that goes right back to the foundations of cosmology. In the legendary Curtis–Shapley debate on the nature of the nebulae, the bright novae that had been observed in nebulae suggested to Shapley (1921) (see Trimble, 1995) that the systems containing them must be nearby. Otherwise, he reasoned, they would have unheard-of luminosities, corresponding to $M = -16$ or brighter. Curtis (1921) countered, concluding, "the dispersion of the novae in spirals and in our galaxy may reach ten magnitudes . . . a division into two classes is not impossible." Curtis missed the opportunity to name the supernovae, but he saw that they must exist if the galaxies are distant. Once the distances to the nearby galaxies were firmly established by the observation of Cepheid variables (Hubble, 1925), the separation of ordinary novae and their extraordinary, and much more luminous super cousins, became clear.

A physical explanation for the supernovae was attempted by Baade and Zwicky (1934). Their speculation that supernova energy comes from the collapse to a neutron star is often cited, and it is a prescient suggestion for the fate of massive stars, but not the correct explanation for the supernovae that Zwicky and Baade studied systematically in the 1930s. In fact, the spectra of all the supernovae that they discovered and followed up in those early investigations were of the distinct, but spectroscopically mysterious, hydrogen-free type that today we call SN Ia. They are not powered by core collapse, but by a thermonuclear flame. Baade (1938)

Dark Energy: Observational and Theoretical Approaches, ed. Pilar Ruiz-Lapuente. Published by Cambridge University Press. © Cambridge University Press 2010.

showed that the luminosities of the supernovae in their program were more uniform than those of galactic novae, with a dispersion of their peak luminosities near 1.1 mag, making them suitable as extragalactic distance indicators. Right from the beginning, supernovae were thought of as tools for measuring the universe.

Nature has more than one way to explode a star. This was revealed clearly by Minkowski (1941) who observed a distinct spectrum for some supernovae, different from those obtained for the objects studied by Baade. SN 1940B had strong hydrogen lines in its spectrum. These are the stars whose energy source we now attribute to core collapse in massive stars. At the time, it seemed sensible to call Baade's original group Type I (SN I) and the new class Type II (SN II). The small dispersion in luminosity for Baade's sample resulted from his good luck in having Zwicky discover a string of supernovae that were all of a single type. SN I are generally less luminous than the galaxies in which they occur. (Introductory texts, and introductory remarks in colloquia concerning supernovae usually get this basic fact wrong.) The SN II are, generally speaking, fainter than SN I and have a larger dispersion in their luminosity. Separating the supernovae, on the basis of their spectra, into distinct physical classes is one way they have become more precise as distance indicators. By the late sixties, Kowal (1968) was able to make a Hubble diagram for 19 SN I. The scatter about the Hubble line for this sample, which reached out to the Coma Cluster of galaxies at a redshift of 7000 km/s was about 0.6 magnitudes. These were photographic magnitudes, obtained with the non-linear detectors of the time, and they contained no correction for absorption by dust in the host galaxies, which we now know is an important source of scatter in the observed samples. But this was a promising step forward.

In 1968, there was plenty of room for improvement in the precision of SN I measurements and in extending the redshift range over which they were studied. As Kowal forecast: "These supernovae could be exceedingly useful indicators of distance. It should be possible to obtain average supernova magnitudes to an accuracy of 5% to 10% in the distances." He also predicted the future use of supernovae to determine cosmic acceleration: "It may even be possible to determine the second-order term in the redshift–magnitude relation when light curves become available for very distant supernovae." The "second-order term" would be the one that indicated cosmic acceleration or deceleration. Along with the Hubble constant (which would require reliable distances from Cepheid variables), this deceleration term was expected to provide an account of cosmic kinematics, and, in the context of general relativity, for the dynamics of the universe, as sketched for astronomers in the classic paper by Sandage (1961).

On the last page of this paper, Sandage worked out the observational consequences of the exponential expansion that would be produced by a cosmological constant. He explicitly showed that you cannot decide between an accelerating

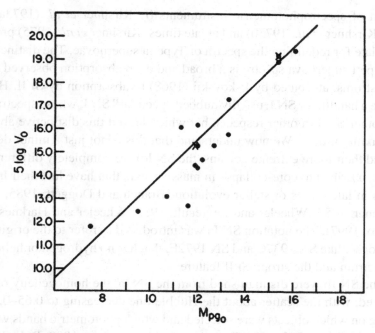

Fig. 5.1. Hubble diagram for 16 individual SN I (dots), plus averages for those in the Virgo and Coma clusters of galaxies (crosses). (From Kowal, 1968.)

universe of this type and the steady-state model (they would both have $q_0 = -1$). Yet, in 1968, the measurement of deceleration was presented by Sandage (1968) as a decisive test between the steady-state model, which predicted acceleration, and Friedmann cosmologies, where matter would produce deceleration. It is possible that, if cosmic acceleration had been discovered earlier, it might have been taken as evidence in favor of the steady-state model. It was the richer physical context of cosmological information, such as the cosmic microwave background, that led to a much different conclusion in 1998.

5.1.1 Classifying supernovae

In 1968, there was ample room for technical improvement in the measurements themselves, a need for a proper account for the effects of dust, and just as important, well into the 1980s the classification scheme for SN I was still incomplete. Core-collapse supernovae were mixed in among the thermonuclear explosions that make up most of the Type I supernovae. As described by Zwicky (1965) and later by Oke and Searle (1974), the definition of a SN I was empirical: it meant that the spectrum resembled the bright supernova SN 1937C as extensively studied by Minkowski (1939). The bright supernova SN 1972E, observed with a new generation of spectrophotometric instruments by Kirshner *et al.* (1973a) in the infrared (Kirshner *et al.,* 1973b) and at late times (Kirshner *et al.*, 1975) provided a rich template for redefining the spectra of Type Ia supernovae. The distinctive feature in Type I supernova spectra is a broad and deep absorption observed at about 6150 angstroms, attributed by Pskovskii (1968) to absorption by Si II. However, there were a handful of SN I, usually dubbed "peculiar" SN I, whose spectra resembled the other SN I in other respects, but which lacked this distinctive absorption line at maximum light. We now understand that this is not just a minor detail: the SN Ib (and their more extreme cousins, the SN Ic) are completely different physical events, ascribed to core-collapse in massive stars that have lost their hydrogen envelopes in late stages of stellar evolution (Branch and Doggett, 1985; Uomoto and Kirshner, 1985; Wheeler and Levreault, 1985; Wheeler and Harkness, 1990; Filippenko, 1997). The notation SN Ia was introduced to refer to the original class of supernovae, like SN 1937C and SN 1972E, that has no hydrogen or helium lines in the spectrum and the strong Si II feature.

Once the SN Ib were distinguished from the SN Ia, the homogeneity of the SN Ia improved, with the scatter about the Hubble line decreasing to 0.65–0.36 mag, depending on which objects were selected and which photometric bands were used (Branch and Miller, 1990, 1993; Tammann and Leibundgut, 1990; Della Valle and Panagia, 1992). This work rested on the assumption that the SN Ia were identical, so that a single underlying template for the light curve (Leibundgut, 1988) could be used to interpolate between the observations of any individual object to determine its apparent brightness at maximum light in the B-band, and put all the objects on a common scale.

5.1.2 SN II as cosmological distance indicators

The idea that supernovae could be used to measure cosmological parameters had more than one component. Another line of work employed Type II supernovae. As pointed out by Kirshner and Kwan (1974), the expanding photospheres of these hydrogen-rich supernovae provide the possibility to measure distances without reference to any other astronomically determined distance. The idea of the expanding photosphere method (EPM) is that the atmosphere was not too far from a blackbody, so the temperature could be determined from the observed energy distribution. If you measure the flux and temperature, that determines the angular size of the photosphere. Since you can measure the temperature and flux many times during the first weeks after the explosion, an observer can establish the angular expansion rate of a supernova. At the same time, absorption lines formed in the expanding atmosphere, from hydrogen and weaker lines that more closely trace the expansion of the photosphere, give the expansion velocity. If you know the angular rate of expansion from the temperature and flux and the linear rate from the shape of the absorption lines, you can solve for the distance to a Type II supernova. The combination of the supernova's redshift and distance allows for a measurement of the Hubble constant that does not depend on any other astronomically determined distance. The departure from a blackbody energy distribution could be computed for a supernova atmosphere, as done by Schmidt, Kirshner, and Eastman (1992), and this held out the prospect of making more precise distance measurements to SN II than had been achieved for SN Ia.

Wagoner (1977) noted that this approach could be extended to high redshift to measure the effects of cosmic deceleration, and also pointed out that the EPM provided an internal test of its own validity: if the distance determined remained the same, while the temperature and the velocity of the atmosphere changed, this was a powerful sign that the measurement was consistent. This was an important point, since the prospects for using galaxies as the principal tracer of cosmic expansion were dimming, due to evidence that the luminosity of a galaxy could easily change over time due to stellar evolution and galaxy mergers. Even the sign of this change was not known for certain. Galaxies might grow brighter over time due to mergers, and they might grow dimmer due to stellar evolution. In either case, unless the effect was carefully calibrated, it could easily swamp the small changes in apparent magnitude with redshift that hold the information on the history of cosmic expansion. Supernovae, though fainter than galaxies, were discrete events that would not have the same set of changes over cosmic time. The use of SN II for cosmology has recently been revived and it promises to provide an independent path to measuring expansion and perhaps even acceleration (Dessart *et al.*, 2008; Poznanski *et al.*, 2008).

For SN II, the expanding photospheres provide a route to distances that can accommodate a range of intrinsic luminosities and still provide accurate distances, because the atmospheres have hydrogen and behave like those of other stars. For SN Ia, the atmospheres are more difficult to analyze, but the hope was simpler: that the physics underlying the explosion of a SN Ia would determine its luminosity. The idea that SN Ia were identical explosions has a theoretical underpinning. In the earliest pictures, the SN Ia were imagined to come from the ignition of a carbon–oxygen white dwarf at the Chandrasekhar mass (Hoyle and Fowler, 1960; Colgate and McKee, 1969). In models of this type, a supersonic shock wave travels through the star, burning it thoroughly into iron-peak isotopes, especially Ni^{56}. Such a standard explosion of a uniform mass would lead to a homogeneous light curve and uniform luminosity, making SN Ia into perfect standard candles. The exponential light curves that suggest an energy input from radioactivity and the late-time spectra of SN Ia, which are made up of blended iron emission lines, were broadly consistent with this picture. Though the simple theoretical idea that SN

Ia are white dwarfs that ignite near the Chandrasekhar mass has been repeated many times as evidence that SN Ia must be perfect standard candles, nature disagrees. Observations show that there is a factor of three range in luminosity from the most luminous SN Ia (resembling SN 1991T) to the least luminous (resembling SN 1991bg). Despite the facts, many popular (and professional!) accounts of SN Ia assert that SN Ia are standard candles because they explode when they reach the Chandrasekhar limit. This is wishful thinking.

5.1.3 Searching for SN Ia for cosmology

Nevertheless, the hope that SN Ia might prove to be good standard candles began to replace the idea that brightest cluster galaxies were the standard candles best suited to measuring the deceleration of the universe. As a coda to his pioneering automated supernova search, Stirling Colgate imagined the way in which a similar search with the Hubble Space Telescope (HST) might find distant supernovae (Colgate, 1979). A more sober analysis of the problem by Gustav Tammann estimated the sample size that would be needed to make a significant detection of deceleration using HST (Tammann, 1979). The result was encouraging: depending on the dispersion of the SN Ia, he found between 6 and 25 SN Ia at $z \sim 0.5$ would be needed to give a 3σ signal of cosmic deceleration. Tammann got the quantities right – it was only the sign of the effect that was wrong.

Unwilling to wait for the advent of the HST, a pioneering group from Denmark began a program of supernova observations using the Danish 1.5-meter telescope at the European Southern Observatory (ESO) (Hansen, Jorgensen, and Norgaard-Nielsen, 1987; Hansen *et al.*, 1989). Their goal was to find distant supernovae, measure their apparent magnitudes and redshifts, and, on the assumption the SN Ia were standard candles, fit for q_0 from the Hubble diagram. This method is described with precision in Chapter 6 in this book by Pilar Ruiz-Lapuente. The difference between $q_0 = 0.1$ and q_0 of 0.5 is only 0.13 mag at redshift of 0.3. At the time they began their work, there was hope that the intrinsic scatter for SN Ia might be as small as 0.3 mag. To beat the errors down by root-N statistics to make a 3σ distinction would take dozens of well-observed supernovae at $z \sim 0.3$.

The Danish group used the search rhythm developed over the decades by Zwicky and his collaborators for finding supernovae. Since the time for a Type Ia supernova to rise to maximum and fall back by a factor of two is roughly one month, monthly observations in the dark of the Moon are the best way to maximize discoveries. Observations made toward the beginning of each dark run were most useful, since that allowed time to follow up each discovery with spectroscopy and photometry. This is the pattern Zwicky established with the Palomar 18-inch Schmidt and which was used for many years by Sargent and Kowal with the

48-inch Schmidt at Palomar (Kowal, Sargent and Zwicky, 1970). It is the pattern used by the Danish group, and all the subsequent supernova search teams until the introduction of dedicated searches like that of Kare *et al.* (1988) and the rolling search led by John Tonry (Barris *et al.*, 2004) that became the model for the recent ESSENCE and SNLS searches.

But there was something new in the Danish search. Photographic plates, which are large but non-linear in their response to light, were replaced by a charge coupled device (CCD). The advantages were that the CCD was much more sensitive to light (by a factor of ~100!) and that the digital images were both linear and immediately available for manipulation in a computer. Fresh data taken at the telescope could be processed in real time to search for new stars, presumably supernovae, in the images of galaxy clusters. The new image needed to be registered to a reference image taken earlier, the two images appropriately scaled to take account of variations in sky brightness, the better of the two images blurred to match the seeing of the inferior image, and then subtracted. The Danish team implemented these algorithms and demonstrated their success with SN 1988U, a SN Ia in a galaxy at redshift 0.31 (Norgaard-Nielsen *et al.*, 1989). Although this group developed the methods for finding distant supernovae in digital data, the rate at which they were able to find supernovae was disappointingly low. Instead of making steady progress toward a cosmologically significant sample at a rate of, say, one object per month, they only found one supernova per year. At this rate, it would take 10 years to beat down the measuring uncertainty and to begin to learn about the contents of the universe. And that was in the optimistic case where the intrinsic scatter of SN Ia was assumed to be small. Instead, the observational evidence was pointing in the opposite direction, of larger dispersion among the SN Ia. Another early effort, carried out by the Lawrence Berkeley Laboratory at the 4-meter Anglo-Australian Telescope had even less luck. Despite building a special-purpose prime focus CCD camera to find supernovae, they reported none (Couch *et al.*, 1991).

5.1.4 SN Ia as standard candles – not!

Starting in 1986, careful observations made with CCD detectors showed ever more clearly that the luminosity and the light curve shapes for SN Ia were not uniform (Phillips *et al.*, 1987). In 1991, two supernovae at opposite extremes of the luminosity scale showed for certain that this variety was real, and needed to be dealt with in order to make SN Ia into effective distance measuring tools. SN 1991bg (Filippenko *et al.*, 1992; Leibundgut *et al.*, 1993) was extremely faint and SN 1991T (Phillips *et al.*, 1992) was extremely bright. Despite hope for a different result, and a theoretical argument why their luminosities should lie in a narrow range, Type Ia supernovae simply are not standard candles: they are known to vary over a factor of three in their intrinsic luminosity. The size of the sample needed to make a cosmological measurement scales as the square of the scatter, so, in 1991, the truly productive thing to harness supernovae for cosmology was not to find more distant supernovae, but to learn better how to reduce the uncertainty in the distance for each object.

Using a set of well-sampled SN Ia light curves with precise optical photometry and accurate relative distances, Phillips (1993) demonstrated a correlation between the shape of a SN Ia light curve and the supernova's luminosity. Supernovae with the steepest declines are the least luminous. More interestingly, even among the supernovae that do not lie at the extremes of the distribution marked by SN 1991T and SN 1991bg, the relation between luminosity and light curve shape provides an effective way to decrease the scatter in the Hubble diagram for SN Ia. Phillips used this correlation to decrease the observed scatter about the Hubble line to about 0.3 mag.

This made the path forward a little clearer. What was needed was a well-run supernova search for relatively nearby supernovae that could guarantee accurate follow-up observations. Mark Phillips, Mario Hamuy, Nick Suntzeff and their colleagues at Cerro Tololo Inter-American Observatory (CTIO) and José Maza and his colleagues at the University of Chile's Cerro Calán observatory worked together to conduct such a search, the Calán–Tololo Supernova Search (Hamuy *et al.*, 1993). The technology was a hybrid of the past and the future – photographic plates were used on the venerable Curtis Schmidt telescope (named in honor of Heber D. Curtis, of the debate cited earlier) at Cerro Tololo to search a wide field (25 square degrees) in each exposure. Despite the drawbacks of photographic plates as detectors, this large field of view made this the most effective search for nearby supernovae. The plates were developed on the mountain, shipped by bus to Santiago, and then painstakingly scanned by eye with a blink comparator to find the variable objects.

The modern part was the follow-up. Since the search area was large enough to guarantee that there would be objects found each month, CTIO scheduled time in advance on the appropriate telescopes for thorough photometric and spectroscopic follow-up with CCD detectors. The steady weather at Cerro Tololo and the dedicated work at Cerro Calán led to a stream of supernova discoveries and a rich collection of excellent supernova light curves. For example, in 1996, the Calán–Tololo group published light curves of 29 supernovae obtained on 302 nights in four colors (Hamuy *et al.*, 1996a). This is what was needed to develop reliable ways to use the supernova light curves to determine the intrinsic luminosity of SN Ia, and to measure the luminosity distance to each object (Hamuy *et al.*, 1996b). The Calán–Tololo Search was restricted to redshifts below 0.1, so it did not, by itself, contain information on the cosmology. However, it provided the data needed to understand how to measure distances with supernovae, and, when used in combination with high-z supernovae, it had the potential to help determine the cosmology.

5.1.5 Dust or cosmology?

However, the accuracy of the distance measurements was compromised by the uncertain amount of dust absorption in the each supernova host galaxy. Two parallel approaches were developed. One, led by Mark Phillips and his colleagues, used the observational coincidence, first noted by Paulina Lira, that the evolution in the color B-V had a very small dispersion at ages from 30 to 90 days after maximum (Phillips *et al.*, 1999). By measuring the observed color at those times, the absorption could be inferred and the true distance measured. The other, based on the same data set, and then later extended through observations at the Whipple Observatory of the Center for Astrophysics (CfA), used an empirical method to find that intrinsically faint supernovae are also intrinsically redder. Since the light curve shape, which was the strongest clue to supernova luminosity, was not greatly affected by absorption, it was possible to determine both the distance and the absorption by dust to each supernova. A formal treatment of the extinction using Bayes' theorem was used to determine the best values and their uncertainty (Riess, Press, and Kirshner, 1996a). This MLCS (multi-color light curve shape) approach was also used to examine whether the dust in other galaxies was the same as dust in the Milky Way (Riess *et al.*, 1996b). While the early indications were that the dust in other galaxies had optical properties that were consistent with those found in the Galaxy, as the samples of supernovae have grown larger and the precision of the measurements has improved, this simple picture is no longer tenable. These early workers recognized that measuring the extinction to individual supernovae was an essential step in deriving reliable information on the cosmology. After all, the dimming due to an accelerating cosmology at redshift 0.5 is only of order 0.2 magnitudes. If instead this dimming were produced by dust like the dust of the Milky Way, the additional reddening would be only 0.07 mag in the B-V color, so good photometry in multiple bands was essential to make reliable inferences on the presence or absence of cosmic acceleration.

5.1.6 Early results

The earliest observations of the Supernova Cosmology Project (SCP) did not take account of these requirements. Their observations of SN 1992bi at $z = 0.458$ were made in only one filter, making it impossible, even in principle, to determine the reddening (Perlmutter *et al.*, 1995). No spectrum for this object was obtained, but it was completely consistent with being a SN Ia. This was a striking demonstration that the search techniques used by the SCP, which resembled those of the Danish team, could reliably detect transient events in galaxies at the redshifts needed to make a cosmologically interesting measurement. The search was carried out with a million pixel CCD camera at the 2.5-meter Isaac Newton Telescope, whose increased speed over the Danish system made it plausible that a supernova could be found in each month's observing. As with the Calán–Tololo search being carried out at low redshift, it was reasonable for the SCP to schedule follow-up observations. The SCP developed the "stretch" method for accounting for the connection between luminosity and light curve shape in the B and V bands. This works very well, but does not, by itself, account for the effects of dust extinction (Goldhaber *et al.*, 2001).

The High-Z Supernova Team (HZT) was formed in 1995 by cooperation between members of the Calán–Tololo group and supernova workers at the Harvard–Smithsonian CfA and ESO. The goal was to apply the new methods for determining the intrinsic luminosity and reddening of a supernova, developed from the low-redshift samples, to objects at cosmologically interesting distances. This required mastering the techniques of digital image subtraction. The first object found by the HZT was SN 1995K, at a redshift of 0.479, which, at that time was the highest yet published (Leibundgut *et al.*, 1996). Observations were obtained in two colors, and the supernova's spectrum showed it was a genuine Type Ia. Leibundgut *et al.* used the observations to show that the light curve for SN 1995K was stretched in time by a factor of $(1 + z)$, just as expected in an expanding universe.

The time-dilation effect had been discussed in 1939 by Olin Wilson (1939), sought in nearby data by Rust (1974), and by Leibundgut (1990). Publications by Goldhaber and his colleagues of the SCP (Goldhaber *et al.*, 1996, 2001) show this effect in their data, though the degeneracy between the light curve shape as analyzed by the "stretch" method and time dilation requires some (quite plausible) constraints on changes in the supernova population with redshift to draw a firm conclusion. Another approach to the same problem uses the evolution of the spectra of SN Ia to show in an independent way that the clocks governing distant supernovae appear to run slower by the factor $(1 + z)$ (Foley *et al.*, 2005; Blondin *et al.*, 2008).

In the mid-1990s, important technical developments improved the ability to discover distant supernovae. At the National Optical Astronomy Observatories, new $2K \times 2K$ CCD systems were implemented at the 4-meter telescopes at Kitt Peak and at Cerro Tololo. In 1997, the Big Throughput Camera (Wittman *et al.*, 1998) became available for general use at the 4-meter telescope at Cerro Tololo. This 16 megapixel camera set the standard for distant supernova searches and was employed by both SCP and HZT as they developed the samples that led to the discovery of cosmic acceleration.

But the path to cosmic acceleration was not smooth or straight. In July 1997, based on seven objects, the SCP published the first cosmological analysis based on supernovae (Perlmutter *et al.*, 1997). Comparing their data from $z \sim 0.4$, most of which was obtained through just one filter, to the nearby sample from Calán–Tololo (Hamuy *et al.*, 1996a) they found a best value for Ω_m of 0.88, and concluded that their results were "inconsistent with Lambda-dominated, low-density, flat cosmologies."

Some theorists had begun to speculate that Λ was the missing ingredient to reconcile the observations of a large value for the Hubble constant (Freedman, Madore, and Kennicutt, 1997), the ages of globular clusters, and a low value for Ω_m in a flat cosmology (Ostriker and Steinhardt, 1995; Krauss and Turner, 1995). If the universe was flat with a total Ω of 1, and had Ω_m of 0.3, then subtraction pointed to a value for Λ of 0.7 and you could match the ages of the globular clusters even if the Hubble constant was significantly larger than previously thought. But the initial results of the SCP pointed in the opposite direction, and their evidence for deceleration threw the cold water of data on these artfully constructed arguments.

The situation began to change rapidly late in 1997. Both teams used the Hubble Space Telescope to observe supernovae that had been found from the ground. The precision of the HST photometry was very good, with the supernova well resolved from the host galaxy thanks to the unique angular resolution of HST. Once the difficult task of accurately connecting the HST photometry to the ground-based work was complete, the observations could be combined to provide additional constraints at the beginning of 1998. For the SCP, there was one additional object from HST, at a record redshift of 0.83. When combined with a subset of the data previously published in July, the analysis gave a qualitatively different answer. In their January 1998 *Nature* paper (submitted on October 7, 1997), the SCP now found "these new measurements suggest that we may live in a low-mass-density universe." There was no observational evidence presented in this paper for cosmic acceleration (Perlmutter *et al.*, 1998). For the High-Z Team, the HST-based sample was larger, with three objects, including one at the unprecedented redshift of 0.97 (Garnavich *et al.*, 1998a). Although the HZT additional sample of ground-based high-redshift observations was meager (just 1995K), using the same MLCS and template-fitting techniques on both the high-z and low-z samples, and augmenting the public low-z sample from Calán–Tololo with data from the CfA improved the precision of the overall result. Taken at face value, the analysis in this paper, submitted on October 14, 1997 and published on January 14, 1998, showed the tame result that matter alone was insufficient to produce a flat universe, and, more provocatively, if you insisted that Λ was zero, and the universe was flat, then the best fit to the data had Ω_m less than 0. This was a very tentative whisper of what, with hindsight, we can now see was the signal of cosmic acceleration.

5.2 An accelerating universe

5.2.1 First results

Both teams had larger samples under analysis during the last months of 1997, and it was not long before the first analyses were published. The HZT, after announcing their results at the Dark Matter meeting in February 1998 (Filippenko and Riess, (1998)), submitted a long article entitled "Observational evidence from supernovae for an accelerating universe and a cosmological constant" to the *Astronomical Journal* on March 13, 1998. This appeared in the September 1998 issue (Riess *et al.*, 1998). It used a sample of 16 high-z and 34 nearby objects obtained by HZT members, along with the methods developed at the CfA and by the Calán–Tololo group to determine distances, absorptions, and their uncertainties for each of these objects. The data clearly pointed to cosmic acceleration, with luminosity distances in the high-z sample 10–15% larger than expected in a low-mass density universe without Λ. The HZT also published a long methods paper (Schmidt *et al.*, 1998) and an analysis of this data set in terms of the dark energy equation of state (Garnavich *et al.*, 1998b).

The SCP, after showing their data at the January 1998 AAS meeting, cautiously warned that systematic uncertainties, principally the possible role of dust absorption, made it premature to conclude the universe was accelerating. They prepared a long paper for publication that showed the evidence from 42 high-redshift objects and 18 low-redshift objects from the Calán–Tololo work. This was submitted to the *Astrophysical Journal* on September 8, 1998 and appeared in June 1999 (Perlmutter *et al.*, 1999). Although the SCP had no method of their own for determining the reddening and absorption to individual supernovae, they showed that the color distributions of their high-z sample and the objects they selected from the Calán–Tololo sample had similar distributions of restframe color, an indication that the extinction could not be very different in the two samples. They also applied the method of Riess, Press, and Kirshner (1996a) to determine the absorption in the cases where they had the required data. The analysis showed, with about the same statistical power as the HZT paper, that the luminosity distances to supernovae clearly favored a picture in which the universe was accelerating.

5.2.2 Room for doubt?

Two important questions about these results soon surfaced.

One was whether the results of the two groups were independent. Some of the machinery for analyzing the data sets, for example, the K-corrections to take account of the way supernova redshifts affect the flux in fixed photometric bands, were based on the same slender database of supernova spectra. Similarly, the low-redshift sample used by the SCP was made up entirely of objects observed by the HZT. The two teams cooperated on observing a few of the high-redshift objects and both teams used the data for those objects. A small number of co-authors showed up on both the HZT and the SCP publications. But the analysis was done independently, most of the high-redshift samples were disjoint, and the astronomical community generally took the agreement of two competing teams to imply that this result was real. But it was the integrity of the results, not the friction of the personalities, that made this work credible.

Another question about the initial results was whether the measured effect – a small, but significant dimming of the distant supernovae relative to nearby ones – was due to cosmology, to some form of dust, or to evolution in the properties of SN Ia with redshift (Aguirre, 1999a,b; Aguirre and Haiman, 2000; Drell, Loredo, and Wasserman, 2000). Aguirre explored the notion that there might be "grey dust" that would cause dimming without reddening. Theoretical difficulties included the limit imposed by using all the available solids, distributing them uniformly, and staying under the limit imposed on the thermal emission from these particles by observations in the far infrared. A direct approach to the possible contribution of dust came from measurements of supernovae over a wider wavelength range – the dust could not be perfectly grey, and a wider range of observations, made with infrared detectors, would reveal its properties more clearly. The earliest application of this was by the HZT (Riess *et al.*, 2000), who observed a supernova at $z = 0.46$ in the restframe I band, with the goal of constraining the properties of Galactic dust or of the hypothetical grey dust. They concluded that the observed dimming of the high-z sample was unlikely to be the result of either type of dust. Much later, this approach was employed by the SCP (Nobili *et al.*, 2005). Dust obscuration, and the relation of absorption to reddening, remains the most difficult problem in using supernova luminosity distances for high-precision cosmology, but the evidence is strong that dust is not responsible for the ~ 0.25 mag dimming observed at $z \sim 0.5$.

A second route to excluding grey dust was to extend observations of SN Ia to higher redshift. If the dimming were due to uniformly distributed dust, there would be more of it along the line-of-sight to a more distant supernova. Due to the discovery of a supernova in a repeat observation of the Hubble Deep Field (Gilliland, Nugent, and Phillips, 1999) and unconscious follow-up with the NICMOS program in that field, Adam Riess and his collaborators were able to construct observations of SN 1997ff at the extraordinary redshift of $z \sim 1.7$ (Riess *et al.*, 2001). In a flat universe with $\Omega_\Lambda \sim 2/3$ and $\Omega_m \sim 1/3$, there is a change in the sign of the expected effect on supernova apparent brightness. Since the matter density would have been higher at this early epoch by a factor $(1+z)^3$, the universe would have been decelerating at that time, if the acceleration is due to something that acts like the cosmological constant. The simplest cosmological models predict that a supernova at $z \sim 1.7$ will appear brighter than you would otherwise expect. Dust cannot reverse the sign of its effect, so these measurements of the light curve of a SN Ia at $z \sim 1.7$ provided a powerful qualitative test of that idea. While the data were imperfect, the evidence, even from this single object, was inconsistent with the grey dust that would be needed to mimic the effect of cosmic acceleration at lower redshift.

Another way of solidifying the early result was to show that the spectra of the nearby supernova of Type Ia, the supernovae at $z \sim 0.5$ that gave the strongest signal for acceleration, and spectra of the most distant objects beyond z of 1 give no sign of evolution. While the absence of systematic changes in the spectra with epoch isn't proof that the luminosities do not evolve, it is a test that the supernovae could have failed. They do not fail this test. The early HZT results by Coil *et al.* (2000) show that, within the observational uncertainties, the spectra of nearby and the distant supernovae are indistinguishable. This approach was explored much later by the SCP (Hook *et al.*, 2005) with consistent results.

5.2.3 After the beginning

By the year 2000, the context for analyzing the supernova results, which give a strong constraint on the combination $(\Omega_\Lambda - \Omega_m)$ soon included strong evidence for a flat universe with $(\Omega_\Lambda + \Omega_m = 1)$ from the power spectrum of the CMB (de Bernardis *et al.*, 2002) and stronger evidence for the low value of Ω_m from galaxy clustering surveys (Folkes *et al.*, 1999). Much later, the Baryon Acoustic Oscillations played this role (Eisenstein *et al.*, 2005). The concordance of these results swiftly altered the conventional wisdom in cosmology to a flat ΛCDM picture. But the concordance of these various methods does not mean that they should lean on each other for support like a trio of drunkards. Instead, practitioners of each approach need to assess its present weaknesses and work to remedy those. For supernovae, the opportunities included building the high-z sample, which was still only a few handfuls, extending its range to higher redshift, augmenting the low-z sample, identifying the systematic errors in the samples, and developing new, less vulnerable methods for measuring distances to supernovae.

5.2.3.1 Building the high-z sample

The High-Z Team published additional data in 2003 that augmented the high-z sample and extended its range to $z = 1.2$ (Tonry *et al.*, 2003). Using the 12K

CCD detector at the Canada–France–Hawaii Telescope and the Suprime-Cam at Subaru 8.2-meter telescope, the HZT then executed a "rolling" search of repeated observations with a suitable sampling interval of 1–3 weeks for 5 months (Barris *et al.*, 2004). This enabled the HZT to double the world's sample of published objects with $z > 0.7$, to place stronger constraints on the possibility of grey dust, and to improve knowledge of the dark energy equation of state. The publication by the SCP of 11 SN Ia with $0.36 < z < 0.86$ included high-precision HST observations of the light curves and full extinction corrections for each object (Knop *et al.*, 2003).

By this point, in 2003, the phenomenon of cosmic acceleration was well established and the interpretation as the effect of a negative pressure component of the universe fit well into the concordance picture that now included results from WMAP (Spergel *et al.*, 2003). But what was not so clear was the nature of the dark energy. Increasing the sample near $z \sim 0.5$ was the best route to improving the constraints on dark energy. One way to describe the dark energy is through the equation of state index $w = p/\rho$. For a cosmological constant, $1 + w = 0$. Back-of-the-envelope calculations showed that samples of a few hundred high-z supernovae would be sufficient to constrain w to a precision of 10%. As before, two teams undertook parallel investigations. The Supernova Legacy Survey (SNLS), carried out at the Canada–France–Hawaii telescope, included many of the SCP team. The ESSENCE program (Equation of State: SupErNovae trace Cosmic Expansion) carried out at Cerro Tololo included many of the HZT. This phase of constraining dark energy is thoroughly described in Chapter 7 in this book by Michael Wood-Vasey.

The SNLS observing program was assigned 474 nights over 5 years at CFHT. They employed the one-degree imager, Megacam, to search for supernovae and to construct their light curves in a rolling search, with a 4 day cadence, starting in August of 2003. In 2006, they presented their first cosmological results, based on 71 SN Ia, which gave a value of $1 + w = -0.023$ with a statistical error of 0.09, consistent with a cosmological constant (Astier *et al.*, 2006).

The ESSENCE program used the MOSAIC II imager at the prime focus of the 4-meter Blanco telescope. They observed with this 64 megapixel camera every other night for half the night during the dark of the moon in the months of October, November, and December for 6 years, starting in 2002. The survey is described by Miknaitis *et al.* (2007) and cosmological results from the first 3 years of data were presented in 2006 (Wood-Vasey *et al.*, 2007). The ESSENCE analysis of 60 SN Ia gave a best value for $1 + w = -0.05$, with a statistical error of 0.13, consistent with a cosmological constant and with the SNLS results. Combining the SNLS and ESSENCE results gave a joint constraint of $1 + w = -0.07$ with a statistical error of 0.09.

We can expect further results from these programs, but the easy part is over. Bigger samples of distant supernovae do not assure improved knowledge of dark energy because systematic errors are now the most important source of uncertainty. These include photometric errors and uncertainties in the light curve fitting methods, and also more subtle matters such as the way dust absorption affects the nearby and distant samples. Collecting large samples is still desirable, especially if the photometric errors are small, but tightening the constraints on the nature of dark energy will also demand improved understanding of supernovae and the dust that dims and reddens them.

5.2.3.2 Extending its range

While the work of Tonry *et al.* (2003) and Barris *et al.* (2004) showed that it was possible, with great effort, to make observations from the ground of supernovae beyond a redshift of 1, the installation of the Advanced Camera for Surveys (ACS) on HST provided a unique opportunity to search for and follow these extremely high-redshift objects (Blakeslee *et al.*, 2003). By enlisting the cooperation of the GOODS (Great Observatories Origins Deep Survey) survey, and breaking its deep exposures of extragalactic fields into repeated visits that formed a rolling search, the Higher-Z Team, led by Adam Riess, developed effective methods for identifying transients, selecting the SN Ia from their colors, obtaining light curves, determining the reddening from IR observations with NICMOS, and measuring the spectra with the grism disperser that could be inserted into the ACS (Riess *et al.*, 2004a,b, 2007). This program has provided a sample of 21 objects with $z > 1$, and demonstrated directly the change in acceleration, the "cosmic jerk", that is the signature of a mixed dark matter and dark energy universe. The demise of the ACS brought this program to a halt. The 2009 servicing mission has restored HST to this rich line of investigation.

5.2.3.3 Augmenting the low-z sample

Both the HZT (which included members of the Calán–Tololo supernova program) and the SCP depended on low-redshift observations of supernovae to establish the reality of cosmic acceleration. The samples at high redshift were assembled, at great effort, and high cost in observing time at the world's largest telescopes because it was clear that these data could shift our view of the universe. The low-z samples require persistence, careful attention to systematic effects, and promised no shift in world view. They have been slower to develop. Two early steps forward were the publication by Riess of 22 BVRI light curves from his Ph.D. thesis at Harvard (Riess *et al.*, 1999), and the publication of 44 UBVRI light curves from the thesis work of Jha (Jha, Riess, and Kirshner, 2007). The U-band observations in Jha's work were especially helpful in analyzing the HST observations of the Higher-Z program, since, for the highest redshift objects observed with HST, most of the observations correspond to ultraviolet emission in the supernova's rest frame. Jha also revised and retrained the MLCS distance estimator that Riess had developed, using this larger data set, and dubbed it MLCS2k2. Recently, Kowalski compiled the "Union" data sample (Kowalski *et al.*, 2008). His work assessed the uncertainties in combining data from diverse sources, and, by applying stringent cuts to the data, provided a set of 57 low-redshift and 250 high-redshift supernovae to derive constraints on dark energy properties. Kowalski noted the imbalance of the low-z and high-z samples and emphasized the opportunity to make a noticeable improvement in the constraints on dark energy by increasing the sample size for the nearby events.

A third Ph.D. thesis at Harvard, by Malcolm Hicken, has just been completed and finally brings the low-redshift sample out of the statistical limit created by our slow accumulation of nearby objects and begins to encounter the systematic limit imposed by imperfect distance estimators. Hicken analyzed the data for 185 SN Ia in 11 500 observations made at the Center for Astrophysics over the period from 2001 to 2008 (see Fig. 5.2). This large and homogeneous data set improves on the Union data set compiled by Kowalski to form the (more perfect) Constitution data set (Hicken *et al.*, 2009a,b; see Fig. 5.3). When Hicken uses the same distance fitter used by Kowalski to derive the expansion history and fits to a constant dark energy, he derives $1 + w = 0.013$ with a statistical error of about 0.07 and a systematic error that he estimates at 0.11. As discussed below, one important contribution to the systematic error that was not considered by Kowalski is the range of results that is produced by employing different light curve fitters such as SALT, SALT2, and MLCS2k2, which handle the properties of dust in different ways.

This CfA work is a follow-up program that exploits the supernova discovery efforts carried out at the Lick Observatory by Alex Filippenko, Weidong Li, and their many collaborators (Filippenko *et al.*, 2001) as well as a growing pace of supernova discoveries by well-equipped and highly motivated amateur astronomers. Since the selection of the Constitution supernova sample is not homogeneous, information extracted from this sample concerning supernova parent populations and host galaxy properties needs to be handled with caution, but it suggests that even after light curve fitting, the SN Ia in Scd, Sd, or irregular galaxy hosts are intrinsically fainter than those in elliptical or S0 hosts, as reported earlier by Sullivan, based on the SCP sample (Sullivan *et al.*, 2003). The idea of constructing a single fitting procedure for supernovae in all galaxy types has proved effective, but it may be missing a useful clue to distinct populations of SN Ia in galaxies that are and are not currently forming stars. There may be a variety of evolutionary paths to becoming a SN Ia that produce distinct populations of SN

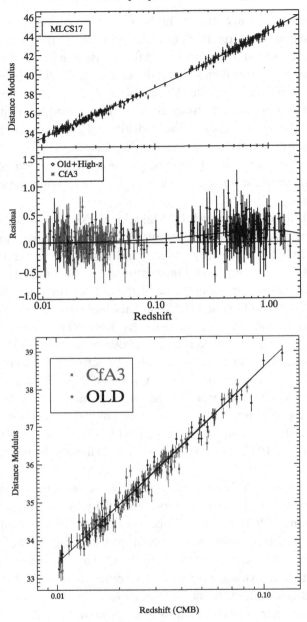

Fig. 5.2. *Top panels*: Hubble diagram and residuals for MLCS17 (which uses $R_V = 1.7$ to minimize the residuals). The new CfA3 points are shown as rhombs and the OLD and high-z points as crosses. *Bottom panel*: Hubble diagram of the CfA3 and OLD nearby SN Ia. (From Hicken *et al.*, 2009a.)

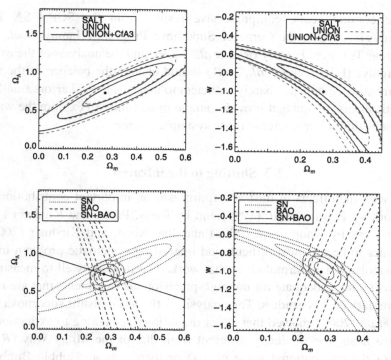

Fig. 5.3. *Left panels*: Today's best constraints from the Constitution data set on Ω_m and Ω_Λ. The lower panel shows the combination of the SN contours with the BAO prior. *Right panels*: Same for w versus Ω_m in a flat universe. (Hicken *et al.*, 2009b.)

Ia in star-forming galaxies that are not exactly the same as the SN Ia in galaxies where star formation ceased long ago (Mannucci *et al.*, 2005; Scannapieco and Bildsten, 2005; Sullivan *et al.*, 2006). Constructing separate samples and deriving distinct light curve fitting methods for these stellar populations may prove useful once the samples are large enough.

A step in this direction comes from the work at La Palma, building up the sample at the sparsely sampled redshift range near $z = 0.2$ (Altavilla *et al.*, 2009). A comprehensive approach to sampling has been taken by the Sloan Supernova Survey (Friedman *et al.*, 2009). By repeatedly scanning a 300 deg^2 region along the celestial equator, the survey identified transient objects for spectroscopic follow-up with excellent reliability and has constructed ugriz light curves for over 300 spectroscopically confirmed SN Ia. With excellent photometric stability, little bias in the supernova selection, and a large sample in the redshift range $0.05 < z < 0.35$, this data set will be a powerful tool for testing light curve fitting techniques, provide a low-redshift anchor to the Hubble diagram, and should result in a more certain knowledge of dark energy properties.

In the coming years, comprehensive results from the SCP's SN Factory (Aldering *et al.*, 2002), the Carnegie Supernova Program (Hamuy *et al.*, 2006), the Palomar Transient Factory (Rau *et al.*, 2009), and the analysis of the extensive KAIT archive (Filippenko *et al.*, 2001) should change the balance of the world's sample from one that is just barely sufficient to make statistical errors smaller than systematic errors, to one that provides ample opportunity to explore the ways that sample selection might decrease those systematic errors.

5.3 Shifting to the infrared

Coping with the effects of dust absorption was an important contribution of the early work by Phillips *et al.* (1999) and by Riess, Press, and Kirshner (1996a), the later work by Knop *et al.* (2003) and Jha, Riess, and Kirshner (2007) and it continues to be the most difficult and interesting systematic problem in supernova cosmology. The formulations that worked sufficiently well to measure 10% effects will not be adequate for the high-precision measurements that are required for future dark energy studies. The analysis of the low-redshift supernova data by Conley *et al.* (2007) showed that either the ratio of reddening to extinction in the supernova hosts was distinctly different from that of the Milky Way ($R_V = 1.7$ instead of the conventional value of 3.1) or there was a "Hubble Bubble" – a zone in which the local expansion rate departed from the global value. As discussed by Hicken *et al.* (2009a), today's larger sample does not show evidence for the Hubble Bubble, but the value of R_V that performs best for MLCS2k2 and for SALT is significantly smaller than 3.1. It seems plausible that the sampling for earlier work was inhomogeneous, with highly reddened objects present only in the nearby region. If the correction for reddening in these cases was not carried out accurately, they could contribute to the illusion of a Hubble Bubble. But the evidence for a small effective value of R_V has not gone away. It seems logical to separate the contribution due to reddening from the contribution that might result from an intrinsic relation between supernova colors and supernova luminosity, as done in MLCS2k2, but the approach of lumping these together, as done by the fitting techniques dubbed SALT and SALT2, also works well empirically (Guy *et al.*, 2005, 2007). In the ESSENCE analysis, the effects of extinction on the properties of the observed sample were carefully considered, and found to affect the cosmological conclusions. Getting this problem right will be an important part of preparing for higher precision cosmological measurements with future surveys.

Fortunately, there is a very promising route to learning more about dust, avoiding its pernicious effects on supernova distances, and deriving reliable and precise measures of dark energy properties. That route is to measure the properties of the

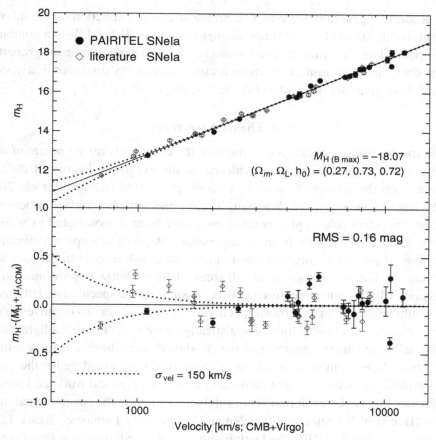

Fig. 5.4. *H*-band SN Ia Hubble diagram. It includes 23 new SN Ia observed with PAIRITEL. (From Wood-Vasey *et al.*, 2008.)

supernovae in the rest frame infrared. As shown in the pioneering work of Krisciunas, Phillips, and Suntzeff (2004), nearby SN Ia in the Hubble flow behave as very good standard candles when measured in near infrared bands (NIR), typically J, H, and K_s. This work has recently been extended by Wood-Vasey *et al.* who used the PAIRITEL system (a refurbished and automated version of the 2MASS telescope) at Mount Hopkins to obtain near-infrared light curves that double the world's sample (Wood-Vasey *et al.*, 2008; Mandel *et al.*, 2009). Even with no correction for light curve shape or dust absorption, the NIR light curves for SN Ia exhibit a scatter about the Hubble line that is typically 0.15 mag. This is comparable to the scatter that is achieved by the output of the elaborate light curve fitters now in use for optical data that correct for the width of the light curve's peak and use the optical colors to infer dust corrections. This means that the SN Ia actually do behave like standard candles – but in the NIR! What's more, the effects of dust absorption generally scale as $1/\lambda$, so the effects of extinction on the infrared measurements should be four times smaller than at the B-band. When combined with optical data, the infrared observations can be used to determine the properties of the dust, and to measure even more accurate luminosity distances. Early steps toward these goals are underway (Friedman *et al.*, 2009).

5.4 The next ten years

Goals for the coming decade are to improve the constraints on the nature of dark energy by improving the web of evidence on the expansion history of the universe and on the growth of structure through gravitation (Albrecht *et al.*, 2006, 2009; Ruiz-Lapuente, 2007; Frieman, Turner, and Huterer, 2008). Supernovae have an important role to play because they have been demonstrated to produce results. Precise photometry from homogeneous data, dust absorption determined with near-IR measurements, and constructing useful subsamples in galaxies with differing star formation histories are all areas where we know improvement in the precision of the distance measurements is possible. More speculative, but plausible, would be the use of supernova spectra in a systematic way to improve the distance estimates. Implementation of statistically sound ways to use the light curves (and possibly spectra) to determine distances should make the results more reliable and robust. What is missing is a level of theoretical understanding for the supernova explosions themselves that could help guide the empirical work, and provide confidence that stellar evolution is not subtly undermining the cosmological inferences (Hoeflich, Wheeler, and Thielemann, 1998; Ruiz-Lapuente, 2004). Large samples from Pan-STARRS, the Dark Energy Survey, and, if we live long enough, from JDEM and LSST will eventually be available. Chapter 8 in this book by Alex Kim makes a persuasive case for the effectiveness of a thorough space-based study of supernovae. Our ability to use these heroic efforts effectively depends on improving our understanding of supernovae as astronomical objects in the context of galaxy formation, stellar evolution, and the physics of explosions. Then we can employ the results with confidence to confirm, or, better yet, to rule out some of the weedy garden of theoretical ideas for the dark energy described in other chapters of this book!

Acknowledgements

Supernova research at Harvard is supported by the US National Science Foundation through grant AST09-07903. I am grateful for the long series of colleagues who have contributed so much to this work. HCO staff member Peter Challis is beyond category. As postdocs: Alan Uomoto, Bruno Leibundgut, Pilar Ruiz-Lapuente, Eric Schlegel, Peter Hoeflich, David Jeffery, Peter Garnavich,

Tom Matheson, Stéphane Blondin, and Michael Wood-Vasey. And as graduate students: Ron Eastman, Chris Smith, Brian Schmidt, Jason Pun, Adam Riess, Saurabh Jha, Marayam Modjaz, Malcolm Hicken, Andy Friedman, and Kaisey Mandel.

References

Aguirre, A. (1999a). *Astrophys. J.* **512**, L19.

Aguirre, A. (1999b). *Astrophys. J.* **525**, 583.

Aguirre, A., and Haiman, Z. (2000). *Astrophys. J.* **532**, 28.

Albrecht, A., Bernstein, G., Cahn, R., *et al.* (2006). [astro–ph/0609591].

Albrecht, A., Amendola, L., Bernstein, G., *et al.* (2009). [arXiv:0901.0721].

Aldering, G., Adam, G., Antilogus, P., *et al.* (2002). *SPIE Conf. Ser.*, **4836**, 61.

Altavilla, G., Ruiz-Lapuente, P., Balastegui, A., *et al.* (2009). *Astrophys. J.* **695**, 135.

Astier, P., Guy, J., Regnault, N., *et al.* (2006). *Astron. Astrophys.* **447**, 31.

Baade, W. (1938). *Astrophys. J.* **88**, 285.

Baade, W., and Zwicky, F. (1934). *Proc. Nat. Acad. Sci. USA* **20**, 254.

Barris, B. J., Tonry, J. L., Blondin, S., *et al.* (2004). *Astrophys. J.* **602**, 571.

Blakeslee, J. P., Tsvetanov, Z. I., Riess, A. G., *et al.* (2003). *Astrophys. J.* **589**, 693.

Blondin, S., Davis, T. M., Krisciunas, K., *et al.* (2008). *Astrophys. J.* **682**, 72.

Branch, D., and Doggett, J. B. (1985). *Astron. J.* **90**, 2218.

Branch, D., and Miller, D. L. (1990). *Astron. J.* **100**, 530.

Branch, D., and Miller, D. L. (1993). *Astrophys. J.* **405**, L5.

Coil, A. L., Matheson, T., Filippenko, A. V., *et al.* (2000). *Astrophys. J.* **544**, L111.

Colgate, S. A. (1979). *Astrophys. J.* **232**, 404.

Colgate, S. A., and McKee, C. (1969). *Astrophys. J.* **157**, 623.

Conley, A., Carlberg, R. G., Guy, J., *et al.* (2007). *Astrophys. J.* **664**, L13.

Couch, W. J., Perlmutter, S., Newburg, H. J. M., *et al.* (1991). *Proc. Astron. Soc. Austral.* **9**, 261.

Curtis, H. D. (1921). *Bull. NRC*, **2**, 194.

de Bernardis, P., Ade, P. A. R., Bock, J. J., *et al.* (2002). *Astrophys. J.* **564**, 559.

Della Valle, M., and Panagia, N. (1992). *Astron. J.* **104**, 696.

Dessart, L., *et al.* (2008). *Astrophys. J.* **675**, 644.

Drell, P. S., Loredo, T. J., and Wasserman, I. (2000). *Astrophys. J.* **530**, 593.

Eisenstein, D. J., *et al.* (2005). *Astrophys. J.* **633**, 560.

Filippenko, A. V., (1997). *Ann. Rev. Astron. Astrophys.* **35**, 309.

Filippenko, A. V., Richmond, M. W., Branch, D., *et al.* (1992). *Astron. J.* **104**, 1543.

Filippenko, A. V., and Riess, A. G. (1998). *Phys. Rep.* **307**, 31.

Filippenko, A. V., Li, W. D., Treffers, R. R., and Modjaz, M. (2001). *Astron. Soc. Pacif. Conf. Ser.* **246**, 121.

Foley, R. J., Filippenko, A. V., Leonard, D. C., Riess, A. G., Nugent, P., and Perlmutter, S. (2005). *Astrophys. J.* **626**, L11.

Folkes, S., Ronen, S., Price, I., *et al.* (1999). *Mon. Not. R. Astron. Soc.* **308**, 459.

Freedman, W. L., Madore, B. F., and Kennicutt, R. C. (1997). In *The Extragalactic Distance Scale*, M. Livio, M. Donahue, and N. Panagia, Eds., Cambridge: Cambridge University Press, p. 171.

Friedman, A. S., Wood-Vasey, M., Mandel, K., *et al.* (CfA Supernova Group and PAIRI-TEL) (2009). *Amer. Astron. Soc. Meet. Abstr.* **213**, 438.06.

Frieman, J. A., Turner, M. S., and Huterer, D. (2008). *Ann. Rev. Astron. Astrophys.* **46**, 385.

Frieman, J. A., Bassett, B., Becker, A., *et al.* (2008). *Astron. J.* **135**, 338.

Garnavich, P. M., Kirshner, R. P., Challis, P., *et al.* (1998a). *Astrophys. J.* **493**, L53.

Garnavich, P. M., Jha, S., Challis, P., *et al.* (1998b). *Astrophys. J.* **509**, 74.

Gilliland, R. L., Nugent, P. E., and Phillips, M. M. (1999). *Astrophys. J.* **521**, 30.

Goldhaber, G., Boyle, B., Bunclark, P., *et al.* (1996). *Nucl. Phys. B Proc. Suppl.* **51**, 12.

Goldhaber, G., Groom, D. E., Kim, A., *et al.* (2001). *Astrophys. J.* **558**, 359.

Guy, J., Astier, P., Nobili, S., Regnault, N., and Pain, R. (2005). *Astron. Astrophys.* **443**, 781.

Guy, J., Astier, P., Baumont, S., *et al.* (2007). *Astron. Astrophys.* **466**, 11.

Hamuy, M., Maza, J., Phillips, M. M., *et al.* (1993). *Astron. J.* **106**, 2392.

Hamuy, M., Phillips, M. M., Suntzeff, N. B., *et al.* (1996a). *Astron. J.* **112**, 2408.

Hamuy, M., Phillips, M. M., Suntzeff, N. B., *et al.* (1996b). *Astron. J.* **112**, 2438.

Hamuy, M., Folatelli, G., Morrell, N. I., *et al.* (2006). *Publ. Astron. Soc. Pacif.* **118**, 2.

Hansen, L., Jorgensen, H. E. and Norgaard-Nielsen, H. U. (1987). *ESO Mess.* **47**, 46.

Hansen, L., Jorgensen, H. E., Norgaard-Nielsen, H. U., Ellis, R. S., and Couch, W. J. (1989). *Astron. Astrophys.* **211**, L9.

Hicken, M., Challis, P., Jha, S., *et al.* (2009a). [arXiv:0901.4787].

Hicken, M., Wood-Vasey, W. M., Blondin, S., *et al.* (2009b). [arXiv:0901.4804].

Hoeflich, P., Wheeler, J. C., and Thielemann, F. K. (1998). *Astrophys. J.* **495**, 617.

Hook, I. M., Howell, D. A., Aldering, G., *et al.* (The Supernova Cosmology Project) (2005). *Astron. J.* **130**, 2788.

Hoyle, F., and Fowler, W. A. (1960). *Astrophys. J.* **132**, 565.

Hubble, E. P. (1925). *Astrophys. J.* **62**, 409.

Jha, S., Riess, A. G., and Kirshner, R. P. (2007). *Astrophys. J.* **659**, 122.

Kare, J. T., Burns, M. S., Crawford, F. S., Friedman, P. G., and Muller, R. A. (1988). *Rev. Sci. Instr.* **59**, 1021.

Kirshner, R. P., and Kwan, J. (1974). *Astrophys. J.* **193**, 27.

Kirshner, R. P., and Oke, J. B. (1975). *Astrophys. J.* **200**, 574.

Kirshner, R. P., Oke, J. B., Penston, M. V., and Searle, L. (1973a). *Astrophys. J.*, **185**, 303.

Kirshner, R. P., Willner, S. P., Becklin, E. E., Neugebauer, G., and Oke, J. B. (1973b). *Astrophys. J.* **180**, L97.

Knop, R. A., Aldering, G., Amanullah, R., *et al.* (The Supernova Cosmology Project) (2003). *Astrophys. J.* **598**, 102.

Kowal, C. T. (1968). *Astron. J.* **73**, 1021.

Kowal, C. T., Sargent, W. L. W., and Zwicky, F. (1970). *Publ. Astron. Soc. Pacif.* **82**, 736.

Kowalski, M., Rubin, D., Aldering, G., *et al.* (2008). *Astrophys. J.* **686**, 749.

Krauss, L. M., and Turner, M. S. (1995). *Gen. Rel. Grav.* **27**, 1137.

Krisciunas, K., Phillips, M. M., and Suntzeff, N. B. (2004). *Astrophys. J.* **602**, L81.

Leibundgut, B. (1988). Ph.D. Thesis, University of Basel.

Leibundgut, B. (1990). *Astron. Astrophys.* **229**, 1.

Leibundgut, B., Kirshner, R. P., Phillips, M. M., *et al.* (1993). *Astron. J.* **105**, 301.

Leibundgut, B., Schommer, R., Phillips, M., *et al.* (1996). *Astrophys. J.* **466**, L21.

Mannucci, F., della Valle, M., Panagia, N., *et al.* (2005). *Astron. Astrophys.* **433**, 807.

Mandel, K. S., Wood-Vasey, W. M., Friedman, A. S., and Kirshner, R. P. (2009). *Astrophys. J.* **704**, 629.

Miller, D. L., and Branch, D. (1990). *Astron. J.* **100**, 530.

Miknaitis, G., Pignata, G., Rest, A., *et al.* (2007). *Astrophys. J.* **666**, 674.

Minkowski, R. (1939). *Astrophys. J.* **89**, 156.

Minkowski, R. (1941). *Publ. Astron. Soc. Pacif.* **53**, 224.

Nobili, S., Amanullah, R., Garavini, G., *et al.* (The Supernova Cosmology Project) (2005). *Astron. Astrophys.* **437**, 789.

Norgaard-Nielsen, H. U., Hansen, L., Jorgensen, H. E., Aragon Salamanca, A., and Ellis, R.S. (1989). *Nature (London)*, **339**, 523.

Oke, J. B., and Searle, L. (1974). *Ann. Rev. Astron. Astroph.* **12**, 315.

Ostriker, J. P., and Steinhardt, P. J. (1995). *Nature (London)*, **377**, 600.

Perlmutter, S., Pennypacker, C. R., Goldhaber, G., *et al.* (1995). *Astrophys. J.* **440**, L41.

Perlmutter, S., Gabi, S., Goldhaber, G., *et al.* (The Supernova Cosmology Project) (1997). *Astrophys. J.* **483**, 565.

Perlmutter, S., Aldering, G., Della Valle, M. *et al.* (1998). *Nature* **392**, 311.

Perlmutter, S., Aldering, G., Goldhaber, G., *et al.* (The Supernova Cosmology Project) (1999). *Astrophys. J.* **517**, 565.

Phillips, M. M. (1993). *Astrophys. J.* **413**, L10.

Phillips, M. M., Phillips, A. C., Heathcote, S. R., *et al.* (1987). *Publ. Astron. Soc. Pacif.* **99**, 592.

Phillips, M. M., Wells, L. A., Suntzeff, N. B., *et al.* (1992). *Astron. J.* **103**, 163.

Phillips, M. M., Lira, P., Suntzeff, N. B., *et al.* (1999). *Astron. J.* **118**, 1766.

Poznanski, D., Butler, N., Filippenko, A. V., *et al.* (2008). [arXiv:0810.4923].

Pskovskii, Y. P. (1968). *Astron. Zh.* **45**, 945.

Rau, A., *et al.* (2009). [arXiv:0906.5355].

Riess, A. G., Press, W. H., and Kirshner, R. P. (1996a). *Astrophys. J.* **473**, 88.

Riess, A. G., Press, W. H., and Kirshner, R. P. (1996b). *Astrophys. J.* **473**, 588.

Riess, A. G., Filippenko, A. V., Challis, P., *et al.* (1998). *Astron. J.* **116**, 1009.

Riess, A. G., Kirshner, R. P., Schmidt, B. P., *et al.* (1999). *Astron. J.* **117**, 707.

Riess, A. G., Filippenko, A. V., Liu, M. C., *et al.* (2000). *Astrophys. J.* **536**, 62.

Riess, A. G., Nugent, P. E., Gilliland, R. L., *et al.* (2001). *Astrophys. J.* **560**, 49.

Riess, A. G., Strolger, L.-G., Tonry, J., *et al.* (2004a). *Astrophys. J.* **600**, L163.

Riess, A. G., Strolger, L.-G., Tonry, J., *et al.* (2004b). *Astrophys. J.* **607**, 665.

Riess, A. G., Strolger, L.-G., Casertano, S., *et al.* (2007). *Astrophys. J.* **659**, 98.

Ruiz-Lapuente, P. (2004). *Astrophys. Space Sci.* **290**, 43.

Ruiz-Lapuente, P. (2007). *Class. Quant. Grav.* **24**, R91 [arXiv:0704.1058].

Rust, B. W. (1974). Ph.D. Thesis, Oak Ridge National Laboratory.

Sandage, A. (1961). *Astrophys. J.* **133**, 355.

Sandage, A. (1968). *The Observatory* **88**, 91.

Scannapieco, E., and Bildsten, L. (2005). *Astrophys. J.* **629**, L85.

Schmidt, B. P., Kirshner, R. P., and Eastman, R. G. (1992). *Astrophys. J.* **395**, 366.

Schmidt, B. P., Suntzeff, N. B., Phillips, M. M., *et al.* (1998). *Astrophys. J.* **507**, 46.

Shapley, H. (1921). *Bull. NRC*, **2**, 171.

Spergel, D. N., Verde, L., Peiris, H. V., *et al.* (2003). *Astrophys. J. Suppl.* **48**, 175.

Sullivan, M., Ellis, R. S., Aldering, G., *et al.* (2003). *Mon. Not. R. Astron. Soc.* **340**, 1057.

Sullivan, M., Le Borgne, D., Pritchet, C. J., *et al.* (2006). *Astrophys. J.* **648**, 868.

Tammann, G. A. (1979). *NASA Conf. Publ.* **2111**, 263.

Tammann, G. A., and Leibundgut, B. (1990). *Astron. Astrophys.* **236**, 9.

Tonry, J. L., Schmidt, B. P., Barris, B., *et al.* (2003). *Astrophys. J.* **594**, 1.

Trimble, V. (1995). *Publ. Astron. Soc. Pacif.* **107**, 1133.

Uomoto, A., and Kirshner, R. P. (1985). *Astron. Astrophys.* **149**, L7.

Wagoner, R. V. (1977). *Astrophys. J.* **214**, L5.

Wheeler, J. C., and Levreault, R. (1985). *Astrophys. J.* **294**, L17.

Wheeler, J. C., and Harkness, R.P. (1990). *Reps. Prog. Phys.* **53**, 1467.

Wilson, O. C. (1939). *Astrophys. J.* **90**, 634.

Wittman, D. M., Tyson, J. A., Bernstein, G. M., *et al.* (1998). *SPIE Conf. Ser.* **3355**, 626.

Wood-Vasey, W. M., Miknaitis, G., Stubbs, C. W., *et al.* (2007). *Astrophys. J.* **666**, 694.

Wood-Vasey, W. M., Friedman, A. S., Bloom, J. S., *et al.* (2008). *Astrophys. J.* **689**, 377.

Zwicky, F. (1965). In *Stars and Stellar Systems*, L. H. Aller and D. B. McLaughlin, Eds., Chicago: University of Chicago Press, p. 367.

6

Dark energy and supernovae

PILAR RUIZ-LAPUENTE

6.1 Introduction

The use of SNe Ia as *calibrated candles* has led to a fundamental discovery: that the rate of the cosmic expansion of the universe is accelerating (Perlmutter *et al.*, 1999; Riess *et al.*, 1998). At present, the data gathered on the expansion rate do not disclose whether the acceleration is due to a component formally equivalent to the cosmological constant introduced by Einstein (1917), whether it is due to a scalar field or other component of the universe unaccounted for so far, or whether we are finding the effective behavior of a theory of a wider scope whose low energy limit slightly departs from general relativity. The presence of an energy component with negative pressure (still undistinguishable from the cosmological constant Λ) and the nature of this new component, commonly termed *dark energy*, is a major challenge in cosmology and in fundamental physics.

The bare Einstein's equations (without cosmological constant) for a universe with a Friedmann–Robertson–Walker (FRW) metric and dust-like matter imply a continuous deceleration of the expansion rate. However, a universe containing a fluid with an equation of state $p = w\rho$ with index $w < -1/3$ overcomes the deceleration when the density of this fluid dominates over that of the dust-like matter. In this context, the cosmological constant, if it is positive and added to the equations, balances the deceleration by acting as a fluid with an equation of state $p = -\rho$.

The cosmological constant was originally in the metric part of Einstein's equations, and alternative metric theories of gravity provide a wealth of terms on that side. When describing a homogeneous and isotropic universe within a FRW model, it is common to come as close as possible to the standard Friedmann–Einstein

Dark Energy: Observational and Theoretical Approaches, ed. Pilar Ruiz-Lapuente. Published by Cambridge University Press. © Cambridge University Press 2010.

equation for the expansion rate by bringing terms to the right-hand side within the so-called "effective index" w_{eff} of the cosmic fluid whose equation of state is $p = w_{\text{eff}}\rho$.

Precise enough observations of SNe Ia extending to redshifts $z \sim 1.5$–2.0 should yield the equation of state of the new component $p = w_{\text{eff}}\rho$. Both the new picture of the universe that is now emerging and the next step in its investigation, that of determining the nature of dark energy, critically depend on the reliability of the calibration of SNe Ia luminosities, which up to now has been purely empirical. While various efforts to build up large consistent samples of SNe Ia covering the z range needed for cosmology are on their way, detailed analyses of the current high-z sample have shed new light on old questions.

6.2 SNe Ia and cosmic expansion

Supernovae enable us to measure the rate of the expansion of the universe along redshift z. They do so by tracing the luminosity distance along z for a standard candle. If the universe is flat, as inferred from WMAP (Komatsu *et al.*, 2009), the d_L relation is simply

$$d_L(z) = cH_0^{-1}(1 + z) \int_0^z \frac{dz'}{H(z')}, \tag{6.1}$$

while in the general case,

$$d_L(z) = cH_0^{-1}(1 + z)|\Omega_K|^{-1/2} \operatorname{sinn}\left(|\Omega_K|^{1/2} \int_0^z \frac{H_0\, dz'}{H(z')}\right), \tag{6.2}$$

where $\operatorname{sinn}(x) = \sin(x)$, x, or $\sinh(x)$ for closed, flat, and open models respectively, where the curvature parameter Ω_K, is defined as $\Omega_K = \Omega_T - 1$ (with $\Omega_T \equiv \Omega_m + \Omega_\Lambda$).

$H(z)$ is the Hubble parameter. For a FRW universe containing matter and a general form of matter–energy X with index of the equation of state w_X, $H(z)$ is given by

$$H^2(z) = H_0^2 \left\{ \Omega_K(1 + z)^2 + \Omega_m(1 + z)^3 + \Omega_X(1 + z)^{3(1+w_X)} \right\}, \tag{6.3}$$

where H_0 is the present value of the Hubble parameter and

$$\Omega_K \equiv \frac{-k}{H_0^2 a_0^2}, \tag{6.4}$$

where k accounts for the geometry and a_0 is the present value of the scale factor,

$$\Omega_m = \frac{8\pi G}{3H_0^2}\rho_M,$$ (6.5)

and Ω_X is the density parameter for dark energy.

When $w(z)$ varies along z, $H(z)$ can be expressed as

$$H(z)^2 = H_0^2 \left\{ \Omega_K(1+z)^2 + \Omega_m(1+z)^3 + \Omega_X \exp\left[3 \int_0^z \frac{1+w(z')}{1+z'}\mathrm{d}z'\right] \right\}.$$ (6.6)

One sees, from (6.1) and (6.6), that d_L contains $w(z)$ through a double integral. This leads to limitations in recovering $w(z)$ if d_L data along z are scarce and with large errorbars.

Data are often given as $m(z)$, which relates to $d_L(z)$ as

$$m(z) = 5\log d_L(z) + [M + 25 - 5\log(H_0)].$$ (6.7)

The quantity within brackets is $\mathcal{M} \equiv M - 5\log H_0 + 25$, the "Hubble-constant-free" peak absolute magnitude of a supernova.

The measurement of $w(z)$ through $d_L(z)$ is the standard approach, though possibilities of measuring directly $H(z)$ have been investigated. We will review the progress done in measuring $w(z)$ since the discovery of dark energy. This advance has been possible thanks to the availability of large samples and the increase in precision of SNe Ia as cosmological distance indicators.

6.3 The d_L test from SNe Ia

6.3.1 Overview

A long path has brought us to the present situation where we can already aim at accuracies of 1% in distance. Back in 1968, a first attempt to draw the Hubble diagram from SNe resulted in a spread of $\sigma \sim 0.6$ mag (Kowal, 1968; see also Chapter 5). Such large scatter was due to the inclusion of other types of supernovae, which were later recognized as being of a different type (SN Ib/c) and characterized by a large scatter in luminosities (Filippenko and Riess, 1998).

SNe Ia are not standard candles either, but they follow a well-known correlation between maximum brightness and rate of decline of the light curve. In 1977, Pskovskii first described the correlation between the brightness at maximum and the rate of decline of the light curve: the brighter SNe Ia have a slower decline of their light curves whereas fainter ones are the faster decliners. A systematic follow-up of SNe Ia confirmed the brightness–decline relation (Phillips, 1993; Phillips *et al.*, 1999). The intrinsic variation of SNe Ia is written as a linear relationship:

$$M_B = M_{B0} + \alpha(\Delta m_{15}(B) - \beta) + 5 \log(H_0/65). \tag{6.8}$$

Δm_{15}, the parameter of the SNe Ia light curve family, is the number of magnitudes of decline in 15 days after maximum. The value of α, as well as the dispersion of that relation, has been evaluated from samples obtained from 1993 to the present. M_{B0} is the absolute magnitude in B for a template SN Ia of β rate of decline. An intrinsic dispersion of 0.11 mag in that law is found when information in more than one color is available.

In fact, the 0.11 mag scatter in the brightness–rate of decline is found when one includes supernovae affected by extinction. The scatter is smaller in environments where dust in the host galaxies of the SNe Ia is a lesser effect (Sullivan *et al.*, 2003).

Additionally, the family of SNe Ia form a sequence of highly similar spectra with subtle changes in some spectral features correlated with the light curve shapes (Leibundgut, 2002; Nugent, Kim and Perlmutter, 2002; Matheson *et al.*, 2005; Hook *et al.*, 2005; Lidman *et al.*, 2005; Blondin *et al.*, 2006; Foley *et al.*, 2008). The multiple correlations allow us to control errors within the family of SNe Ia.

The parameter Δm_{15} requires the maximum in the light curve of the supernova to have been observed. This is not always possible, and therefore the various SN collaborations have formulated the brightness–decline correlation in different ways. The High-Z SN Team uses the full shape of the light curve with respect to a template, a method referred to as MLCS2k2 (Riess, Press and Kirshner, 1995a; Jha, Riess and Kirshner, 2007; Kirshner in Chapter 5 of this book) (see Fig. 6.1). The Supernova Cosmology Project collaboration, on the other hand, has introduced the *stretch factor, s,* as a parameter to account for the brightness–decline relationship (Perlmutter *et al.*, 1998, 1999; Goldhaber *et al.*, 2001). The stretch parameter fits the SN Ia light curve from premaximum to up to 60 days after maximum. Thus, even if the maximum brightness phase is not observed, the whole light curve provides the decline value. More recently, other ways to measure the decline have been proposed (Tonry *et al.*, 2003; Barris *et al.*, 2004; Clocchiatti *et al.*, 2006; Conley *et al.*, 2008). The degree to which the brightness–rate of decline correlation reduces the scatter in the Hubble diagram is shown in Fig. 6.2.

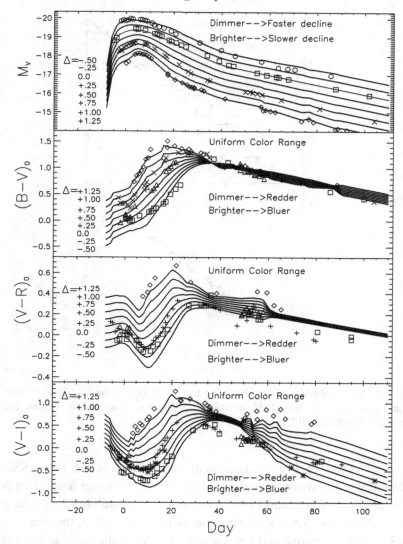

Fig. 6.1. The correlation maximum brightness–rate of decline of SNe Ia as described through the SNe Ia light curve shapes by Riess, Press and Kirshner (1995a).

Measuring the evolution of the rate of expansion of the universe does not require knowledge of its present value, H_0. Large projects have been devoted to the determination of H_0 by attempting the absolute calibration of M_B from nearby SNe Ia that occur in galaxies with Cepheid-measured distances (Freedman *et al.*, 2001; Sandage *et al.*, 2006). Controversy persists, though the most recent values quoted by the two collaborations using this method are now close (in the range

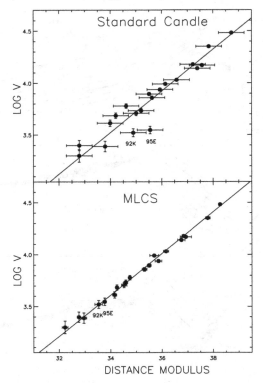

Fig. 6.2. The Hubble diagram for SNe Ia before (top panel) and after (bottom panel) correction of the width–absolute brightness relationship. (From Riess, Press and Kirshner, 1995a.)

65–75 km s^{-1} Mpc^{-1}). From a theoretical angle, the mere fact that a SN Ia is an explosion of a white dwarf bounds very efficiently the lower and upper possible values of H_0 (Ruiz-Lapuente, 1996). The emission at late phases of SNe Ia gives a precise value of H_0, since such emission is very sensitive to the density profile, the mass of the supernova and the amount of ^{56}Ni synthesized in the explosion (Ruiz-Lapuente, 1996). Those quantities are easily measurable from the spectra. The value favored by this method is $H_0 = 68 \pm 6$ (syst) km s^{-1} Mpc^{-1}. At present the classical measured value of H_0 is affected by error in the Cepheid distance ladder. Riess (2007) has proposed to improve the measurements by simplifying the ladder. His method aims at reducing the error in the measurement to 1% in H_0. This digression made, we emphasize the *nuisance role* of the absolute magnitude of SNe Ia when it comes to dark energy cosmology. The apparent magnitude of supernovae along z gives us the cosmology, m_B, as (Perlmutter *et al.*, 1999; Knop *et al.*, 2003)

$$m_B = \mathcal{M} + 5 \log \mathcal{D}_{\mathcal{L}}(z; \Omega_m, \Omega_\Lambda) - \alpha(s - 1), \qquad (6.9)$$

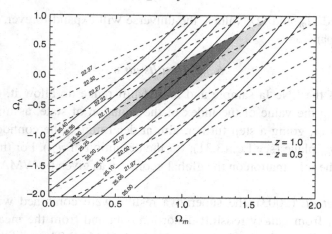

Fig. 6.3. Constraints in the Ω_m–Ω_Λ plane from two SNe Ia (one at $z = 1.0$ and another at $z = 0.5$), according to Goobar and Perlmutter (1995). SNe Ia at different z allow us to discriminate between possible universe models.

where s is the stretch value for the supernova, $\mathcal{D}_\mathcal{L} \equiv H_0 d_L$ is the "Hubble-constant-free" luminosity distance, d_L is given by (6.1) and $\mathcal{M} \equiv M_B - 5 \log H_0 + 25$ is the "Hubble-constant-free" B-band peak absolute magnitude of an $s = 1$ SN Ia with true absolute peak magnitude M_B. Similar use of the nuisance parameter in their light curve fitting is made in Riess *et al.* (2004) and Riess *et al.* (2007).

In the first determinations of dark energy cosmology, there were four parameters in the fit: the matter density parameter Ω_m, the cosmological constant density parameter Ω_Λ, as well as the two nuisance parameters, $\mathcal{M}$ and α (either α itself, used by the SCP, or some equivalent parameter used by other collaborations). The four-dimensional (Ω_m, Ω_Λ, $\mathcal{M}$, α) space, explored as a grid, yields χ^2 and $P \propto e^{-\chi^2/2}$ of the fit of the luminosity distance equation to the peak B-band magnitudes and redshifts of the supernovae. After normalization by integrating over the "nuisance" parameters, the confidence regions in the Ω_m–Ω_Λ plane can be obtained.

This use of the magnitude–redshift relation $m(z)$ as a function of Ω_m and Ω_Λ with a sample of high-z SNe Ia of different z, is a very powerful pointer to the allowed values in the Ω_m–Ω_Λ plane (Goobar and Perlmutter, 1995: see Fig. 6.3; Garnavich *et al.*, 1998). By 1998, such use yielded important results, implying that $\Omega_\Lambda > 0$ at the 3σ confidence level. For a flat universe ($\Omega_T = 1$), the results from the Supernova Cosmology Project meant $\Omega_m = 0.28^{+0.09}_{-0.08}(\text{statistical})^{+0.05}_{-0.04}$ (systematic), and the High-Z Supernova Team obtained for a flat universe $\Omega_m = 0.24 \pm 0.1$. The resulting picture of our universe is that about 20–30% of its density content is in matter and 70–80% in cosmological constant. According to the allowed Ω_m and Ω_Λ values, the universe will expand forever, accelerating its rate of expansion.

6.3.2 Results on w

If the size of the SNe Ia sample and its redshift range do allow it, one can aim at determining the value of the index of the equation of state, assuming it to be a constant w or, going a step further, aim at determining the confidence regions on the $w_0 - w_a$ plane (see Eq. (3.31), for the definition of w_a). For this latter purpose, using the information on the global geometry as given by WMAP provides a helpful prior.

In Knop *et al.* (2003), the supernova results were combined with measurements of Ω_m from galaxy redshift distortion data and from the measurement of the distance to the surface of last scattering from WMAP. The confidence regions provided by the measurement of the distance to the surface of last scattering and the SNe Ia confidence regions cross in the $\Omega_m - w$ plane, thus providing very good complementarity. This work considers an averaged w (given the range in z of the SNe Ia observed). From the fitting assuming w constant, the value is $w = -1.06^{+0.14}_{-0.21}$(statistical) $^{+0.08}_{-0.08}$(identified systematics).

The analysis of the SNLS (Astier *et al.*, 2006) also aims at an average value of w, allowed by the z range of this survey. Their 71 high-z SNeIa give $w = -1.023 \pm 0.090$(statistical) ± 0.054(identified systematics). Recently, the results from ESSENCE, using 60 SNe Ia centered at $z = 0.5$ point to a similar result with $w = -1.05^{+0.13}_{-0-12}$(statistical) ± 0.13(systematics) (Wood-Vasey *et al.*, 2007a). In this last work, the supernovae from various collaborations are joined with two separate light curve fitting approaches giving a consistent result for the value of w, assuming it to be constant.

High-z SNe Ia beyond $z > 1$ enable us to have a first look at possible evolution of $w(z)$. The first campaign by Riess *et al.* (2004) gave the results in the $w_0 - w'$ plane, using the Taylor expansion for $w(z)$ as $w(z) = w_0 + w'z$. The best estimates for w_0 and w' are $w_0 = -1.36^{+0.12}_{-0.28}$ and $w' = 1.48^{+0.81}_{-0.98}$. Those results (Riess *et al.*, 2004; Riess *et al.*, 2007) exclude a fast evolution in the equation of state, and therefore they rule out a number of modified gravity proposals. Moreover, the results are confirmed in Wood-Vasey *et al.* (2007a) where the available SNe Ia at large z, which can constrain w_0 and w_a, are incorporated to test with a joined sample encompassing a wide range of z. Along these lines, the most consistent attempt to create a sample including all sets gathered by the various collaborations has resulted in the Union set of the Supernova Cosmology Project (Kowalski *et al.*, 2008). This sample consists of 307 SNe Ia whose light curves are fitted with a single method. The value for the parameter is $w = -0.969^{+0.059}_{-0.063}$(stat) $^{+0.065}_{-0.066}$(syst),

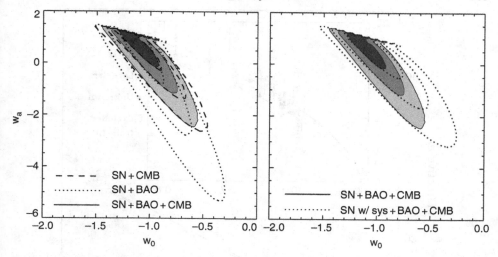

Fig. 6.4. Results on w_0–w_a from the Union sample. 68.3%, 95.4% and 99.7% confidence level contours on w_0 and w_a for a flat universe. *Left*: The Union SN sample was combined with CMB and BAO constraints. *Right*: Combination of SNe, CMB and BAO data with and without systematic uncertainties. (From Kowalski *et al.*, 2008.)

assuming w constant. The diagram w_0–w_a has been explored as well with the Union set (see Fig. 6.4).

Significant limits have been placed on the evolution of $w(z)$. Now the potential of the method relies on an improved control of systematic effects. In this area, there has been substantial progress in recent years.

6.3.3 Results on systematic uncertainties

Huge advances have taken place in controlling the uncertainties due to extinction by dust, the universality of the supernova properties, the control of lensing by dark matter and the observational process leading to the construction of the Hubble diagram of SNe.

6.3.3.1 Dust

The study with the HST of the host galaxies of the SNe Ia sample of the Supernova Cosmology Project allowed us to subclassify the galaxy hosts of the P99 sample (Perlmutter *et al.*, 1999). It was found that, at high z, early-type galaxies show a

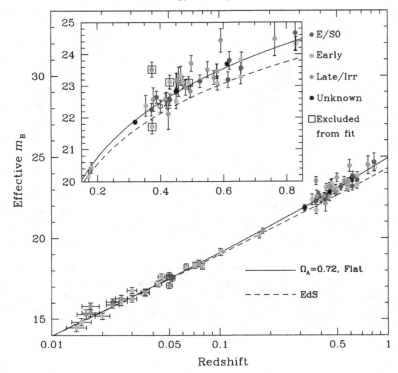

Fig. 6.5. The stretch-corrected SNe Ia Hubble diagram plotted according to the type of the host galaxy. Residuals from the adopted cosmology ("fit–C" of Perlmutter *et al.*, 1999: $\Omega_m = 0.28$ and $\Omega_\Lambda = 0.72$) are small in E/S0 galaxies. (From Sullivan *et al.*, 2003.)

narrower dispersion in SNe Ia properties than late-type galaxies, as they do at low z (Branch, Romanishin and Baron, 1996).

This result is very encouraging as it supersedes previous dispersion values obtained from samples of supernovae in all galaxy types. Supernovae in dust-rich environments, such as spiral galaxies, are more affected by extinction. The supernova magnitudes are corrected regularly for Galactic extinction as in Schlegel, Finkbeiner and Davis (1998), but the correction for dust residing in the host galaxy or along the line of sight is often not included in the standard fit, as it would require extensive color information that might not be available for high-z SNe Ia. When it is included, it is still subject to the uncertainty in the extinction law (Jha, Riess and Kirshner, 2006).

Let us, for instance, consider the correction for dimming by dust as entering in the term A_B, the magnitudes of extinction in the B-band. We take, for instance, a supernova of a given stretch s, and we place its m_{eff} (the magnitude of the

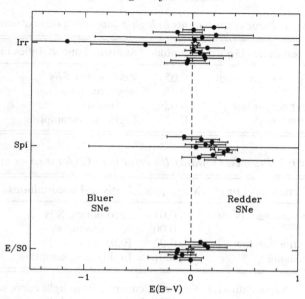

Fig. 6.6. The SN rest-frame "color excess" E(B−V) plotted as a function of host galaxy type. Error bars are are derived from R and I photometry for each SN. (From Sullivan *et al.*, 2003.)

equivalent supernova of $s = 1$) in the Hubble diagram. We might use, as in Perlmutter *et al.* (1999), Knop *et al.* (2003):

$$m_{\text{eff}} = m_B + \alpha(s - 1) - A_B. \tag{6.10}$$

The extinction in the canonical B-band of the spectrum (A_B) is determined by the extinction coefficient R_B in that band. Any uncertainty in this coefficient, due to variation of dust properties, leads to uncertain estimates of A_B. R_B is of the order of 4.1 ± 0.5 as the properties of dust might change going to high z. If we have an error of 0.1 in the observable $E(B - V)$ due to dust extinction, it is amplified as

$$A_B \equiv R_B \times E(B - V). \tag{6.11}$$

The present way of dealing with SNe Ia, including information in several photometric bands of the spectrum where extinction coefficients (R_V, R_I) are lower and one gets smaller values in the corresponding A_V and A_I, allows the dust problem to be controlled. In fact, the coefficients R_B, R_V and R_I are also determined by methods giving distances to SNe Ia. In a sample of 133 nearby SNe Ia, it is found (Jha, Riess and Kirshner, 2006; Conley *et al.*, 2007) that the dust in their host galaxies is well described by a mean extinction law with $R_V \simeq 2.7$. In all

Table 6.1. *Error contributions to high-z supernova distances in 1998*

Systematic uncertainties (1σ)	Magnitude	Statistical uncertainties (1σ)	Magnitude
Photometric system zero point	0.05	Zero points, S/N	0.17
Evolution	<0.17	K-corrections	0.03
Evolution of extinction law	0.02	Extinction	0.10
Gravitational lensing	0.02	Light curve sampling	0.15

Table 6.2. *Prospects towards 1% error in $d_L(z)$ for a space mission*

Systematic uncertainties (1σ)	Magnitude	Statistical uncertainties (1σ)	Magnitude
Photometric system zero point[a]	0.01	Zero points, S/N	0.00*
Evolution	0.00*	K-corrections	0.00*
Evolution of extinction law	<0.02	Extinction	0.00*
Gravitational lensing	<0.01	Light curve sampling	0.00*

[a] For prospects in a space mission, the statistical error from light curve sampling of SNe Ia is reduced due to the large number of SNe Ia data per redshift bin.
* 0.00* means values of the order of 10^{-3}. The statistical errors are for the redshift bin.

this, the level at which SNe Ia are treated, including how many bands are used, becomes critical. The use of SNe Ia without taking color information into account leads to a dispersion of 0.17 mag. When taking color information into account, the limit goes down to 0.11 mag. Limiting the effect of extinction, as in SNe Ia in ellipticals, should bring down the dispersion. This finding is motivating the ongoing SN searches in clusters of elliptical galaxies (Perlmutter, 2005). In environments where the effect of dust is controlled, such as SNe in clusters rich in elliptical galaxies, the systematic effect introduced by dust extinction can be better controlled. There, one can also control the evolution of dust properties.

Alternatively, it has been found by Wood-Vasey *et al.* (2007b) that SNe Ia are good standard candles in the near infrared, showing an intrinsic dispersion uncorrected for extinction or light curve shape of 0.14 mag. This opens the possibility of reducing systematic effects in reddening by adding NIR observations to the optical. However, this requires going to longer wavelengths in the observation of high-z SNe Ia, an approach that could soon be adopted with JWST (James Webb Space Telescope).

In Table 6.1, the level of accuracy in the first uses of SNe Ia back in 1998 is shown, with the contributions for each systematic and statistical error. The expected improvement from a future space mission able to control the systematics in the best possible way is given in Table 6.2. It is foreseen that several sources of systematic errors will become negligible within the next decade.

6.3.3.2 Evolution or population drift

This question resides in knowing whether SNe Ia properties evolve with z or not. Much has been done along this path. Nowadays, while we know that dust is systematically unavoidable, several studies confirm the universality of the light curve properties of SNe Ia. The study of a significant sample of SNe Ia at various z by the Supernova Cosmology Project reveals that the rise times to maximum of high-z SNe Ia are similar to the low-z SNe Ia (Aldering, Knop and Nugent, 2000; Goldhaber *et al.*, 2001). The same is found in the SNLS sample (Conley *et al.*, 2006). Statistical evaluation is so far consistent with no difference between the low-z and high-z samples.

Moreover, the possibility of the existence of an extra parameter in the maximum brightness–rate of decline relation has been carefully examined. Among possible influences, metallicity was one of the obvious ones to consider. Examining supernovae in galaxies with a gradient of metallicity, it is found (Ivanov, Hamuy and Pinto, 2000) that there is no evidence for metallicity dependence as an extra parameter in the light curve correlation of SNe Ia. This result comes from an analysis of the SNe Ia light curves of a sample of 62 supernovae in the local universe. The SNe Ia belong to different populations along a metallicity gradient. This check has also been done at high-z (Quimby *et al.*, 2002) using 74 SNe Ia ($0.17 < z < 0.86$) from the SCP sample. No significant correlation between peak SN Ia luminosity and metallicity is found.

Often we address the question of the spread of the SNe Ia samples in galaxies of different types: of how many fast SNe Ia versus intermediate or slow decliners are found in the various morphological types. Spirals at low or moderate redshifts should encompass all ranges of variation in the SNe Ia properties since they contain populations with a wide spread in age. The cosmological SN collaborations find that in those samples the one-parameter correlation does give a good description of the variation of the SNe Ia light curves: there is no residual correlation after the light curve shape correction (Goldhaber *et al.*, 2001). Kowalski *et al.* (2008) have derived the same cosmological results using the supernova Union set when taking SNe Ia with stretch $s < 0.96$ and supernovae with stretch $s > 0.96$. This repeated result suggests that the stretch correction takes away the spread of SNe Ia light curves.

One can go a step further and test the SN Ia objects themselves. This has been attempted in an intermediate-z sample and it is a feasible project at all z (Ruiz-Lapuente *et al.*, 2002; Ruiz-Lapuente, 2006). The physical diagrams of SNe Ia at intermediate z and at low z appear to be similar. Those diagrams reveal the composition, velocity gradient and radiation properties of SNe Ia. Looking at the inside of the SNe Ia is the best way to test whether they are the same at all z. Such a step forward is giving definitive results that reaffirm the validity of the SNe Ia method. The progress in this domain is summarized in Tables 6.1 and 6.2.

6.3.3.3 K-corrections

There is another element yet to be introduced here: it deals with the K-correction. The K-correction is necessary to calculate the apparent magnitude in a y filter band of a source observed in an x filter. As we are comparing the observed flux of supernovae in a given spectral band along the expansion history, we need to take into account that the photons of the supernova have shifted to redder wavelengths and spread over a different spectral wavelength range and the observed flux is different from that at emission:

$$m_Y(z) = m_X(z = 0) + K_{xy}(z), \qquad (6.12)$$

where

$$K_{xy} = 2.5 \log \left\{ (1+z) \frac{\int F(\lambda) S_x(\lambda) \, d\lambda}{\int F\left[\lambda/(1+z)\right] S_y(\lambda) \, d\lambda} \right\} + \mathcal{Z}_y - \mathcal{Z}_x, \qquad (6.13)$$

where $F(\lambda)$ is the source spectral energy distribution (a SN in this case), and $S_x(\lambda)$ is the transmission of the filter x. The $\mathcal{Z}_y - \mathcal{Z}_x$ term accounts for the different zero points of the filters (Kim, Goobar and Perlmutter, 1996). At present, this term amounts to an uncertainty of 0.02 mag in the overall budget of SNe Ia as distance indicators, but it can be decreased to 0.01 mag.

The comparisons of supernovae with fine time sequences of spectra along all phases available are turning the uncertainty in the K-correction into a decreasingly small figure. The large amount of spectral data on SNe Ia, and the good correlation with photometric properties, can indeed make this uncertainty very small.

6.3.3.4 Gravitational lensing of SNe Ia

Gravitational lensing causes dispersion in the Hubble diagram for high redshift sources. There was an early concern on the systematics and bias that this effect could cause in the determination of the cosmological parameters using SNe Ia (Amanullah, Mörtsell and Goobar, 2003). It has been shown through the study of lensing in the highest z SNe Ia sample (Jonsson et al., 2006) that the magnification distribution of SNe Ia matches very well the expectations for an unbiased sample, i.e., their mean magnification factor is consistent with unity. The effect can be very well controlled by studying the galaxy fields (Jonsson et al., 2006).

Moreover, SNe Ia should show negligible systematics caused by gravitational lensing when having a few SNe Ia per redshift bin (Wood-Vasey *et al.*, 2007a). This means that in future missions this effect should not have any impact on limiting the accuracy of $w(z)$.

6.3.3.5 The limit of SNe Ia as calibrated candles

The better we deal with the light curve and spectral information of SNe Ia, the lower becomes the dispersion in the Hubble diagram. If SNe Ia have a systematic error in their use (independently from evolution), this is likely well below the accuracy to which we would like to determine distances (1%). We have not yet hit that physical limit. Using multi-epoch spectral information and accurate light curves in various filters should enable us to test the power of these candles and their limiting irreducible error.

6.3.3.6 The path towards a unified data set

Supernovae collected over the years show the effect of observations treated with various degrees of accuracy. The dispersion around the best fit reflects this fact. For instance, some samples had information on only two colors and at very few epochs, making it difficult to control reddening, or supernovae had no spectrum confirming that they were SNe Ia. The new samples will naturally have a higher degree of accuracy in magnitude and extinction for each supernova than previous samples. The ones with poorer information have been placed (Riess *et al.*, 2004, 2007) into the "silver" category (as opposed to the "gold" category). However, even within the gold sample the requirements have changed: the latest definition of the gold sample demands higher quality standards than in Riess *et al.* (2004). In some of these SNe Ia lists, distance moduli of supernovae are assembled, but they have been obtained through procedures that deal differently with dust extinction. The Hubble diagram resulting from those compilations will naturally show a large dispersion. In Nesseris and Perivolaropoulos (2007), it is argued that a few supernovae from an early High-Z SN Search sample are pulling $w(z)$ towards evolution. The exploration done with an independent method by Riess *et al.* (2004) showed that the central value varied, within the 1σ confidence region, according to the light curve and reddening fitter of those SNe Ia (Ruiz-Lapuente, 2006). Fortunately, supernovae enable us to compare results obtained with different light curve fitting approaches. The photometry published by the various collaborations can be brought together into the Hubble diagram for the SNe Ia using a consistent procedure. This is done in Wood-Vasey *et al.* (2007a) and Kowalski *et al.* (2008), for samples including the latest data from the Higher-Z collaboration.

6.4 Testing the adequacy of a FLRW metric

In most of the above discussion on dark energy, the FLRW metric is taken as the right metric for our universe. Modern cosmology supports the concept that the universe is highly isotropic on average on large scales. The main evidence comes from the statistical isotropy of the cosmic microwave background (CMB) radiation. There are also cosmological observations supporting a matter distribution homogeneous on large scales, of the order of $100\,h^{-1}$ Mpc. Given the observed structures of matter in voids and filaments, it is worth considering whether departures from the homogeneous and isotropic universe could alter the present discussion; in particular, the possibility that there is no need for dark energy, but instead, that the observations are the result of an effect of inhomogeneity.

As pointed out in Bonvin, Durrer and Kunz (2006), large samples of supernovae will soon allow us to test the luminosity distance relation in various directions and redshift bins. One can consider the luminosity distance d_L as a function of direction **n** and redshift z. The direction-averaged luminosity distance is

$$d_L^{(0)}(z) = \frac{1}{4\pi} \int d\Omega_{\mathbf{n}}\, d_L(z, \mathbf{n}) = (1 + z) \int_0^z \frac{dz'}{H(z')}. \tag{6.14}$$

This is the usual value retrieved by the different SN collaborations. But its directional dependence $d_L(z, \mathbf{n})$ can be a test of the validity of the isotropy assumption. These authors propose expanding the luminosity distance in terms of spherical harmonics, to obtain the observable multipoles $C_l(z)$. This idea was also investigated in Riess *et al.* (1995b), where the motion of our Local Group with respect to the CMB was measured. Results in Bonvin, Durrer and Kunz (2006) are in agreement with those in Riess *et al.* (1995b).

Moreover, we can consider using the dispersion in the magnitude of SNe Ia in each direction **n** to put constraints on anisotropic models.

In redshift space, the dispersion of the magnitude in the observed Hubble diagram can serve to discard some inhomogeneous models. Simple inhomogeneous models predict a higher spread in $m(z)$ and also a non-linear behavior.

The anisotropic and inhomogeneous models have to confront present and future values on the SNe Ia dispersion at each redshift interval z_{bin} and in every direction **n**:

$$\frac{\sigma(d_L(z_{\text{bin}}, \mathbf{n}))}{d_L(z_{\text{bin}}, \mathbf{n})} = \frac{\ln(10)}{5} \sigma(m(z_{\text{bin}}, \mathbf{n})). \tag{6.15}$$

The scaling solutions of the regionally averaged cosmologies can simulate the presence of a quintessence field or other negative pressure component. But any backreaction from inhomogeneity and anisotropy would show in the Hubble diagram in distinct ways. The tests of the backreaction hypothesis are starting to give observational results (Larena *et al.*, 2008). Brouzakis, Tetradis and Tzavara (2008) address the question of whether the luminosity distance of a light source in a background of voids surrounded by denser regions might increase enough to simulate the effect caused by dark energy. They find that the luminosity distance arising from void domination is too small to account for the supernova data. The relative increase at a redshift around 1 is of the order of a small percentage while the required increase is around 30%. On the other hand, Alnes and Amarzguioui (2008) inspect whether an off-center observer in a spherically symmetric inhomogeneous universe is compatible with the supernova data. They find restrictions in the possible displacement of the off-center observer. For these tests and others, the new generation of projects surveying thousands of supernovae with all-sky coverage can offer definitive answers.

6.5 Next decade experiments

The aim of discriminating between various possibilities for dark energy places stringent requirements on the use of standard candles. Within the two decades from the discovery (1998) to the gathering of space mission data, it should become feasible to greatly reduce the errors, as mentioned before, and this would reflect into an error of only 1–2% in $d_L(z)$.

Errors much bigger than this 1% error in d_L result in the impossibility to distinguish between models (see Fig. 6.7). Even at the level of 1% in $d_L(z)$ one finds some degeneracy (Maor and Brustein, 2003). Space projects aimed at determining the nature of dark energy are counting on a systematic error in the method of around 0.02 mag, i.e., 1% in the distance (Kim *et al.*, 2004).

A different question is whether from the ground, within the planned experiments at work in the present decade, one can converge towards that limit.

These questions have been investigated in detail by using Monte Carlo simulations including the systematic errors in $m(z)$ (or $d_L(z)$) to be found in the evaluation of the present value of the equation of state of dark energy w_0 and its first derivative with respect to the scale factor w_a.

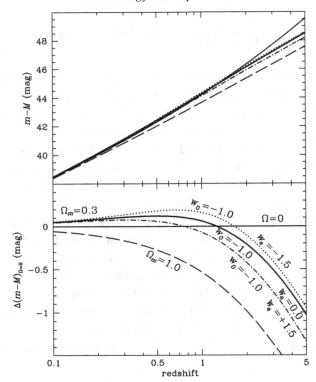

Fig. 6.7. The cosmological constant case (bold line) is compared with evolving models close to $w = -1$, i.e., a model with $w_0 = -1.0$ and $w_a = -1.5$ (short dashed line) and a model with $w_0 = -1.0$ and $w_a = +1.5$ (dash-dotted line). Only very accurate measurements will be able to differentiate between different equations of state.

Missions such as the originally proposed SNAP (SuperNova Acceleration Probe) mission have been specifically designed to gather about 2000 SNe Ia up to $z \sim 1.7$. As a single mission gathering supernova data with the same instrumentation, it is aimed at eliminating possible photometric offsets and includes high-precision photometry out to 1.7 micrometers.

With such a large number of SNe Ia, statistical errors in every redshift bin of $\Delta z = 0.1$ become negligible. The error is dominated by the systematics. The total error in each redshift bin is given by

$$\sigma_{\text{bin}} = \sqrt{\frac{\sum \sigma_{\text{st}}^2}{N_{\text{bin}}} + \sigma_{\text{sys}}^2}, \tag{6.16}$$

with expected systematic error

$$\sigma_{sys} = 0.02 \, z/z_{max}. \tag{6.17}$$

Tables 6.1 and 6.2 give a progress evaluation on the control of high-redshift supernova distance uncertainties.

Ground-based projects will not be able to achieve such accuracy. They aim at a reduction of systematic errors to 0.04–0.05 mag. Much of what will be feasible from now until the next space era of determination of the equation of state depends on the use in the best possible way of all the information already gathered. Spectra gathered so far can be exploited further than has been done up to now, in order to reduce systematic errors.

In the coming years, a lot can be done to improve the present dispersion of the Hubble diagram of SNe Ia and aim at a scatter of 1% in d_L. The above limit is only reachable if all the systematic effects are dealt with as well as possible.

6.6 Tested dark energy models

As new samples of SNe Ia have become available, it has been possible to test various dark energy models.

A candidate for dark energy is a scalar field ϕ that does not couple with matter (see Chapter 1 by Uzan). This candidate, called *quintessence*, has the feature of avoiding effects on variations of constants and other effects already restricted by experimental evidence. There is already a reconstruction of what could be the potential of the minimally coupled field $V(\phi)$ obtained from the SNLS SNe Ia (Li, Holz and Cooray, 2007). Another reconstruction of $V(\phi)$ using SNe Ia from various collaborations and considering Padé approximant expansions of the potential is found in Sahlén, Liddle and Parkinson (2006). By now, many $V(\phi)$ proposals have ben discarded since the equation of state between $z = 0$ and $z = 1.5$ does evolve very slowly, as shown by the high-z SNe Ia.

The case of non-minimally coupled fields has been analyzed as well. Scalar–tensor proposals of the type $f(R)$ have been tested against supernovae samples. While in some cases the models are compatible with supernova data, in other cases it has been possible to exclude many candidates (Capozziello, Carloni and Trosi, 2003; Mena, Santiago and Weller, 2006; Ruiz-Lapuente, 2007).

Davis *et al.* (2007) have tested a range of dark energy models using the ESSENCE data. They find that the extradimensional DGP braneworld model gives a worse fit to the data than ΛCDM. Rubin *et al.* (2008) have recently examined a wide range of models using the Union supernova sample. This sample also shows that the DGP gravity model does not achieve an acceptable fit to the combined data, even allowing for a spatial curvature parameter. The areas of intersection of

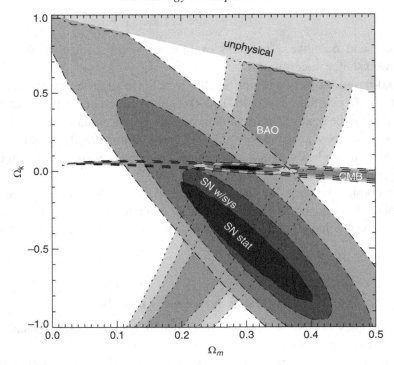

Fig. 6.8. The DGP model gives a poor goodness of fit to the data with $\Delta\chi^2 = 2.7$ relative to the flat ΛCDM model. (Figure from Rubin *et al.*, 2008.)

any pair of probes are distinct from the other pairs, indicating that the full data disfavor the braneworld model (see Fig. 6.8).

An interesting model has been proposed recently in the area of coupled dark energy. It is the growing neutrino model by Amendola *et al.* (2007) and Wetterich (2007), which couples the dark energy field to massive neutrinos. The model brings the scalar field to near cosmological constant behavior when the neutrinos go non-relativistic. The mass of the neutrino grows due to the coupling, the value today being larger than at high z. There are two free parameters in this model: the neutrino mass or density $\Omega_\nu = m_\nu(z=0)/(30.8\,h^2\,\text{eV})$ and the early dark energy Ω_e (Fig. 6.9). It is found that the 95% confidence level limit on the neutrino mass is $2.1(h/0.7)^2$ eV.

These examples, extracted from the wide range of dark energy models tested, illustrate how prolific is the supernova test of dark energy. Obviously, any dark energy candidate has to be confronted with other tests such as the growth of structure. In the next section we address the need for complementary probes.

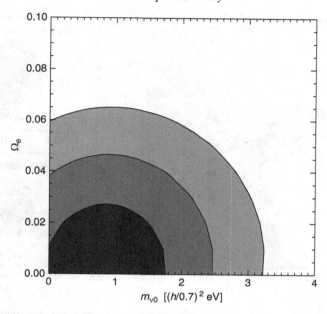

Fig. 6.9. Test of the growing neutrino model with the Union supernova sample (Rubin *et al.*, 2008). A 10% prior on the total linear growth (or the mass variance σ_8) is added to the SNe Ia data.

6.7 Complementarity

The SNe Ia test of $w(z)$ benefits from using information on Ω_m obtained by some other method. This can come from weak lensing or from large scale structure (Heavens, Kitching and Taylor, 2006; Tegmark *et al.*, 2006; Percival *et al.*, 2007a). Baryon acoustic oscillations (BAO) can also provide a measurement of distance along z, in this case angular distances $d_A(z)$, and of $E(z) = H(z)/H_0$. As expressed in Eisenstein *et al.* (2005), one can measure the parameter A which is related to the angular distance of the BAO scale.

For any geometry, one has

$$A = \frac{\sqrt{\Omega_m}}{E(z)^{\frac{1}{3}}} \left[\frac{1}{z\sqrt{|\Omega_K|}} \text{sinn} \left(\sqrt{|\Omega_K|} \int_0^z \frac{dz'}{E(z')} \right) \right]^{\frac{2}{3}}. \tag{6.18}$$

A first measurement has been obtained (Eisenstein *et al.*, 2005) giving $A = 0.469 \pm 0.017$, where the redshift at which the acoustic scale has been measured is $z = 0.35$. A second measurement has been obtained recently at $z = 0.2$ (Percival *et al.*, 2007b). This measurement combines 2dFGRS and SDSS main galaxy samples. The results of the combined BAO measurements are found to be

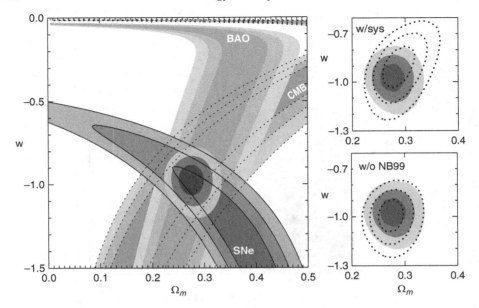

Fig. 6.10. Confidence contours for w, assuming a constant value, and Ω_m for a flat universe using the SN Union sample (Kowalski *et al.*, 2008). 68.3%, 95.4% and 99.7% confidence level contours. The large plot shows the individual and the combined constraints from CMB, BAO, and the Union SN set. The upper right plot shows the effect of including systematic errors. The lower right plot illustrates the impact of the SCP Nearby sample from 1999.

roughly consistent with that predicted by supernova data for ΛCDM cosmologies. Forcing the cosmological model to be flat and assuming constant w, this method measures $w = -1.004 \pm 0.0889$ after combining with the SNLS data and including WMAP measurements of the apparent acoustic horizon angle in the CMB.

Combining BAO and SNe Ia distances, one can aim to trace dark energy throughout a high-z range. The value of $w(z = z_{\text{rec}})$ is provided by WMAP and Planck. In between, other methods such as studying the growth of large scale structure (LSS) obtained from Lyα clouds and clusters could depict the growth factor of structure along z. Considering the information at high-z derived from LSS, SNe Ia, and CMB should lead to learning about early dark energy.

6.8 Future prospects

The observational determination of w and its variation with z has already started: while the Supernova Cosmology Project is enlarging the sample of high-z SNe Ia, the SNFactory[1] is providing an anchor in the low-z domain with hundreds of SNe Ia below $z = 0.2$. ESSENCE pursues the goal of obtaining ~ 200 SNe Ia in the z interval 0.15–0.75 over a five year period. It has already issued the first results of the project. About a thousand SNe Ia discoveries are expected to come from the Supernova Legacy Survey, now giving preliminary results on the equation of state. The very high-z range has been targeted by the Higher-Z Team and the Supernova Cosmology Project, aimed at supernovae that can test the epoch of the deceleration of the universe. To add to this the intermediate-z searches by the Sloan Digital Sky Survey and the very nearby ones by a number of collaborations such as the European Supernova Collaboration (ESC), the Carnegie Supernova Project (CSP), and a number of other surveys should greatly enhance the knowledge of SNe Ia and reinforce their cosmological use. In the coming years the Panoramic Survey Telescope and Rapid Response System (Pan-STARRS) will survey the whole sky and find thousands of supernovae per year, while the Large Synoptic Survey Telescope (LSST) will provide all-sky coverage but to higher redshifts. Those projects will allow us to test the acceleration of the universe in different directions in the sky and test our FLRW metric.

Ultimately, to unveil the nature of dark energy, ground-based programmes are limited in accuracy and scope. It seems unavoidable to go to a fully devoted mission from space, such as the Joint Dark Energy Mission (JDEM), to achieve an improved level of accuracy in the cosmological measurements, and to be able to discriminate among possible candidates for dark energy.

From all the above, one sees that the next decades will be filled with SNe Ia data and the possibility to discard competing dark energy models will increase. At the moment, w seems to be near -1, but a wide range of z has not yet been explored to test the evolution of dark energy. Thus, we expect the coming years to bring a definite understanding of what lies behind the observed acceleration of the expansion of the universe.

References

Aldering, G., Knop, R., and Nugent, P. (2000). *Astron. J.* **119**, 210.
Alnes, H., and Amarzguioui, M. (2007). *Phys. Rev. D* **75**, 023506.
Amanullah, R., Mörtsell, E., and Goobar, A. (2003). *Astron. Astrophys.* **397**, 819.
Amendola, L., Baldi, M., and Wetterich, C. (2007). [arXiv:0706.3064].

---

[1] http://SNFactory.lbl.gov
http://www.ctio.noao.edu/essence
http://www.cfht.hawaii.edu/Science/CFHTLS
http://www.sdss.org
http://www.mpa-garching.mpg.de/ rtn
http://www.ociw.edu/CSP
http://pan-starrs.ifa.hawaii.edu
http://www.lsst.org

Astier, P., *et al.* (The SNLS Collaboration) (2006). *Astron. Astrophys.* **447**, 31 [astro-ph/0510447].

Barris, B. J., *et al.* (The High Z SN Search Collaboration) (2004). *Astrophys. J.* **602**, 571.

Blondin, S., *et al.* (The ESSENCE Collaboration) (2006). *Astron. J.* **131**, 1648.

Bonvin, C., Durrer, R., and Kunz, M. (2006). *Phys. Rev. Lett* **96**, 191302.

Branch, D., Romanishin, W., and Baron, E. (1996). *Astrophys. J.* **465**, 73.

Brouzakis, N., Tetradis, N., and Tzavara, E. (2007). *JCAP* **0702**, 013.

Capozziello, S., Carloni, S., and Trosi, A. (2003). *Recent Res. Dev. Astron. Astrophys.* **1**, 625 [astro-ph/0303041].

Clocchiatti, A., *et al.* (The High Z SN Search Collaboration) (2006). *Astrophys. J.* **642**, 1.

Conley, A., *et al.* (The SNLS Collaboration) (2006). *Astron. J.* **132**, 1707.

Conley, A., Carlberg, R. G., Guy, J., *et al.* (2007). *Astrophys. J.* **664**, L13.

Conley, A., *et al.* (2008). *Astrophys. J.* **681**, 482.

Davis T. M., *et al.* (2007). *Astrophys. J.* **666**, 716 [astro-ph/0701510].

Einstein, A. (1917). *Sitzungsber. Preuss. Akad. Wiss.* **142**.

Eisenstein, D. J., *et al.* (2005). *Astrophys. J.* **633**, 560.

Filippenko, A. V., and Riess, A. G. (1998). *Phys. Rep.* **307**, 31.

Foley, R. J., *et al.* (2008). *Astrophys. J.* **684**, 68.

Freedman, W. L., *et al.* (2001). *Astrophys. J.* **553**, 47.

Garnavich, P. M., *et al.* (1998). *Astrophys. J.* **509**, 74.

Goldhaber, G., *et al.* (The Supernova Cosmology Project) (2001). *Astrophys. J.* **558**, 359.

Goobar, A., and Perlmutter, S. (1995). *Astrophys. J.* **450**, 14.

Heavens, A. F., Kitching, T. D., and Taylor, A. N. (2006). *Mon. Not. R. Astron. Soc.* **373**, 105.

Hook, I. M., *et al.* (The Supernova Cosmology Project) (2005). *Astron. J.* **130**, 2788.

Ivanov, V. D., Hamuy, M., and Pinto, P. A. (2000). *Astrophys. J.* **542**, 588.

Jha, S., Riess, A. G., and Kirshner, R. P. (2007). *Astrophys. J.* **659**, 122.

Jonsson, J., Dahlen, T., Goobar, A., Gunnarson, C., Mortsell, E., and Lee, K. (2006). *Astrophys. J.* **639**, 991.

Kim, A. G., Goobar, A., and Perlmutter, S. (1996). *Publ. Astron. Soc. Pacif.* **108**, 190.

Kim, A. G., Linder, E. V., Miquel, R., and Mostek, N. (2004). *Mon. Not. R. Astron. Soc.* **347**, 909.

Knop, R. A., *et al.* (The Supernova Cosmology Project) (2003). *Astrophys. J.* **598**, 102.

Komatsu, E., *et al.* (2009). *Astrophys. J. Suppl.* **180**, 330.

Kowal, C. T. (1968). *Astron J.* **73**, 1021.

Kowalski, M., *et al.* (The Supernova Cosmology Project) (2008). *Astrophys. J.* **686**, 749.

Larena, J., Alimi, J.-M., Buchert, T., Kunz, M., and Corasaniti, P.-S. (2008). [arXiv:0808.1161].

Leibundgut, B. (2002). *Ann. Rev. Astron. Astrophys.* **39**, 67

Li, C., Holz, D., and Cooray, A. (2007). *Phys. Rev. D* **75**, 3503.

Lidman, C., *et al.* (The Supernova Cosmology Project) (2005). *Astron. Astrophys.* **430**, 843.

Maor, I., and Brustein, R. (2003). *Phys. Rev. D* **67**, 103508.

Matheson, T., *et al.* (The ESSENCE Collaboration) (2005). *Astron. J.* **129**, 2352.

Mena, O., Santiago, J., and Weller, J. (2006). *Phys. Rev. Lett.* **96**, 1103.

Nesseris, S., and Perivolaropoulos, L. (2007). *JCAP* **02**, 25 [astro-ph/0611572].

Nugent, P., Kim, A., and Perlmutter, S. (2002). *Publ. Astron. Soc. Pacif.* **114**, 803.

Percival, W. J., *et al.* (2007a). *Astrophys. J.* **657**, 51.

Percival, W. J., Cole, S., Eisenstein, D., Nichol, R. C., Peacock, J. A., Pope, A. C., and Szalay, A. S. (2007b). *Mon. Not. R. Astron. Soc.* **381**, 1053.

Perlmutter, S. (2005). *HST Proposal-GO 10496.*

Perlmutter, S., *et al.* (The Supernova Cosmology Project) (1998). *Nature (London)* **391**, 51.

Perlmutter, S., *et al.* (The Supernova Cosmology Project) (1999). *Astrophys. J.* **517**, 565.

Phillips, M. M. (1993). *Astrophys. J.* **413**, L105.

Phillips, M. M., *et al.* (1999). *Astron. J.* **118**, 1766.

Pskovskii, Y. P. (1977). *Soviet Astron.* **21**, 675.

Quimby, R., *et al.* (The Supernova Cosmology Project) (2002). *Bull. Amer. Astron. Soc.* **201**, 2305.

Riess, A. G. (2007). *HST proposal GO 10802.*

Riess, A. G., Press, W. H., and Kirshner, R. P. (1995a). *Astrophys. J.* **438**, L17.

Riess, A. G., Press, W. H., and Kirshner, R. P. (1995b). *Astrophys. J.* **445**, L91.

Riess, A. G., *et al.* (The High-Z SN Search Collaboration) (1998). *Astron. J.* **116**, 1009.

Riess, A. G., *et al.* (The Higher-Z Team) (2004). *Astrophys. J.* **607**, 665.

Riess, A. G., *et al.* (The Higher-Z Team) (2007). *Astrophys. J.* **659**, 122

Rubin D., *et al.* (2008). [arXiv 0807.1108].

Ruiz-Lapuente, P. (1996). *Astrophys. J.* **465**, L83 [astro-ph/9604044].

Ruiz-Lapuente, P. (2006). Invited review at *Bernard's Cosmic Stories*, Valencia, June 2006.

Ruiz-Lapuente, P. (2007). *Class. Quant. Grav.* **24**, 91.

Ruiz-Lapuente, P., *et al.* (2002). ITP2002 on Ω *and* Λ *from Supernovae, and the Physics of Supernova Explosions.*

Sahlén, M., Liddle, A. R., and Parkinson, D. (2007). *Phys. Rev. D.* **75**, 3502 [astro-ph/0610812].

Sandage, A., *et al.* (2006). *Astrophys. J.* **653**, 843.

Schlegel, D. J., Finkbeiner, M., and Davis, M. (1998). *Astrophys. J. Suppl.* **500**, 525.

Sullivan, M., *et al.* (The Supernova Cosmology Project) (2003). *Mon. Not. R. Astron. Soc.* **340**, 1057.

Tegmark, M., *et al.* (2006). *Phys. Rev. D* **74**, 123507.

Tonry, J. L., *et al.* (The High-Z SN Search Collaboration) (2003). *Astrophys. J.* **594**, 1.

Wetterich, C. (2007). *Phys. Lett. B* **655**, 201.

Wood-Vasey, W. M., *et al.* (The ESSENCE Collaboration) (2007a). *Astrophys. J.* **666**, 694 [astro-ph/0701041].

Wood-Vasey, W. M., *et al.* (2007b). [arXiv:0711.2068].

7

The future of supernova cosmology

W. MICHAEL WOOD-VASEY

The mysterious acceleration of the expansion of the universe remains as tantalizing as ever. This accelerated expansion is attributed to a dark energy in the universe that accounts for ~70% of the total energy density at the present epoch. There are several clear promising avenues to pursue in the next decade to reveal the nature of dark energy. The standard complement of the four methods currently being discussed in observational cosmology are: weak lensing; cluster abundances; baryon acoustic oscillations; and the originator of much of the recent excitement, Type Ia supernovae.

Type Ia supernovae (SNe Ia) are powerful probes of the kinematics of the universe because they are visible with optical telescopes to at least two-thirds of the lifetime of the universe, during which it has expanded by a factor of 2.5 (i.e., out to a redshift of $z \sim 1.5$). During this latest doubling, the expansion of the universe transitioned from decelerating (Riess et al., 2004b) to accelerating (Riess et al., 1998; Perlmutter et al., 1999). Further work (Knop et al., 2003; Tonry et al., 2003; Barris et al., 2004; Astier et al., 2006; Wood-Vasey et al., 2007) confirmed this picture of an accelerating universe whose kinematics are consistent with a constant vacuum energy density that permeates the universe. This remarkable result remains unexplained and this era of transition from acceleration to deceleration continues to be the epoch of central interest in attempts to elucidate the nature of dark energy. In this chapter I will focus on SNe Ia as cosmological probes and present the current challenges and opportunities for SN Ia cosmology in the next decade.

Dark Energy: Observational and Theoretical Approaches, ed. Pilar Ruiz-Lapuente. Published by Cambridge University Press. © Cambridge University Press 2010.

Our current understanding indicates that dark energy is without form or structure. However, a key aspect of the testing in the next decade will be searching for evidence of non-uniformity of dark energy either in time or in space. Searches for a change of the equivalent dark energy physical density over time are a strong motivation for exploring the universe at redshifts beyond $z \sim 1$. For certain cosmological probes, such as SNe Ia and weak lensing, this redshift range naturally leads one to consider observations from space for dramatically improved sensitivity in the infrared (to which the optical slight has been shifted) and stability of the instrumental apparatus. The billion-dollar question is twofold: (a) does dark energy vary with cosmic time or space, and (b) is a space-based mission necessary to determine this?

7.1 Current results from SN Ia cosmology

I have spent most of my recent time working on the ESSENCE supernova survey: a six-year project using the 4-meter Blanco telescope at Cerro Tololo Interamerican Observatory (CTIO) to discover and study 200 SNe Ia with $0.15 < z < 0.75$. This project has involved an international collaboration of scientists from five separate countries on four continents and many nights of time on the biggest optical telescopes (the Keck Observatory, the Very Large Telescope of the European Southern Observatory, the pair of Gemini telescopes) to spectroscopically confirm and measure redshifts for these hundreds of SNe Ia. Our competition during this time has been the SuperNova Legacy Survey (SNLS), primarily based in Canada and France with significant partnerships with institutions in the UK and the USA. In 2006 and 2007 each group published their respective papers (Astier *et al.*, 2006; Wood-Vasey *et al.*, 2007) in which they reported on new sets of intermediate redshift SNe Ia. These papers emphasized the increasingly important consideration of systematic effects in measuring the luminosity distance to SNe Ia and, in particular, any pathological evolution of such systematics with redshift that could mimic the precise cosmological measurements to tease out the nature of dark energy. Each group found that, when compared to the existing sample of nearby SNe Ia (Hamuy *et al.*, 1996; Riess *et al.*, 1999), the new set of luminosity distances from SNe Ia were well explained by a model of the universe where dark energy is equivalent to a cosmological constant (Λ) and the balance of the energy density of the universe is cold dark matter (CDM). This new "standard" cosmology model is referred to as the ΛCDM concordance cosmology and is in agreement with all of the major cosmological measurements: the cosmic microwave background, the large-scale distribution of galaxies in the universe, the initial balance of elements of the universe from Big-Bang nucleosynthesis, the angular correlation of galaxies from baryon acoustic oscillations, and the luminosity distance measurements from SNe Ia.

Both ESSENCE and SNLS are now finishing up their observations and have accumulated respective samples of ~200 and ~500 SNe Ia. New improvements in analysis techniques and understanding of the intrinsic nature of SNe Ia and their host galaxy environments are being developed to achieve the ultimate goal of a 10% measurement (including both systematic and statistical uncertainties) of a constant equation-of-state of dark energy. But before these intermediate-redshift surveys can realize their full potential, the sample of low-redshift SNe Ia will need to grow to roughly match their high-redshift counterparts in total number. Currently there ~50 well-observed nearby SNe Ia in the linear Hubble flow (i.e., sufficiently distant from our local neighborhood so that their redshift–distance relation is not hopelessly confused by the individual velocities of galaxies and clusters with respect to the overall Universal expansion). Since the basic measurement of SN Ia cosmology is a comparison of the apparent brightness of nearby and distant SNe Ia, we would ideally obtain roughly equal numbers of each. Upcoming data releases from the KAIT project at UC Berkeley, the nearby supernova follow-up program at the Harvard–Smithsonian Center for Astrophysics, and the Carnegie Supernova Project led by the Carnegie Observatories should add another ~200 nearby SNe Ia to the published literature. In addition, the Nearby Supernova Factory led by Lawrence Berkeley National Laboratory will provide detailed spectro-photometry for at least 100 SNe Ia in the next couple of years. Half of these new nearby SNe Ia will immediately be useful in our large-scale cosmological measurement of the universe. The other new survey soon to release its first results is the very successful Sloan Digital Sky Survey II Supernova Survey, which found over 400 confirmed SNe Ia during its two-year run. These SNe Ia are concentrated in the range of redshifts $0.1 < z < 0.2$ and will both bring great reassurance in connecting the lowest-redshift and intermediate-redshift SN Ia samples and also provide a wealth of information of SN Ia properties and correlations with the photometric and spectroscopic properties of their host galaxies.

While the ΛCDM model is very successful at explaining virtually all existing cosmological observations, the current SN Ia data (cf., Astier *et al.*, 2006; Riess *et al.*, 2007; Wood-Vasey *et al.*, 2007) allow for variation of the dark energy density at $z > 1$, as shown in Fig. 7.1, and quantitatively for an arbitrary parameterization of w in Fig. 7.2. Further constraints on the time variability of dark energy from

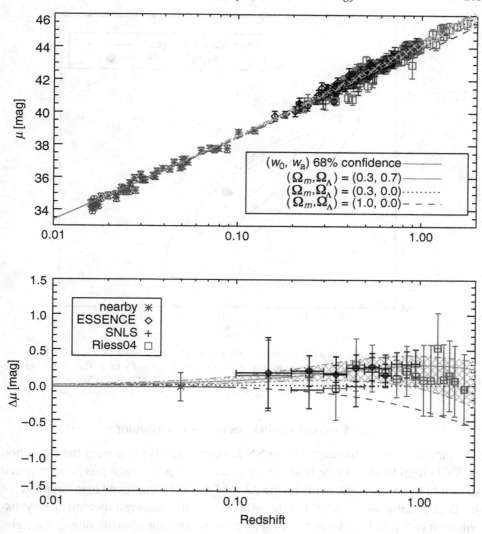

Fig. 7.1. The global SNe Ia constraints on the variation of the dark energy equation-of-state parameter $w = P/(\rho c^2)$ in a flat universe. The current SNe Ia data (shown here binned) provide less constraint on w for $z > 0.9$, as indicated by the range of allowed $w = w_0 + w_a(1 - a)$ models as shown in Fig. 7.2. (Adapted from Wood-Vasey *et al.*, 2007, Fig. 14.)

SN Ia cosmology will come from increased samples at the highest ($z > 1$) and lowest redshifts ($z < 0.1$) because (a) this range maximizes the contrast and (b) the range $0.2 < z < 0.8$ has been well sampled by the current generation of supernova surveys.

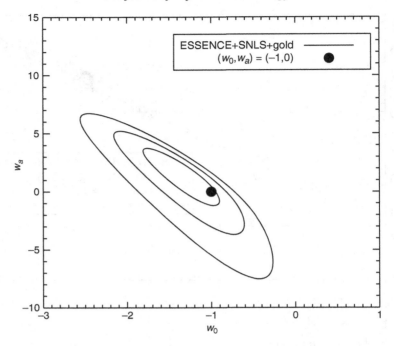

Fig. 7.2. Constraints on a simplistic $w = w_0 + w_a(1 - a)$ model as a local expansion of a non-constant w around the current epoch $(a = 1)$ in a flat universe assuming $\Omega_m = 0.27 \pm 0.03$. (From Wood-Vasey *et al.*, 2007, Fig. 13.)

7.2 Current challenges in SN Ia cosmology

The largest current challenge facing SN Ia cosmology is measuring the extinction of SN Ia light by dust in the host galaxy to the small percentage precision required to enable improved constraints on the ΛCDM model or improved measurements of w. Traditionally astronomers use the reddening of the observed spectral energy distribution (often in broad-band colors) to infer the amount of extinction of the light. In our own Milky Way galaxy, we have observed that the relationship between the amount of color change (reddening) in observed light from a source and the amount of extinction in a given passband can be expressed by a particular model (Cardelli *et al.*, 1989; O'Donnell, 1994) that is generally referred to as the CCM law. This law parameterizes a relationship between the extinction in the V band and the color "excess", $E(B - V)$, by a parameter R_V

$$A_V = R_V\, E(B - V).$$

Analogous relationships can be written down for the relationship between the extinction in other passbands and the excess $B - V$ color, and the CCM model makes a prediction for the value of these R coefficients; e.g., if $R_V = 3.1$ then $R_H = 0.5$. CCM found that, for their sample of stars, the average extinction through our galaxy could be modeled by $R_V = 3.1$. However, the extinction laws for the Large and Small Magellanic Clouds differ from Milky Way extinction, and, in general, we have few good measurements of the extinction–color relationship in other galaxies. The very thorough dust map of Schlegel *et al.* (1998) confirmed this average property for our galaxy but also revealed clear variations in R_V from ~ 2 to 5 along specific lines of sight, particularly at higher values of A_V.

We do not know the appropriate equivalent R_V to use or whether or not the R_X formalism even works, nor the distribution of dust in host galaxies along the line of sight between the supernova and our observations, nor how this distribution or extinction law is likely to evolve with cosmic time. We will need to improve our understanding of all of these aspects to begin to have confidence in our overall measurements to the small percentage level necessary to distinguish between the current competing models for dark energy. Progress toward this goal is quite likely to come from the impending data releases from the nearby SN Ia projects along with complementary UV data from the *Swift* satellite and NIR observations from ground-based telescopes. With wavelength coverage from 200 to 3000 nm, we will be able to parameterize the behavior of dust and extinction in our local universe, develop a sample sufficient to identify the dependence on this dust distribution as a function of host galaxy time and use the diversity of types and formation history of nearby galaxies to understand their high-redshift analogs.

7.3 Local inhomogeneities

The current SN Ia cosmological results are based on a comparison of the apparent brightness of nearby SNe Ia at redshifts <0.1 with more distant SNe Ia at $0.3 < z < 1.5$. The unexpected faintness of SNe Ia at $0.3 < z < 0.5$ was taken as the first sign of the existence of dark energy and moved us to adopting the standard ΛCDM concordance model of the universe. As we have explored this model in more detail, it has become clear that it is time to expand on the simple model of a homogeneous and isotropic universe, which is known as the Friedmann–Lemaître–Robertson–Walker framework (FLRW). While we observe that our everyday experience is clearly neither isotropic nor homogeneous, on larger scales of a few hundred megaparsecs the statistical properties of the universe do indeed appear to be homogeneous and isotropic. Several questions immediately occur to us when considering these assumptions of homogeneity and isotropy.

The structure of the universe exhibits correlation on scales larger than 200 mega-parsecs. Does this correlation affect our interpretation of the propagation of light from standard candles? Temporal evolution and spatial averaging do not commute. Does this non-commutation lead to different overall expansion behavior than a simple perturbation on FLRW? Does the Copernican principle hold? Do we inhabit a special place in the universe?

The first question is beginning to be addressed quantitatively (cf., Hui and Greene, 2006; Cooray and Caldwell, 2006) and incorporated into the cosmological analysis. Theorists still argue about whether the so-called "back-reaction" can lead to significantly different expansion behavior once perturbations from FLRW become non-linear. The Copernican question has definite philosophical overtones regarding our place in the universe but should be ultimately testable. We can easily measure the dipole moment of our relative motion with respect to the cosmic microwave background, but it is very challenging to measure a monopole moment. Thus our actual constraints on living in a significant local over- or under-density in the universe are more limited than one might assume. There have been some possible indications of such a local Hubble bubble on the scale of $\sim$100 megaparsecs in some analyses of the magnitude–redshift relation of nearby SNe Ia (Jha *et al.*, 2007). However, other analyses have shown that this effect is intimately related to our lack of understanding of the intrinsic color vs. host galaxy dust behavior of SNe Ia (Conley *et al.*, 2007).

On a larger scale, one can wonder if all of the nearby SNe Ia are within a special region. If we inhabited a local under-density with a radius of a giga-parsec such that the effective local expansion rate (the Hubble parameter) were different from the universal value by on the order of 10%, we could be incorrectly interpreting the SN Ia data as implying a universal accelerated expansion rather than just revealing the local increased expansion rate of our local region within, e.g., an Einstein–de Sitter (EdS) universe (flat, $\Omega_m = 1$, $\Omega_\Lambda = 0$) (for a review, see Célérier, 2007). One can see this inherent degeneracy as something that naturally arises because of our finite look back time in the universe. Because we can only observe our local volume out to 3 billion light-years, we cannot observe other regions in the universe within their comoving past 3 billion years that are outside our light cone. Thus we are fundamentally limited in our ability to distinguish a universal acceleration that began 3 billion years ago from a local acceleration within the nearest 3 billion gigaparsecs. However, the ΛCDM model also deviates from an EdS model at earlier times in the expansion of the universe. An EdS model predicts that the universe has always been decelerating, whereas the ΛCDM model predicts a decelerating phase above a redshift of $z > 0.8$ that then transitions to an accelerating phase at more recent times ($z \sim 0.5$). Thus the high-redshift SNe Ia ($z > 1$) provide an equally valid comparison point as the low-redshift SNe Ia traditionally used for the luminosity distance–redshift comparison.

The Hubble Space Telescope (HST) has been productive in the past decade in discovering SNe Ia with $z > 1$ (Dahlen *et al.*, 2004; Riess *et al.*, 2004a,b, 2007; Strolger *et al.*, 2004; Kuznetsova *et al.*, 2008), which is the effective limit in discovering SNe Ia from the ground because of the difficulties in observing in the infrared through the Earth's atmosphere. At the time of publication, the cosmological analysis of these HST-discovered SNe Ia (Riess *et al.*, 2004b, 2007) was touted as conclusively demonstrating that grey-dust models that attempt to explain the dimming of distant SNe Ia through inter-galactic dust that dims irrespective of optical wavelength would have to be significantly fine-tuned to explain the dimming and then brightening of SNe Ia at $z > 1$ with respect to a $\Omega_\Lambda = 0$ cosmology. While this remains true, these observations also provide the aforementioned high-redshift comparison point to demonstrate that the universe transitioned from a decelerating phase to an accelerating phase between redshifts of $z \sim 1.2$ and $z \sim 0.2$ without reference to the nearby SNe Ia or any potential local large-scale Hubble bubble. However, the statistical significance of this result is only at the $\sim 2\sigma$ level and will need approximately an additional 50 SNe Ia at $z > 1$ to conclusively rule out an EdS model of the universe using only SNe Ia at $z > 0.2$.

From the lower-redshift end, the results of the SDSS supernova survey are eagerly awaited to fill in the region for $0.1 < z < 0.25$ with at least 300 high-quality SNe Ia. This data will connect the existing and future samples of low-redshift SNe Ia with intermediate-redshift SN Ia such as those from the ESSENCE and SNLS surveys. Any large-scale inhomogeneity will reveal itself as a significant deviation in the luminosity distance–redshift relationship from a ΛCDM cosmology in the region of $0.03 < z < 0.3$.

My prediction is that the SDSS results will confirm the ΛCDM picture while teaching us much about systematic uncertainties in measuring luminosity distances to SNe Ia. These systematic uncertainties will motivate detailed analyses of the imminent data releases of nearby SNe Ia observed from the UV through the optical and NIR to resolve the questions of intrinsic color, dust, and possible additional parameters that correlate with and help further standardize the luminosity distances to SNe Ia. While the proper treatment of correlations and inhomogeneities of our real universe versus the simplistic smooth FLRW model will provide interesting discussions in the theoretical literature, in the end it is very difficult to believe that any argument for effective expansion of regions within an EdS universe does not have an empty universe as its limiting case. Thus, while it is possible that a more detailed treatment will reveal necessary improvements in how cosmological luminosity distances are interpreted, the observed accelerated expansion of the universe is a real property of the dynamical universe and is almost certainly a result of some innate generic property involving new physics in the vacuum, in gravity, or in particle physics.

7.4 Future cosmological results from SNe Ia

As we obtain larger samples of SNe Ia, we will begin to use them to measure properties of the galaxies and the distribution of galaxies. In this section I will focus on two upcoming ground-based survey missions, Pan-STARRS and LSST, that will change the nature of supernova cosmology and observational astronomy in general. I should preface this section by recognizing my glaring omission of the SkyMapper and Dark Energy Survey projects in the following discussions. SkyMapper (scheduled for 2009) will be analogous to Pan-STARRS1 but in the southern hemisphere rather than the northern, while the Dark Energy Survey (currently scheduled to begin in 2010) will fall midway between Pan-STARRS1 and LSST in both schedule and capabilities. Space-based supernova cosmology missions such as SNAP and DESTINY will focus on the measurement of the overall dynamics of the universe and are covered in more detail elsewhere in this book.

The Large Synoptic Survey Telescope (LSST) is a 6.7-meter telescope[1] that will begin observation in Chile in the middle of the next decade. With a 10-square-degree camera, LSST will scan the entire visible sky every three to four nights. This rapid, all-sky coverage will enable the discovery and study of millions of supernovae during the 10-year planned LSST mission. This incredible number of supernovae will allow new types of investigations that exploit the standard brightness of SNe Ia.

The Pan-STARRS project in Hawaii is paving the way toward the new era of SN Ia studies that will be enabled by LSST. Currently consisting of a single 1.8-meter telescope (Pan-STARRS1) with a 7-square-degree field of view, this project will eventually consist of four telescopes (Pan-STARRS4) that will provide similar rapid-cadence coverage of the entire visible sky. The Pan-STARRS approach to all-sky time-series monitoring trades off the expense and sensitivity of a large aperture for the more flexible configuration of multiple identical telescopes that will have the ability to simultaneously monitor phenomena in multiple filters and on different time scales. The present incarnation, Pan-STARRS1, is currently in commissioning at Haleakala on the Hawaiian island of Maui. In addition to the exciting science enabled by this project, Pan-STARRS1 will serve as a prototype mission for both Pan-STARRS4 and LSST.

---

[1] The primary mirror is an annulus that is 8.4 meters in outer diameter, but, with the central region taken up with the third mirror, the effective area of the primary mirror is equivalent to a 6.7-meter telescope.

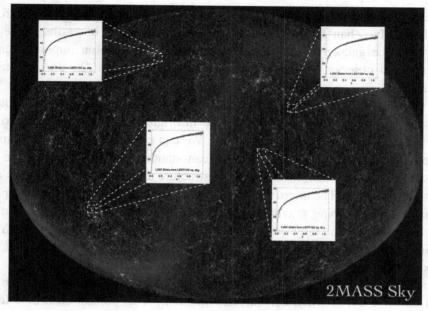

Fig. 7.3. The LSST will discover 10 SNe Ia per square degree per year. Observing the entire visible 30 000 square-degree sky from its site in Cerro-Pachón, the
LSST will find 3 000 000 SNe Ia over the ten years of the survey. These SNe Ia
will allow for cosmological tests equivalent to the current sample of 1000 SNe Ia
in each of 3000 separate 10-square-degree lines-of-sight. This ensemble of angular samples on the sky will allow for sensitive tests of isotropy in the universe
and analogous binning of the sample into redshift shells will test homogeneity
on large scales. (The background all-sky image is from the Two-Micron All Sky
Survey, http://www.ipac.caltech.edu/2mass/.)

In the era of LSST we will begin to use SNe Ia in a correlated statistical sense.
When measuring the overall expansion of the universe we currently treat SNe Ia on
an individual basis. However, with hundreds of thousands of SNe Ia, we will start
having to deal with correlations in rates and brightness from large-scale structure
and lensing. These will simultaneously be sources of uncertainty for the overall
measurement and projects measuring more complicated cosmological parameters:
σ_8, dark matter versus luminous matter bias, baryon acoustic oscillation, dynamics
and correlations within clusters.

We will be able to make measurements of the dynamical evolution of the
universe along 3000 lines of sight to test isotropy, or across shells in redshift
to test homogeneity, or more sophisticated combinations to test the correlations
imprinted on the universe from the era of inflation immediately after the Big Bang.
Figure 7.3 visually conveys this ability to measure the luminosity distance–redshift
relation through different regions of the universe. It will in fact be possible to

make a SN Ia-based measurement of the baryon acoustic oscillation with the LSST SNe Ia (Zhan *et al.*, 2008). This measurement will be entirely independent from the galaxy-based techniques, and serve as both an important check on the BAO observations and a method to constrain both Ω_m and Ω_Λ using only SNe Ia.

If the individual calibration uncertainty of SN Ia luminosity can be reduced from the current 15% in brightness to 5% in brightness, we will be able to use SNe Ia to probe the structure of clusters. We can currently use them to explore our own local cluster and the more we can standardize the distances, the farther out we can use them to measure the structure and dynamics of clusters on a cosmological scale. The internal dynamics of clusters will be a specific astrophysical extreme of probing gravitationally bound systems that will allow us to use SNe Ia to connect the large-scale dynamic behavior of the universe through any medium-scale inhomogeneities down to the astrophysical scale of gas, dust, and galaxies.

There is the potential for LSST to provide probes of smaller scales in our local universe out to when our universe was half of its present size. SNe Ia are sensitive to the underlying stellar populations because supernovae are an end result of stellar evolution. In interesting results (e.g., Sullivan *et al.*, 2006b), the rates of SNe Ia are found to be sensitive to two different populations: one short-lived, of perhaps a few hundred million years old; and one long-lived, at perhaps a few billion years old. Thus SNe Ia are in fact sensitive to the star-formation rate (SFR). Core-collapse SNe are completely sensitive to the SFR because their progenitors are stars younger than 100 Myr. However, SNe Ia are significantly brighter (by a factor of 2–10 from the average population of core-collapse SNe) and thus are visible through much more of the universe for any given survey.

In an interesting coincidence that must say something about the SFR and galactic evolution, the SNe Ia rate traces B-band luminosity. Non-star-forming galaxies are red because they have older stars. Star-forming galaxies are blue because they have young hot stars, but also red because they have many stars in the dusty regions that are correlated with star formation. It is very intriguing that these factors balance sufficiently well that the SN Ia rate traces the B-band luminosity (Filippenko and Li, 2008).

A key feature of SNe Ia is that these with wider rest-frame lightcurves are found in galaxies with more star formation and these with narrower rest-frame lightcurves are found in galaxies with less star formation. The fact that after width–luminosity correction there is no difference between these samples is another issue that prompts interesting questions and suggests important connections to the nature of SNe Ia. Thus, SNe Ia can be used as probes of the SFR both in an overall sense and in the individual galaxy-by-galaxy, or, more likely, cluster-by-cluster sense. Perhaps for large clusters, one could have enough SNe Ia over the 10-year span of LSST to trace out the history of star formation in a separate way from the current light of the stars. Cross-correlating the SN Ia rate with other tracers of recent and old star formation will be a fruitful direction of inquiry. In fact, combining these measurements of the SFR will allow for detailed modeling of a cluster by constraining both the star-formation history and thus the supernova history for a cluster. This information will lead to relatively well-informed models on the likely amount of supernova activity over the Hubble time and thus the supernova feedback and population of the intra-cluster medium. In fact, it is possible and likely that collisions and mergers are dominant contributors to star formation, but these processes might dominate only at certain epochs in the cluster formation, which will be somewhat different, although possibly correlated through the accompanying bursts of star formation, from the supernova history.

One very interesting question is whether long-delay and short-delay SNe Ia are both generated in the same mechanisms of star formation or if they come from different methods. That is, is there a phase delay between the short- and long-duration progenitors/events, or are they unassociated? If there is a correlation then a particular burst of star formation at one epoch and an accompanying production of short-delay SNe Ia would be associated with another rise 3 billion years later (or whatever the long-delay time (and its width) turn out to be).

For some of these tests it is not that SNe Ia will become the preferred way of making the measurement, it is that they will become an interesting cross-check that will sometimes be subject to different systematics, and certainly different selection effects than other methods.

SNe Ia will continue to be powerful cosmological yardsticks as we begin to zoom in from the current gross measurements of the universe to exploring the dynamics and perturbations on smaller scales and correlations on all scales to test the fundamental application of a homogeneous and isotropic model (FLRW) to the real universe with structure, motion, and light.

7.5 Final musings

Supernovae have captured our attention for centuries with their own explosive intensity and their fundamental reminder of the impermanence of existence. In the past decade, they have drawn renewed focus for their amazing power to reveal new surprises about the nature of the universe. In the next decade they will fulfill their potential to make precision measurements of the dark energy currently accelerating the expansion of the universe. While the field has grown from its precocious youth, much work is being done to take SNe Ia to higher levels of precision and finer scales of the universe. The upcoming generation of wide-field time-domain surveys of the sky will allow us to exploit millions of SNe Ia to study the history, growth, and fundamental nature of our universe.

References

Astier, P., Guy, J., Regnault, N., *et al.* (2006). *Astron. Astrophys.* **447**, 31.

Barris, B. J., Tonry, J. L., Blondin, S., *et al.* (2004). *Astrophys. J.* **602**, 571.

Cardelli, J. A., Clayton, G. C., and Mathis, J. S. (1989). *Astrophys. J.* **345**, 245.

Célérier, M.-N. (2007). [astro-ph/0702416].

Conley, A., Carlberg, R. G., Guy, J., Howell, D. A., Jha, S., Riess, A. G., and Sullivan, M. (2007). *Astrophys. J. (Lett.)*, **664**, L13.

Cooray, A., and Caldwell, R. R. (2006). *Phys. Rev. D* **73**, 103002.

Dahlen, T., Strolger, L.-G., Riess, A. G., *et al.* (2004). *Astrophys. J.* **613**, 189.

Filippenko, A. V., and Li, W. (2008). In *Aspen Workshop on Supernovae.*

Hamuy, M., Phillips, M. M., Suntzeff, N. B., Schommer, R. A., Maza, J., and Aviles, R. (1996). *Astron. J.* **112**, 2391.

Hui, L., and Greene, P. B. (2006). *Phys. Rev. D* **73**, 123526.

Jha, S., Riess, A. G., and Kirshner, R. P. (2007). *Astrophys. J.* **659**, 122.

Knop, R. A., Aldering, G., Amanullah, R., *et al.* (2003). *Astrophys. J.* **598**, 102.

Kuznetsova, N., Barbary, K., Connolly, B., *et al.* (2008). *Astrophys. J.* **673**, 981.

O'Donnell, J.E. (1994). *Astrophys. J.* **422**, 158.

Perlmutter, S., Aldering, G., Goldhaber, G., *et al.* (The Supernova Cosmology Project) (1999). *Astrophys. J.* **517**, 565.

Riess, A. G., Filippenko, A.V., Challis, P., *et al.* (1998). *Astron. J.* **116**, 1009.

Riess, A. G., Filippenko, A. V., Li, W., and Schmidt, B.P. (1999). *Astron. J.* **118**, 2668.

Riess, A. G., Strolger, L., Tonry, J., *et al.* (2004a). *Astrophys. J. (Let.)* **600**, L163.

Riess, A. G., Strolger, L.-G., Tonry, J., *et al.* (2004b). *Astrophys. J.* **607**, 665.

Riess, A. G., Strolger, L.-G., Casertano, S., *et al.* (2007). *Astrophys. J.* **659**, 98.

Schlegel, D. J., Finkbeiner, D.P., and Davis, M. (1998). *Astrophys. J.* **500**, 525.

Strolger, L.-G., Riess, A. G., Dahlen, T., *et al.* (2004). *Astrophys. J.* **613**, 200.

Sullivan, M., Le Borgne, D., Pritchet, C. J., *et al.* (2006). *Astrophys. J.* **648**, 868.

Tonry, J. L., Schmidt, B. P., Barris, B., *et al.* (2003). *Astrophys. J.* **594**, 1.

Wood-Vasey, W. M., Miknaitis, G., Stubbs, C. W., *et al.* (2007). *Astrophys. J.* **666**, 694.

Zhan, H., Wang, L., Pinto, P., and Tyson, J. A. (2008). *Astrophys. J. (Let.)* **675**, L1.

8

The space advantage for measuring dark energy with Type Ia supernovae

ALEX KIM

8.1 Introduction

The observed accelerated expansion of the universe (Riess *et al.*, 1998; Perlmutter *et al.*, 1999) is in conflict with longstanding expectations of cosmology. The dynamics of the universe is not dominated by the self-attraction of its contents as expected from "normal" gravity: in that case the rate of expansion would have been decelerating. In response to the observations, a slew of theories have been proposed to describe the repulsive force responsible for the acceleration. One possibility is that most of the energy content in the universe has, unexpectedly, a negative equation of state. Alternatively, there may be a problem with our standard conception of gravity: in the dimensionality of space, the existence of a geometric cosmological constant, or a breakdown of general relativity. The source of the observed accelerated expansion, whatever its underlying cause, is commonly referred to as "dark energy".

The universe's expansion history is charted using brightness and redshift measurements of distant Type Ia supernovae (SNe Ia), which serve as standard candles. The bulk of the data used to discover the accelerating universe and provide the first coarse measurements of the dark energy parameters was obtained from ground-based observatories. As interest now shifts toward identifying the physics responsible for the acceleration, more distant supernovae and more accurate and precise distances as inferred from light curves and spectra are required. The stability of a space-based platform allows for an experiment that accurately and precisely maps the expansion history and probes dark energy models.

Dark Energy: Observational and Theoretical Approaches, ed. Pilar Ruiz-Lapuente. Published by Cambridge University Press. © Cambridge University Press 2010.

8.1.1 Cosmology theory

A homogeneous and isotropic universe has a geometry described by the Robertson–Walker metric where the kinematics are encoded in the scale factor $a(t)$. The luminosity distance of an astronomical source, d_L, is associated with the bolometric energy flux passing by an observer today (t_0) on the surface of a sphere centered at the source. The source redshift, $z \equiv a(t_0)/a(t_{emit}) - 1$, measures the relative scale of the universe between when the light was measured and emitted. Luminosity distance and redshift are related through the Hubble parameter, $H(z) \equiv \dot{a}/a$, by

$$d_L(z) = (1+z)Rf\left(\frac{1}{R}\int_0^z \frac{dz'}{H(z')}\right), \tag{8.1}$$

where

$$f(x) = \begin{cases} \sinh(x) & \text{for an open universe,} \\ x & \text{for a flat universe,} \\ \sin(x) & \text{for a closed universe.} \end{cases} \tag{8.2}$$

R is the proper radius of the universe today. The $(1+z)$ factor accounts for the reduced flux due to the time dilation at the source as seen by an observer, and the loss of photon energy due to the redshifting. The rest of the equation is the geometric measure of the sphere's radius. The bolometric energy flux depends on the source redshift, the geometry of spacetime, the size of the universe, and the Hubble parameter.

Physics fixes the dynamical evolution of $a(t)$; given a theory of gravity and the properties of the constituents of the universe, we can predict the behavior of $H(z)$. For general relativity and a universe filled exclusively with non-relativistic matter and "dark energy",

$$H^2(z) = H_0^2\left(\Omega_m(1+z)^3 + \Omega_R(1+z)^2 + \Omega_{de}e^{3\int_0^z (1+w(x))d\ln(1+x)}\right), \tag{8.3}$$

where H_0 is the Hubble constant, Ω_m and Ω_{de} are today's normalized energy densities of non-relativistic matter and dark energy respectively ($\Omega_X \equiv \frac{8\pi G\rho_X}{3H^2}$), and $\Omega_R \equiv (RH_0)^{-2}$. The dark energy contribution to the dynamics is characterized by its equation of state $w(z) = p/\rho$. The theoretical luminosity distance, Eq. (8.1), can be rewritten in terms of the dynamical parameters:

$$d_L(z) = \frac{(1+z)}{H_0\Omega_R^{1/2}} f\left(\Omega_R^{1/2}\int_0^{z'} \frac{dz'}{h(z')}\right), \tag{8.4}$$

with

$$h^2(z') = \Omega_{NR}(1+z')^3 + \Omega_R(1+z')^2 + \Omega_{de}e^{3\int_0^{z'}(1+w(x))d\ln(1+x)}. \tag{8.5}$$

The luminosity distance is often recast as the distance modulus $\mu \equiv 5 \log \frac{d_L}{10\,\text{pc}}$.

The ratio of luminosity distances at different redshifts $d_L(z_1)/d_L(z_2)$ is not explicitly dependent on the Hubble constant but remains sensitive to the remaining cosmological parameters. We therefore can rely on relative, not absolute, distances to measure the dynamical parameters. The dependence of relative distances at different redshifts on Ω_m and Ω_{de} was explored by Goobar and Perlmutter (1995) who demonstrated (using supernova magnitudes normalized to a nearby sample) that isocontours of constant distance modulus in the $\Omega_m - \Omega_{de}$ plane have different slopes at different redshift, as shown in the top plot of Fig. 8.1 for $z = 0.5$ and $z = 1.0$. The intersecting contours demonstrate that distance measurements at different redshifts provide leverage in making a simultaneous measurement of Ω_m and Ω_{de}.

The acceleration of the universe's expansion today is cast as the deceleration parameter (a remnant of pre-1998 expectations) $q_0 \sim -\frac{\ddot{a}a}{\dot{a}^2}|_0$. The sign of q_0 is opposite to $\ddot{a}$ and therefore a negative q_0 corresponds to an accelerating universe. A dataset of luminosity distances with corresponding redshifts can be used to measure the kinematics. However, the observable d_L depends on integrals over the scale factor. Practically speaking, recovering a from derivatives of discretely sampled noisy data is difficult, particularly when the uncertainty of each measurement propagates into large slope uncertainties (Daly and Djorgovski, 2004; Wang and Tegmark, 2005; Rapetti *et al.*, 2007). It is thus common to adopt a model for the expansion history, use data at all redshifts to constrain the model, and then express the deceleration parameter in terms of the model parameters; for a homogeneous and isotropic universe under the influence of general relativity $q_0 = \Omega_m/2 + (1 + 3w)\Omega_{de}$.

The current acceleration of $a(t)$ was discovered and remains best constrained by mapping distances as a function of redshift. As seen in Fig. 8.1 the luminosity distance at redshift z is sensitive to $\alpha \Omega_m - \Omega_{de}$ where α varies from 1 to 2 for $0.5 < z < 1$ (Goobar and Perlmutter, 1995): practically the same expression as q_0. Other probes of cosmology are not as sensitive to q_0, clusters are more sensitive to Ω_m and the cosmic microwave background sensitive to Ω_R, or $\Omega_m + \Omega_{de}$.

8.1.2 Type Ia supernovae as distance indicators

The luminosity distance is defined so that it can be associated with a physical observable and gives a clear connection between theoretical predictions and measurements. Observationally, the luminosity distance is measured through the ratio of the bolometric energy luminosity, L, and the bolometric flux, f, through

$$d_L = \sqrt{\frac{L}{4\pi f}}. \tag{8.6}$$

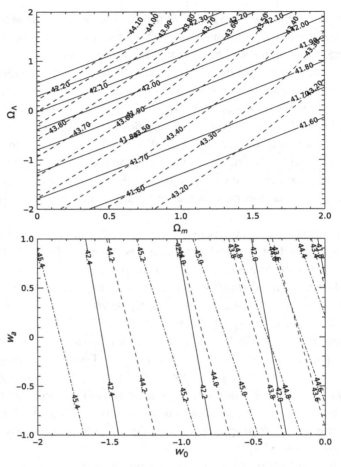

Fig. 8.1. Isocontours of μ for fixed redshifts. The top plot shows the $\Omega_m - \Omega_\Lambda$ plane for a cosmological constant dark energy and steps of 0.1 mag, and bottom plot shows the $w_0 - w_a$ plane for an $\Omega_m = 0.3$, $\Omega_{de} = 0.7$ cosmology with 0.2 steps in mag. The solid and dashed curves are for distances at $z = 0.5$ and 1.0 respectively; the bottom plot has an additional dash-dot curve for $z = 1.5$. The intersecting contours on the top plot indicate that luminosity distance measurements over a range of redshift can be used to determine Ω_m and Ω_{de}. The contours' redshift dependence is subtle in the $w_0 - w_a$ plane indicating that accurate and precise measurements are necessary to constrain these parameters.

Associating the theory (Eq. (8.1)) with observation (Eq. (8.6)) provides a direct probe of the universe's geometry and expansion history, or the dynamical parameters of a specific cosmological model.

The broad-band energy magnitude used in astronomy for the measure of flux is defined as $m \equiv -2.5 \log f + Z$, where Z is a constant set by the magnitude system.

The absolute magnitude, which is a measure of luminosity, is the magnitude of the source as seen at a distance of 10 pc, $M \equiv -2.5 \log \left(\frac{L}{4\pi (10 \, \text{pc})^2} \right) + Z$.

Luminosities of astronomical objects are generally poorly determined whereas relative distances or magnitude differences, which are not dependent on L, are more robustly measured. Fortunately, absolute distances are not required for determining the dynamical parameters.

The familiar definition of magnitude just described doesn't quite apply to measured data as observer measurements are neither bolometric nor of energy. Observations are generally filtered, detectors are photon counters, and standard magnitude systems, such as Johnson–Cousins (Johnson and Morgan, 1953; Cousins, 1976) and that of the Sloan Digital Sky Survey, fundamentally based on three subdwarf stars (Smith *et al.*, 2002), are count-based. In practice, observations are compared to theoretically predicted magnitudes that incorporate the band transmission with the object spectral energy distribution (SED) (Kim *et al.*, 1996), and either convert observed to flux magnitude (e.g. via the flux of α Lyr: Hayes, 1985) or work with "photon distance" rather than luminosity distance (Nugent *et al.*, 2002).

The brightness and homogeneity of Type Ia supernovae (SNe Ia) give their luminosity distance measurements an important role in cosmology. The Hubble Key Project linked Cepheid and SN Ia distances to measure the Hubble constant (Freedman *et al.*, 2001), while measurements of the relative distances between low- and high-redshift supernovae have demonstrated the accelerating expansion of the universe (Riess *et al.*, 1998; Perlmutter *et al.*, 1999) and continue to provide the strongest constraints on dark energy properties (Astier *et al.*, 2006). Reaching a peak $M_B \sim -19.5$, SNe Ia can become as bright as their host galaxies and have been seen out to $z = 1.7$. In optical wavelengths, they have an observed magnitude dispersion of ~ 0.3 mag, and after further subclassification and dust correction can have their dispersion reduced to ~ 0.1 mag. There remains the prospect for further narrowing of the dispersion as supernova colors, near-ultraviolet (UV), and near-infrared (NIR) emissions continue to be better characterized and associated with luminosity.

The basic theoretical understanding of SNe Ia is based on their observational properties. Their homogeneous brightness and their occurrence in both early and late galaxies indicate that these SNe must be standardized explosions whose progenitors can be associated with both young and old stellar populations. The triggered thermonuclear explosion of a C/O white dwarf in a binary system (Hoyle and Fowler, 1960; Arnett, 1969) naturally satisfies these conditions. A white dwarf supported by electron degeneracy pressure increases in mass and density through accretion from a companion star. There is a range in latent times necessary to grow the white dwarf from its initial mass up to its mass at explosion. Thermonuclear burning commences when the carbon ignition density is reached, providing a standardized condition for the triggering of the explosion. The burning of $\sim M_{\odot}$ of material to nuclear statistical equilibrium powers the explosion that disrupts the star. The delayed detonation model (Khokhlov, 1991) provides a phase of preexpansion due to deflagration to account for the observed presence of intermediate mass elements (Nomoto *et al.*, 1976), a subsequent transition to detonation burns a fraction of the star to ^{56}Ni. Alternatively, the detonation may be initiated by a buoyant bubble of burnt fuel that emerges from the surface and converges with an explosion at the antipode (Plewa *et al.*, 2004). The energy seen by observers originates from the decay chain of ^{56}Ni $\rightarrow$ ^{56}Co $\rightarrow$ ^{56}Fe in the expanding ejecta.

Despite the great advances in the modeling and numerical simulation of supernova explosions, the numerous degrees of freedom that describe the progenitor initial condition are unconstrained and simulations produce a broad range of events not seen in nature. Currently, theory provides qualitative explanations for the empirical supernova relations on which cosmology analysis is based.

Although SNe Ia are remarkably homogeneous, there are non-trivial differences in their light curves, spectra, and absolute magnitudes. The heterogeneity in absolute magnitude imposes a limit on how well an individual object's distance can be measured; the averaging of many measured distances is necessary to deliver an arbitrarily small uncertainty in true distance for a specific redshift interval. Fortunately, much of the diversity is accounted for when SNe Ia are considered as a multi-parameter family. The heterogeneity in different observables is correlated, with much of the diversity in absolute magnitude, light-curve shapes, colors, and spectral line ratios accounted for by a single parameter (Nugent *et al.*, 1995; Hamuy *et al.*, 1996b; Riess *et al.*, 1996). There is evidence for a second independent parameter identified through a correlation between absolute magnitude and color (Tripp and Branch, 1999; Guy *et al.*, 2007). (In astronomy parlance, color refers to the difference between magnitudes of different bands.) New data already in hand may identify even more intrinsic supernova parameters. Each parameter can be thought to be associated with an important degree of freedom in the supernova explosion.

The pioneering work in high-redshift supernova cosmology was done at ground-based observatories. In 1988, a $z = 0.31$ supernova was discovered using the 1.5-meter Danish telescope at La Silla (Norgaard-Nielsen *et al.*, 1989). The Supernova Cosmology Project made their first supernova discovery in 1992 at the Isaac Newton Telescope (Perlmutter *et al.*, 1995) and further discoveries ensued there and at the Kitt Peak 4-meter (Perlmutter *et al.*, 1997). Mass harvesting of supernovae became possible with the wide-field 2048^2 pixel prime-focus CCD camera and the 4×2048^2 pixel Big Throughput Camera at the Cerro Tololo Inter-American

Observatory 4-meter telescope. The series of searches run with these instruments provided the bulk of supernovae used in the measurement of the accelerated expansion of the universe (Riess *et al.*, 1998; Perlmutter *et al.*, 1999). Ground-based searches continue to improve the measurement of the expansion history of the universe and of the properties of the dark energy responsible for its acceleration (Astier *et al.*, 2006; Wood-Vasey *et al.*, 2007a). The large field of view of the Canada–France–Hawaii Telescope (CFHT) imager enabled the next significant advance in search strategy. The Megacam prime-focus imager covers one square degree, so that each of its exposures can contribute light-curve photometry for tens of active supernovae out to $z \sim 0.8$. The Supernova Legacy Survey (SNLS) (Astier *et al.*, 2006) uses the Megacam in a rolling search mode of operation: by regularly reimaging a fixed field in multiple bands with a cadence of several days, SNLS automatically builds multi-band light curves for all supernovae within the field without the need for specially targeted observations. The multiplex advantage of wide-field cameras is critical for the efficient collection of supernova data.

8.1.3 Observing supernovae from space

The data quality possible from space observations was first demonstrated with the serendipitous Hubble Space Telescope (HST) imaging of SN 1996cl and then the targeted follow-up of SN 1997ap (Perlmutter *et al.*, 1998). Specifically, the diffraction-limited seeing and avoidance of the Earth's atmosphere provide high-resolution imaging of supernovae and their host galaxies at observer-frame NIR wavelengths without bright sky emission and absorption. HST followup of ground-discovered supernovae and discovery from space data themselves has provided better signal-to-noise, color measurements, and redshift coverage than has been possible from ground data alone, improving distance measurements of supernovae at moderate redshifts ($z \sim 0.5$) and extending the Hubble diagram to $z > 0.8$ (Garnavich *et al.*, 1998; Riess *et al.*, 2001, 2004, 2007; Knop *et al.*, 2003; Clocchiatti *et al.*, 2006).

Just as with the ground, the next major advance in space-based supernova surveys will come with wide-field imaging. The HST optical cameras, WFPC2, ACS, and WFC3, have fields of view of $\sim$10–20 square arcminutes whereas at least a quarter square degree is necessary for multiplexed discovery and follow-up of a rolling search targeting $z < 2.0$. The primary imager for the JWST (the next-generation successor to HST) will cover infrared wavelengths and have only a $2 \times 2.16 \times 2.16 \, \text{arcmin}^2$ field of view. Although this instrument is ideal for studying extremely distant supernovae or the restframe infrared bands of individual supernovae, it cannot efficiently collect SN-frame optical photometric time series of large numbers of events with $z < 2$. This limitation has spawned the development of alternative space missions tailored for supernova-cosmology and wide-field surveys.

The optical detectors appropriate for space missions are enabled by the development of CCD's designed for use at vertex detectors at particle accelerators (Holland *et al.*, 2003). Made from high-resistivity material, the CCDs are thick and fully depleted, and so have high quantum efficiency up to and beyond 1 μm. The devices are p-channel, shown to be more resistant to the damaging radiation from space (Bebek *et al.*, 2004). For NIR wavelengths, R&D done for the Supernova/Acceleration Probe (SNAP) mission have led to technological improvements in HgCdTe 1.7 μm cutoff detectors that nearly double the quantum efficiency while retaining low noise (Brown *et al.*, 2006). Longer wavelengths are accessible with 2.5 μm HgCdTe technology commonly used in astronomy.

The concept for SNAP (Aldering *et al.*, 2004) was first developed in 1998 at Lawrence Berkeley National Laboratory. In November 2003, the US Department of Energy and NASA announced the Joint Dark Energy Mission (JDEM), a space-based program to study dark energy, to be supported by the two agencies. There are three experiments that engaged in Advanced Mission Concept Studies for JDEM. The fact that all three, SNAP (Aldering *et al.*, 2004), Destiny (Benford and Lauer, 2006), and Advanced Dark Energy Physics Telescope (ADEPT)[1] proposed to use supernovae (in conjunction with other cosmological probes) to measure the dynamics of the universe testifies to the strength of the technique and recognizes the need for space observations to advance the science.

8.2 Connecting supernova measurements to cosmology

To appreciate the importance of the data quality afforded by space-based observations, it is important to know which data are taken and how they are used to deduce distances and probe cosmology.

Much of the data comes in the form of time-series multi-band photometry and/or spectroscopy at the position of each supernova. The set of supernova photometry in a single band is referred to as a light curve. The wavelength coverage is set by the capabilities of the observatory, typically spanning optical and NIR bands where most of the supernova flux resides. Additional data for the photometric calibration standards, the transmission functions of the optical components, and the standard star flux in physical units are necessary to calibrate the supernova data. Independent measurements of the Milky Way extinction are also used.

To be included in the cosmology-analysis sample, the supernovae are typed: SNe Ia are nominally defined by the absence of hydrogen and the presence of

---

[1] www7.nationalacademies.org/ssb/mtg_2_ADEPT.pdf

silicon in their spectra. Though peculiar events such as SN 2002ic (Wood-Vasey, 2002; Hamuy *et al.*, 2003) and SN 2005gj (Aldering *et al.*, 2006; Prieto *et al.*, 2007) demonstrate that this definition does not tag thermonuclear supernovae with 100% efficiency, it does appear to provide a pure sample. This identification is unambiguous with a spectrum with sufficient signal-to-noise that covers $\sim$6200 Å, the location of the ubiquitous SiII feature and (the absent) Hα; for the purposes of this article we assume that precision supernova-cosmology surveys have unambiguous typing. The homogeneity of SNe Ia does allow typing derived from light curves, spectra covering alternative wavelengths, or from host-galaxy morphology (Howell *et al.*, 2005; Blondin and Tonry, 2007). This typing can produce impure samples since other backgrounds, particularly SNe Ibc, can resemble SNe Ia at some phases and wavelengths. The population characteristics of these interlopers are not well quantified and would introduce a poorly constrained source of systematic error in a cosmology analysis.

The physical processes that occur to photons along their trajectory determine the fluxes that are observed. The time-dependent luminosity of each supernova establishes the intensity of photons at the start of their journey. Wavelength-dependent dust extinction attenuates the photons as they make their way out of the host galaxy. Gravitational lensing due to mass inhomogeneities focuses or defocuses photon bundles during their travel through the intergalactic medium. The changing scale factor redshifts photon wavelengths and sets the surface area of the spherical shell centered at the supernova that intersects the observer. Milky Way dust attenuates the photons during their travel within our own galaxy. Finally, the photons make their way through the complete optical chain of the site, telescope, and camera before they are detected and take the form of measured data.

We present a series of equations to represent all the data used in the cosmology analysis and how they are interpreted. The equations have data to the left of the equal sign and the model interpretation of those data to the right. All model parameters are expressed as p except the cosmological and dark energy parameters that retain their standard notation.

The many algorithms in use to analyze supernova data (Phillips *et al.*, 1999; Wang *et al.*, 2003; Guy *et al.*, 2007; Jha *et al.*, 2007) can be abstractly viewed as modeling a photometric or spectroscopic datum, $m_{ij}^{\alpha} \sim -2.5 \log\left(\frac{\text{counts}_{ij}^{\alpha}}{\sec}\right)$, through a parameterized model of the physical processes responsible for the incoming flux:

$$m_{ij}^{\alpha} = M(t_i; F_{ij}, p_z^{\alpha}, \vec{p}_{SN}^{\alpha}) + p_{ij,\text{res}}^{\alpha}$$
$$+ A(t_i; F_{ij}, p_z^{\alpha}, \vec{p}_{\text{dust}}^{\alpha}, \vec{p}_{SN}^{\alpha}) + p_{\text{lensing}}^{\alpha} + \mu^{\alpha}$$

$$+ A_{MW}(t_i; F_{ij}, p_z^{\alpha}, \vec{p}_{MW}^{\alpha}, \vec{p}_{SN}^{\alpha})$$
$$+ Z(F_{ij}(M^{\star}, \vec{p}_F), M^{\star}(\vec{p}_{\star})) + m_{gal,ij}^{\alpha}(\vec{p}_{gal}^{\alpha}), \tag{8.7}$$

where α is the supernova index, i is the temporal index, j is the wavelength resolution-element index, and t is the date of observation. The first $M + p_{ij,res}^{\alpha}$ term is the intrinsic luminosity of the supernova, where M is model prediction and $p_{ij,res}^{\alpha}$ is the residual offset from the model for object α. The A term is the extinction due to host-galaxy dust, and $p_{lensing}^{\alpha}$ is the magnification from gravitational lensing. The term of ultimate interest, μ^{α}, is the distance modulus to the supernova. Milky Way dust extinction is given by A_{MW}. The calibration zeropoint term Z is the expression of the throughput and standard star flux in physically based magnitude units. The underlying host-galaxy flux is modeled as $m_{gal,ij}^{\alpha}(\vec{p}_{gal}^{\alpha})$.

The other variables that appear in the equation are described in detail in the following paragraphs, but it is important to point out that the determination of the μ's and all other "nuisance" astrophysical and instrumental parameters are interrelated through Eq. (8.7).

Additional data are necessary to break the coupling and improve the marginalized uncertainties in μ. Independent measurement of the supernova redshifts z_k^{α} are directly associated with the true supernova redshifts, p_z^{α}, through

$$z_k^{\alpha} = p_z^{\alpha}. \tag{8.8}$$

We assume that the supernova host galaxies' peculiar velocities are negligible compared to the Hubble bulk flow, true for those distant supernovae that need observations from space.

Calibration data are used to model the band transmissions. The transformation from observer magnitudes to a physical system depends on the transmission functions and the flux of the fundamental standard star that defines the magnitude system. The best estimate of the jth band's transmission function at epoch i is denoted as F_{ij} and its model parameters $\vec{p}_F$. *In situ* measurement of standard stars and individual components in the optical chain (denoted by c_k) help constrain the transmission function model:

$$c_k = F_{ij}(M^{\star}(\vec{p}_{\star}), \vec{p}_F). \tag{8.9}$$

The calibration of the fundamental standard stars, though generally not done by the supernova observer, also involves data $m_k^{\star}$ and a model for the stars' flux $M^{\star}$ and its parameters $\vec{p}_{\star}$. For completeness we write

$$m_k^{\star} = M^{\star}(\vec{p}_{\star}). \tag{8.10}$$

We now discuss each of the terms in Eq. (8.7). The phase and wavelength-dependent absolute magnitude M is usually based on an underlying model for the evolving SED as a function of the supernova parameters $\vec{p}_{SN}^{\alpha}$. The SED's are typically generated from an independent supernova data training set. The parameters $\vec{p}_{SN}^{\alpha}$ include the date of explosion and those chosen to describe intrinsic supernova diversity, typically in the form of light-curve shape and color (Tripp and Branch, 1999; Guy *et al.*, 2007). The number of independent parameters used to label the SED's should increase as the SN Ia class is better characterized. The color and morphology of the host galaxy may also provide data that constrain $\vec{p}_{SN}^{\alpha}$. A flux measurement is associated with the model's restframe absolute magnitude using $\vec{p}_{SN}$ to select the appropriate SED and integrating over the band transmission F blueshifted by p_z.

Even after subclassification SNe Ia are not perfectly homogeneous. The deviation of an individual supernova's SED from the model for a particular epoch and band is given by $p_{ij,res}^{\alpha}$. The prior on its distribution is obtained from the residuals of the training set data used in the building of the SED template, e.g. the "error snakes" of SALT2 (Guy *et al.*, 2007) and MLCS2k2 (Jha *et al.*, 2007). Without the inclusion of a prior for these parameters, the model would be underconstrained.

Magnitude extinction due to host-galaxy dust is denoted as A and specified by its parameters $\vec{p}_{dust}^{\alpha}$. The Cardelli *et al.* (1989) extinction law with its parameters A_V and R_V is commonly used. The predicted extinction is constructed by using the values of $\vec{p}_{SN}$ and p_z to select the appropriate redshifted SED at the correct supernova phase. The extinction law for the $\vec{p}_{dust}$ is applied to the spectrum and the result is integrated over by the band transmission F. As a consequence, the magnitude extinction for a broad-band light curve can vary due to the time-evolving supernova SED.

The term $p_{lensing}^{\alpha}$ is the magnification due to gravitational lensing caused by mass inhomogeneities in the path of the photons of supernova α (Holz, 1998). For a single supernova, the fluxes at all wavelengths and phases incur the same magnification. Without prior information on the lensing distribution the model would be unconstrained, except that it is known that the mean magnification in fluence (but not magnitude!) units over all sky positions at a given redshift must be one.

The term of cosmological interest, the distance modulus μ, can either be considered as a model parameter in itself or as the functional prediction of luminosity distance for a specific cosmological model with parameters (as in Eq. (8.4)) using the relation $\mu = 5 \log \frac{d_L}{10\,pc}$. For the standard parameterization of dark energy in terms of its present equation of state w_0 and slope $w_a \sim -dw/da|_{z=1}$, the distance to each supernova is expressed as $\mu^{\alpha} = \mu(p_z^{\alpha}, H_0, \Omega_m, \Omega_{de}, w_0, w_a)$. In the former case, the set of μ's and its accompanying covariance matrix can be used as input to compute the likelihood of any cosmological model.

The extinction due to Milky Way dust, A_{MW}, is constrained by maps of dust tracers (Burstein and Heiles, 1982; Schlegel *et al.*, 1998). Absorption uncertainties in directions of low Galactic extinction have no appreciable effect on cosmological measurements, and so are not further considered in this article.

The calibration-model zeropoints Z transform observer units to a magnitude system based on physical units, cgs for example. The zeropoint flux is constructed by convolving the transmission function F with the fundamental standard star SED $M^\star$. F is generally parameterized by the efficiency at a set of wavelengths, or more coarsely in terms of its various moments. $M^\star$ is often parameterized as emission at a set of wavelengths, or by the parameters of the theoretical radiative transfer model upon which the standard star's flux is based.

The spatially varying host-galaxy flux is constrained by deep observations well before or after the supernova explosion. The galaxy contribution at the supernova position varies with seeing and is thus time-dependent.

A supernova cosmology analysis can proceed with fitting the model parameters on the right sides of Eqs. (8.7), (8.8), (8.9), and (8.10) to the data on the left. Priors on the probability distribution functions (pdf's) of $p^\alpha_{\text{lensing}}$ and $p^\alpha_{ij,\text{res}}$ are included in the fit. Uncertainties in μ or the cosmological parameters are given after marginalization over the other parameters. In practice, the analysis is broken into several steps; for example the calibration solution is usually decoupled from light-curve fits that may be decoupled from the cosmological parameter fitting. However, as the statistical uncertainties decrease with growing supernova samples, it is important to recognize that the cosmological and dark energy parameters are fundamentally coupled to parameters associated with SN Ia subclassification, dust, the flux of the fundamental standard star, and the transmission function of each optical element.

Systematic uncertainties in the analysis arise when the models for the intrinsic supernova SED, dust, and calibration, or the priors (e.g. probability distribution function for magnitude residuals from the SN model, lensing magnification) inadequately reflect reality.

8.3 Type Ia supernova homogenization and diversity

The homogenization of the broad range of SN Ia behavior and quantification of the remaining random residuals, as denoted by $M + p^\alpha_{ij,\text{res}}$ in Eq. (8.7), are critical elements in the interpretation of the data. The intrinsic properties of supernovae drive observing requirements by establishing which wavelengths and phases must be measured for SN Ia subclassification and reduction in intrinsic flux dispersion.

Supernovae are best understood at phases and wavelengths where they emit most of their light. Figure 8.2 (Hsiao *et al.*, 2007) shows the template SEDs 10 days

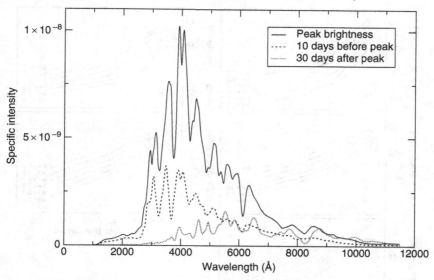

Fig. 8.2. The SEDs of a normal SN Ia 10 days before, 30 days after, and at *B* peak brightness from the spectral time series of Hsiao *et al.* (2007).

before, 30 days after, and at *B* peak brightness. During this time window, when the supernova is brightest and most easily detected, the bulk of the supernova light is emitted in optical wavelengths. There is little emission at wavelengths below 0.26 μm, the flux increases at longer wavelengths despite the line-blanketing of iron-group elements. The emission peaks in the optical, evolving from 0.35 to 0.55 μm from 10 days before peak to 30 days after peak. The drop in emission at redder wavelengths into the NIR is not blackbody but empirically follows a power law.

The supernova time evolution over broad bands is shown in Fig. 8.3 (Jha *et al.*, 2007). Initially the light curves have a power-law increase powered by the decay of ^{56}Ni in the optically thick expanding ejecta. The increase in brightness peters out as diffusion loss matches the radioactive input at peak brightness. The flux drops as the supernova cools and becomes optically thin. In optical bands, a plateau phase occurs 20–30 days after peak with a shallow exponential decline corresponding roughly to the ^{56}Co lifetime. (See Arnett, 1982, for an analytic description of early light-curve evolution.) Redder bands exhibit a second peak due to the ionization evolution of iron group elements in the ejecta (Kasen, 2006).

SNe Ia are not perfectly homogeneous but their diversity appears to be constrained by a small number of degrees of freedom. The observable most often used to parameterize supernovae is the time-scale of the *B* and *V* light curves spanning [−10, 30] days relative to peak brightness (Perlmutter *et al.*, 1999;

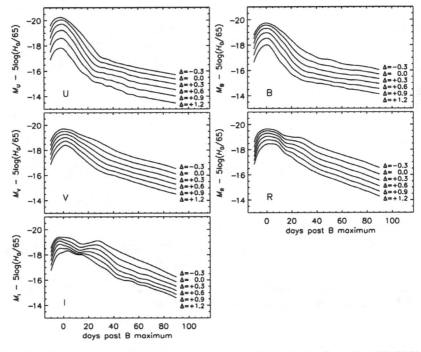

Fig. 8.3. *U*, *B*, *V*, *R*, and *I* light curves for the SN Ia family from the MLCS2k2 model. Δ is the intrinsic supernova parameter of this model. (Figure from Jha *et al.*, 2007.)

Phillips *et al.*, 1999; Jha *et al.*, 2007). Objects with broad light curves tend to be intrinsically bright. The light-curve shape during this phase is correlated (perhaps degenerate) with other observables such as the spectroscopic line ratios of the $\lambda 3650$, 3900 CaII, $\lambda 5650$ SII, and $\lambda 6355$ SiII features at peak brightness (Nugent *et al.*, 1995; Bongard *et al.*, 2006).

There is growing belief that supernova colors indicate intrinsic supernova heterogeneity independent of light-curve shape (Tripp and Branch, 1999; Conley *et al.*, 2008). It is difficult to distinguish between colors intrinsic to the supernova versus those due to dust, particularly in optical wavelengths. It remains inconclusive whether the spread in the UV colors is inconsistent with being purely due to dust (Fig. 3 in Guy *et al.*, 2007; Conley *et al.*, 2008; Nobili and Goobar, 2008).

The heterogeneity in supernova flux has been quantified by its variance (Guy *et al.*, 2007) and its temporal and wavelength covariance (James *et al.*, 2006; Jha *et al.*, 2007). After model corrections, it is the low ~ 0.1 RMS achieved in the *B* and *V* bands at peak brightness that make SNe Ia excellent calibrated candles. The largest variance occurs in the UV and at the rising and plateau phases.

These are the wavelengths and phases most sparsely represented in the training sets, so may become better modeled with the addition of more data.

Supernova *J* and *H* peak magnitudes without extinction or light-curve shape correction are found to have similar RMS (~ 0.15) as those of corrected optical magnitudes (Krisciunas *et al.*, 2004a; Wood-Vasey *et al.*, 2007b). Smaller intrinsic dispersion in the infrared is predicted by Kasen (2006). Use of NIR wavelengths could provide a viable alternative way to measure distances, although getting accurate observations of the low supernova emission at these long wavelengths is challenging at any redshift.

Since some of the dispersions probably have as-of-yet undiscerned correlation with observables, the search for new independent supernova parameters continues to be the subject of active research. Benetti *et al.* (2005) find that SNe Ia can be split into distinct groups through their light-curve shapes and velocity evolution of the $\lambda 6355$ SiII absorption. Theoretical modeling predicts that the specific initial conditions of the supernova progenitor system (e.g. metallicity, white dwarf C/O ratio, central density) have observable consequences in UV flux, colors, light-curve rise times and plateau levels, and spectral velocities (Höflich *et al.*, 2003). With the large supernova datasets now being analyzed, we anticipate learning about new correlations in the near future. Current ambiguity in whether there remains an unidentified non-random component to the SN magnitude dispersion results in experiments making different subjective choices of how to track supernova heterogeneity.

There is diversity in the polarization and hence the geometry of SNe Ia (Leonard *et al.*, 2005; Wang *et al.*, 2007). The effect of the explosion asymmetry seems consistent with the random magnitude dispersion and these data are prohibitively difficult to obtain for supernovae at cosmological distances, so we do not consider polarization further in this article.

8.4 Why space?

8.4.1 Potential for further discovery

Although nature may have conspired to doom cosmologists to making increasingly tighter constraints but never "proving" that the cosmological constant is responsible for the accelerating universe, theory tells us that there is real potential for discovery within the reach of supernova experiments. The goal is to distinguish between an accelerated expansion due to the cosmological constant or due to new physics in the form of modified gravity or fields beyond the Standard Model. The expected distance modulus as a function of redshift for a cosmological constant and a suite of other dark energy models are calculated in Weller and Albrecht (2002) and shown in Fig. 8.4. Their results show that these models

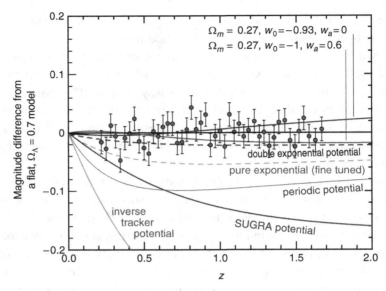

Fig. 8.4. The difference between the expected Hubble relation for a flat, $\Omega_\Lambda = 0.7$ universe and those for a suite of dark energy models. The data points are a realization of simulated SNAP supernovae binned in groups of 50. (Based on the figure in Weller and Albrecht (2002).)

exhibit differences in the predicted Hubble relations over the full redshift range $0 < z < 1.8$ down to 0.02 mag. There is a range in the shapes of the predicted Hubble relation; different models may have similar curves in one redshift window but differ in another. This diversity in the slope of the Hubble relation occurs over the entire plotted redshift range. Supernovae at higher redshifts $z > 2$ are not sensitive to dark energy as normal matter dominates the dynamical behavior in the early universe (Aldering *et al.*, 2007). We conclude that supernovae over a broad redshift range extending out to $z \sim 1.8$ with distances known to <0.02 mag and redshifts known to $<0.01 \, (\mathrm{d}\mu/\mathrm{d}z)^{-1}$ provide leverage in disentangling dark energy models.

In terms of the empirical dark energy parameters w_0 and w_a (or equivalently w_0 and $w' \equiv -\mathrm{d}w/\mathrm{d}\ln a|_{z=1}$), Caldwell and Linder (2005) show that scalar-field quintessence models fall into two broad categories. In "thawing" models the field evolution initiated recently so that in the early universe $w \sim -1$ and it is only now that the equation of state evolves to a less negative number. In the "freezing" scenario, the field is dynamic at early times but slows down as it begins to dominate the energy content of the universe: initially, $w > -1$ and then evolves toward $w \sim -1$. The difference in sign of the evolution of w results in the two classes occupying different regions in the w_0–w' plane, as shown in Fig. 8.5. The two classes are separated by an empty region with constant w; for a field to fall here it must be

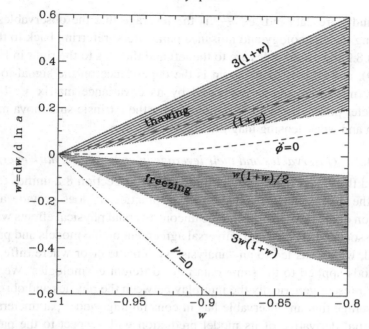

Fig. 8.5. The phase space $w'-w$ possesses distinct regions corresponding to thawing and freezing behavior. In between, the field evolution would need to be finely tuned to coast, $\ddot{\phi} \approx 0$, neither accelerating nor decelerating. The upper long-dashed line shows the degeneracy surface of observations that can measure only the averaged or assumed constant equation of state. (From Linder, 2007.)

finely tuned to have its second time derivative $\ddot{\phi} = 0$ over many dynamical Hubble times.

Improved supernova data *can* distinguish between interesting dark energy models and hone in on their parameters. However, existing surveys are nearing their systematic limit and planned ground surveys will provide only differential improvement in data quality. We need to determine the important properties of the survey that provide the next leap forward in cosmology measurement accuracy.

8.4.2 Survey resolving power

In Section 8.2 we see that measuring dark energy parameters relies on the determination of a suite of other nuisance parameters that describe things like calibration, dust, and SN subtype. The interplay between all the parameters and the ultimate accuracy in the ones that describe the cosmology can be assessed by projecting the parameter covariance matrix U, which is calculated to first order by

$$(U^{-1})_{ij} \approx \sum_{kl} \frac{\partial f_k}{\partial p_i} (V^{-1})_{kl} \frac{\partial f_l}{\partial p_j} + (U^{-1}_{\text{priors}})_{ij}, \qquad (8.11)$$

where k and l are data indices. $\frac{\partial f_k}{\partial p_i}$ is the leverage that the observables have in constraining the cosmology and nuisance parameters; referring back to the model of Section 8.2 the observables are to the left and the f_k's to the right in Eqs. (8.7), (8.8), (8.9), and (8.10). Also shown is the dependence on the signal-to-noise of the survey measurements as expressed by its covariance matrix V. U_{priors} are the parameter priors, particularly important for the intrinsic supernova magnitude dispersion and weak-lensing magnification parameters.

8.4.2.1 Observables and their leverage in constraining dark energy

We remind the reader that the monolithic model in Section 8.2 unifies (and complicates) the distinct analysis steps used in practice. It does provide a uniform presentation of the observables and the uncontroversial physical effects with which they are associated. There is no universal agreement on the models and parameters to be used, which has led to an "analysis" systematic error where different analysis methods applied to the same data infer different cosmologies. We therefore attempt to present generically the interplay between the physics and observables.

The leverage that an observable has in constraining model parameters is given by the partial derivative of its model prediction with respect to the parameters: $\partial f / \partial p$. We now take the cosmology, redshift, supernova, dust, transmission, and standard star parameter sets in turn, and describe their interplay with observables and each other.

We begin with the cosmological and dark energy parameters, either represented by the set of all supernova μ's, or alternatively by the set $\{H_0, \Omega_m, \Omega_{\text{de}}, w_0, w_a\}$. Only the predicted supernova fluxes depend on distance modulus, and the μ parameters require the observation of precision fluxes of supernovae in the range $0 < z < 1.8$ as discussed in Section 8.4.1. Supernova distances do provide resolution in terms of the w_0–w_a dark energy parameters though not as acutely as for $\Omega_m - \Omega_{\text{de}}$ as seen in the bottom plot in Fig. 8.1.

The other axis of the Hubble diagram, the supernova redshifts $\vec{p}_z$, is most trivially constrained by redshift measurements (Eq. (8.8)) but can also be accessed through a supernova photometric redshift ($\partial f_{ij\alpha}/\partial p_z^\alpha$ in Eq. (8.7)). There is recent interest in determining supernova redshifts from photometric data for future imaging surveys that may have insufficient coordinated spectroscopic followup. In such a scenario, an independent measure z^α may be available from the photometric redshift of the host galaxy. Kim and Miquel (2007), however, show that redshifts derived from even excellent light curves and galaxy photo-z's are poorly determined and are strongly correlated with μ. A spectroscopic (or extremely accurate galaxy photometric) redshift must therefore be provided.

The intrinsic supernova parameters affect observed brightness, most prominently through their specific prediction of the absolute magnitude. The sensitive

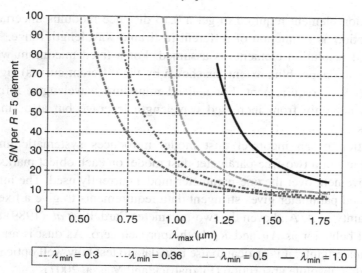

Fig. 8.6. The S/N per resolution element required to get a 0.15 distance modulus uncertainty after dust correction as a function of different restframe wavelength range. The same S/N for all resolution elements is assumed. The different curves correspond to $[0.3 - \lambda_{max}]$, $[0.36 - \lambda_{max}]$, $[0.5 - \lambda_{max}]$, and $[1.0 - \lambda_{max}]$.

mission targets observations at epochs and wavelengths that cover light-curve and spectral features that distinguish supernova subclasses. As discussed in Section 8.3, the most commonly used signatures to subtype SNe Ia are light-curve and spectral features extending from the blue to the defining SiII feature, 0.4–0.63 µm, from −10 to +30 days relative to peak brightness. Newer work and theory hint that additional diversity is encoded in the flux at 0.3–0.4 µm. Figure 12 in Höflich *et al.* (2003) provides quantitative estimates of correlations between SN parameters and absolute magnitude. Future surveys should be flexible enough to provide spectroscopy and well-sampled light curves over at least SN-frame optical wavelengths in anticipation of any new observables being identified as being important for SN Ia subclassification. Analyses that do not subtype can have a systematically imprecise model for M and thus are susceptible to "evolution" systematics if the subtype distribution evolves with redshift.

The dust parameters have an observational effect on the set of relative supernova fluxes measured over a broad wavelength range. Dust extinction is wavelength dependent, with blue light more heavily absorbed than red. When an object with known intrinsic color is obscured by dust, the difference between observed and expected colors quantifies the absolute amount of extinction. With several colors, the absorption properties of the dust can be determined. A is frequently given by the two-parameter dust model of Cardelli *et al.* (1989). The importance of the wavelength coverage is demonstrated in Fig. 8.6, which shows the S/N

per resolution element required to get a 0.15 distance modulus uncertainty after dust correction as a function of different restframe wavelength range. The same S/N for all resolution elements is assumed. For a fixed maximum wavelength in the optical, deeper blue extent (as shown by the different curves) significantly relaxes the S/N requirement. For a fixed minimum wavelength, infrared data give gains not only from increased wavelength leverage but also smaller dust corrections.

Application of an incorrect dust model introduces systematic uncertainty; clearly the use of a two-free-parameter dust model on each object marks a significant improvement over the one-parameter model currently used. The introduction of the second parameter gives stringent data requirements to give a fixed statistical uncertainty in A. And even the two-parameter Cardelli *et al.* (1989) model has unphysical behavior as A_V and R_V both approach zero. As dust is an important source of uncertainty, it is critical to use new data to test how well optical and NIR colors predict absolute absorption (Fitzpatrick and Massa, 2007).

The calibration parameters that describe the band transmissions and the flux of the fundamental standard of the magnitude system are most directly constrained by calibration-specific observations (Eqs. (8.9) and (8.10)). We make only a few comments on the measurements necessary for the calibration program.

The system transmission F characterizes the combined throughput of the entire optical chain for each individual exposure: from the atmosphere, optical telescope assembly, filter, detector, and electronics. The parameters $\vec{p}_F$ must describe all the wavelength, temporal, and spatial variability in the transmission. In ground observations, the atmospheric transmission varies in time, with extinction contributions from Rayleigh scattering, scattering off of aerosols, and molecular absorption from O_2, O_3, and H_2O. The angular and temporal correlations in atmospheric composition are currently not well known. Next-generation calibration programs are aiming for 1% relative photometry with simultaneous atmospheric monitoring (Stubbs and Tonry, 2006). It remains to be seen how well the absolute transmission for a specific position in the sky for a specific exposure can be determined for a ground-based observatory. A space-borne telescope in a thermally stable environment has transmissions that are temporally stable.

The standard star flux $M^\star$ ties observer counts to a physical system. The best standards used today come from models of the white dwarf stars G191B2B, GD153, and GD71, providing flux calibrations from 0.4 to 1.8 μm that are reliable to $\sim$1% in the optical and to $\sim$5% in the NIR (Bohlin, 2007). As the fluxes are based on models, many of the wavelength-dependent errors are correlated.

The function Z in Eq. (8.7) gives the observed signal of $M^\star$ as viewed through F_{ij}. The common contribution of $M^\star$ and F_{ij} to all modeled supernova fluxes produces strongly correlated uncertainties in μ, $\vec{p}_F$, and $\vec{p}_\star$. A constant wavelength-independent magnitude offset in $M^\star$ and Z results in a constant shift of all μ^α; it is therefore the color and not absolute calibration that is of importance for the study of dark energy.

Calibration (along with dust extinction) is named as the major limiting source of uncertainty in the results from two large ground-based surveys (Astier *et al.*, 2006; Wood-Vasey *et al.*, 2007a), indicating that the calibration data c_k and $m_k^\star$ contribute the limiting source of uncertainty in the determination of μ. Calibration, through its coupling to extinction corrections, dominates supernova uncertainties in programs with 1000 or more objects out to $z = 1.7$. Any supernova cosmology experiment, ground- or space-based, must actively seek to improve this fundamental calibration (Deustua, 2007). One possibility is to use the standard candle nature of SNe to improve the determination of flux of the fundamental standard (Kim and Miquel, 2006).

8.4.2.2 *Parameter priors*

The parameters for the magnitude deviation from the supernova model, $\vec{p}_{ij,\text{res}}^\alpha$, and the gravitational lensing magnification, $p_{\text{lensing}}^\alpha$, are unconstrained by the supernova dataset and must be set by priors to constrain the cosmology fit.

Supernova heterogeneity is described by the pdf of the residual scatters from the SED model training set. The pdf's of all and a subset of SNe Ia at redshift z are related by $P(\vec{p}_{\text{res}}^\alpha | z) = \int d\vec{p}_{\text{SN}} P(\vec{p}_{\text{res}}^\alpha | \vec{p}_{\text{SN}}, z) P(\vec{p}_{\text{SN}} | z)$. One reason that subclassification is important in using SNe Ia as distance indicators is the reduction in pdf variance, for example the magnitude dispersion at peak brightness drops from ~ 0.3 to ~ 0.15 after distinguishing light-curve shapes. Secondly, the pdf's are redshift dependent so any redshift evolution in $P(\vec{p}_{\text{SN}} | z)$ can bias Hubble diagrams that use all supernovae. This type of error, commonly referred to as the supernova-evolution systematic error, is no longer a source of concern when supernovae are properly subclassified.

MLCS2k2 (Jha *et al.*, 2007) and SALT2 (Guy *et al.*, 2007) provide residual magnitude dispersions along with their SED models that cover U to I wavelengths. The dispersion is smallest in the BVR-bands from $[-5, 15]$ days relative to peak brightness, meaning either that the model parameterization best captures variability, and/or supernovae have little intrinsic random dispersion in this temporal and wavelength range. Of course, these results do not have the final say as to where SNe are best standardizable. The regions of high dispersion tend to be at wavelengths and phases poorly populated by the training data set. In the NIR for example, the relatively small data set indicates that supernovae are very homogeneous in JHK bands even without subclassification (Krisciunas *et al.*, 2004b).

The expected pdf of $p^\alpha_{\text{lensing}}$ is a function of cosmology and can be calculated via simulation and used as a prior for the lensing of each individual supernova (Holz and Liender, 2005). The dispersion and the offset of the mode from zero increase with redshift, due to the asymmetry between over- and under-densities and the increasing number of lenses in the intervening space between observer and source. Although the pdf is non-Gaussian, the conservation of energy ensures that these distributions have a mean of zero, meaning that no biases will be incurred as long as the pdf is fairly sampled, e.g. through independent lines of sight.

Direct calculation can provide an estimate of magnification by assigning mass halos to the intervening galaxies and galaxy clusters detected in the images (Gunnarson *et al.*, 2006), as has been done for several high-redshift objects (Riess *et al.*, 2001; Jönsson *et al.*, 2006). It is not yet clear whether the reduction in expected magnitude dispersion justifies potential biases induced by the predictions, and whether the slightly more accurate mass maps made possible with the small and low surface brightness galaxies in space images provide a real advantage in making lensing corrections.

8.4.2.3 Data covariance matrix and a survey figure of merit

The designer of a supernova mission has flexibility in customizing the data covariance matrix V (introduced in Eq. (8.11)) to best measure the dynamical effects of dark energy. The signal-to-noise requirements of the fundamental data are chosen to satisfy two conditions. Firstly, the distance modulus uncertainty deduced for each individual object should have similar contributions from the intrinsic magnitude dispersion and measurement uncertainty. Secondly, the anticipated systematic uncertainties dictate how large the sample should be: the dark energy parameter uncertainties of the survey should be barely dominated by external limitations.

The data the observer has most control over are the signal-to-noise of the flux measurements (Eq. (8.7)) and the redshift determination (Eq. (8.8)). The SN Ia subtyping strategy is the other important choice to be made by the experiment designer. Calibration measurements of band transmission and standard star flux do not directly influence the supernova survey, and the detailed discussion on this complex field falls outside the scope of this chapter. The following references provide a sampling of strategies and observations being used to improve telescope calibrations and fundamental calibrators (Stubbs and Tonry, 2006; Bohlin, 2007; Deustua, 2007; Kaiser *et al.*, 2007; Padmanabhan *et al.*, 2008).

As shown in the previous section, restframe multi-channel optical data of supernovae out to $z \sim 1.8$ that extend redwards to 0.63 μm and beyond from at least -10 to 30 days from peak brightness can discriminate SN Ia subtypes, foreground dust, and ultimately dark energy models. It is also at these wavelengths and phases that SNe Ia are most powerful as distance indicators, i.e. where the dispersion in the distribution function of $\vec{p}_{res}$ is small. There is no commonly accepted set of signal-to-noise requirements of particular features adopted by the JDEM candidates. However, the subject of this article, the relative efficiency of ground and space observatories in generating these data, can be calculated independently of the signal-to-noise (or covariance matrix V) requirement in the sky-noise limit. To this end, we present a figure of merit for dedicated observatories running rolling supernova searches: the number of supernovae whose light curves can be satisfactorily observed over the desired phase of evolution. In addition to the dependence on etendue (telescope collecting area $\times$ camera solid-angle) there are terms for the sky brightness, seeing, and the continuous viewing zone of discovered supernovae. As a rolling search, exposure times are not tuned to specific objects in the field but are set to obtain the data most difficult to obtain, generally off-peak measurements at the high redshift limit.

In a rolling search, the desired temporal sampling of the supernova establishes the cadence at which fields are revisited. During that period there is a time duration T_C available for exposure integration at an observatory.

The per-exposure signal-to-noise requirements vary over the phase of the supernova; the exposure time t_E on a field for each visit must satisfy the most demanding signal-to-noise requirement for all phases and redshifts. Assuming a Poisson sky-dominated background and simple aperture photometry,

$$t_E = \frac{(S/N)^2 f_{sky} \pi \sigma^2}{\epsilon A f_{SN}^2}, \tag{8.12}$$

where f_{sky} is the sky surface brightness, ϵ the throughput efficiency, A the telescope aperture area, σ seeing, and S/N and f_{SN} are the signal-to-noise and supernova flux corresponding to the most stringent observing requirement.

The number of fields that can be observed in a cycle is then

$$N_F = T_C/t_E = \frac{T_C \epsilon A f_{SN}^2}{(S/N)^2 f_{sky} \pi \sigma^2}. \tag{8.13}$$

During one cycle of the observing cadence N_F fields can be observed. However, it is possible that only a subset of those fields will still be visible some months later when further light-curve points are needed to characterize the supernova. At the major ground observatories, only the regions around the equatorial poles are visible throughout the year but they lie at extremely high airmass. Surveys typically have target fields uniformly distributed in RA with declination limited in range of the latitude of the observatory. These fields are visible neither for the full night nor

Table 8.1. *Variables used in the SN Survey figure of merit*

Ω	Field of view
T_L	Time the SN must be followed after explosion
T_V	Time over the year the field is monitored
T_C	Available exposure time during one cadence cycle
A	Telescope aperture area
f_{sky}	Sky flux
σ	PSF FWHM

over a full year. On a given night all accessible fields are observed, but the early fields are only taken to follow old objects and not to discover new transients that cannot be followed up.

The fraction of fields on which a new supernova can be observed over the extent of its light curve is given as F so that the number of useful discovery fields is $F N_F$. If T_V is the amount of time in a year that a field is visible and T_L is the extent of time over which the supernova must be followed, then $F = (T_V - T_L)/T_V$.

The number of useful supernovae discovered is then proportional to the field of view Ω and the number of useful fields:

$$N_{SN} \propto \frac{\Omega (1 - T_L/T_V) T_C \epsilon A f_{SN}^2}{(S/N)^2 f_{sky} \pi \sigma^2}. \tag{8.14}$$

The variables and parameters in this equation are summarized in Table 8.1.

The relative yield of surveys with similar observing requirements and instrumentation but run at different observatories can be compared by comparing their ratio of Eq. (8.14):

$$\frac{N_{SN,1}}{N_{SN,2}} = \frac{\Omega_1 (1 - T_L/T_{V,1}) T_{C,1} \epsilon_1 A_1 f_{sky,2} \sigma_2^2}{\Omega_2 (1 - T_L/T_{V,2}) T_{C,2} \epsilon_2 A_2 f_{sky,1} \sigma_1^2}. \tag{8.15}$$

Several factors in Eq. (8.15) directly demonstrate the advantages of observations from space. Ground-based observatories collect data only at night and are susceptible to poor weather. Small correlated patches of poor weather leave detrimental gaps in all objects being actively monitored. We conservatively adopt $T_{C,s}/T_{C,g} \sim 2$ as the advantage that the space observatory (s) has in the available time to collect science data relative to the ground (g).

To capture the rise and plateau phases, a supernova must be monitored from explosion to two months post explosion in its restframe, i.e. $T_L = 2(1 + z)$ months. A space observatory can easily point toward the ecliptic poles whose line of sight never passes near the Sun; neglecting the edge effects of the finite duration of the survey, the fields accessible in space have $F \to 1$. Almost all optical observatories are located at latitudes where the ecliptic poles lie at high airmass $\chi = 2-3$. Observations at high airmass are bad primarily because the seeing degrades with airmass, $\sigma \propto \chi^{0.6}$, and atmospheric extinction increases as well. At the Paranal Observatory in Chile, the south ecliptic pole is visible for the full night over the entire year at $\chi = 2.4$. By shifting the targeted declination toward the equator, a field can be observed at lower airmass but for fewer hours within a night and fewer months over a year. At Paranal a field with the latitude of the site is visible with $\chi < 1.5$ for at least one hour over <8 months and four hours over <5 months of the year. Anticipating the long exposure times necessary for the highest redshift supernova, we consider a ground-targeted field that is continuously visible for 4 hours for 6 contiguous months (some observations are taken with $\chi > 1.5$). Then the ratio of discovered supernovae to those that get adequate light-curve coverage is $\frac{(1 - T_L/T_{V,g})}{(1 - T_L/T_{V,s})} = 3/(2 - z)$.

We can compare the zodiacal background and gray sky emission at wavelengths that correspond to 5500 Å for $z = 1.5$, which corresponds to the restframe V band. For the observer $\lambda = 13\,750$ Å, $f_s/f_g = 0.004$ taking zodiacal light at high ecliptic latitude and for half-moon sky illumination.

The optics and instrumentation of a space telescope with a mirror diameter D provide a diffraction-limited seeing of $\sigma_s = 1.44\lambda/D = 1.28\lambda/\sqrt{A} = 0.13\left(\frac{\lambda}{5000\,\text{Å}}\right)$ $\left(\frac{1\,\text{m}^2}{A}\right)^{1/2}$ arcsec. Using the measured seeing distribution from Subaru at Mauna Kea of $\sqrt{\langle\sigma_g^2\rangle} = 0.74\left(\frac{\lambda}{5000\,\text{Å}}\right)^{-0.2}\chi^{0.6}$ arcsec, we get

$$\frac{\sigma_s^2}{\sigma_g^2} = 0.031\left(\frac{\lambda}{5000\,\text{Å}}\right)^{2.4}\left(\frac{1\,\text{m}^2}{A}\right)\chi^{-1.2}. \qquad (8.16)$$

Assigning the same throughput for both missions, $\epsilon_s/\epsilon_g = 1$, we find the relative survey figure of merit for surveys measuring the SN-frame V band at $z \sim 1.5$ with $\langle\chi^{1.2}\rangle^{1/1.2} = 1.5$ to be

$$\frac{N_{\text{SN,s}}}{N_{\text{SN,g}}} = 3.0 \times 10^4\left(\frac{A_s}{1\,\text{m}^2}\right)\frac{A_s\Omega_s}{A_g\Omega_g}. \qquad (8.17)$$

For a ground survey to be competitive, it must have an etendue $A\Omega > 10^5$ times larger than a 2-meter space telescope.

To provide perspective, the Large Synoptic Survey Telescope (LSST) (Sweeney, 2006) is the proposed next-generation imaging survey telescope with largest etendue, with an effective 6.5 meter aperture and a 10 square degree field of view (though with no plans for a NIR camera). The proposed space mission SNAP (Aldering *et al.*, 2004) is a 2-meter telescope with a 0.7 square degree imager. LSST has a factor 100 larger etendue than SNAP, falling way short of making it competitive with a canonical space-based supernova survey. It should be emphasized that the figure of merit for the 6.5-meter ground telescope improves and exceeds that of the 2-meter space telescope when the most challenging observations move to shorter wavelength, i.e. for lower-redshift surveys.

Supernova redshift is the other observable to be obtained by the survey. This is best done with spectroscopic measurement of either the supernova or its host galaxy. When the surveys are spectroscopic, as is the case for Destiny and ADEPT, the redshift measurement requirement is superfluous and the survey figure of merit Eq. (8.14) applies. If the spectrum is observed with triggered followup, Eq. (8.14) without the $(1 - T_L/T_V)$ term is appropriate. The redshift measurement is usually superseded by other spectroscopy requirements, e.g. SNAP requires supernova/host galaxy spatial resolution for supernova subtyping and spectrophotometry. Then the ratios used for the calculation of relative ground and space figure of merit apply. If the spectrum is only used to obtain redshift, adaptive optics can be used from the ground. In this case $\sigma_s^2/\sigma_g^2 = A_g/A_s$ and $\epsilon_s/\epsilon_g \approx$ (Strehl ratio)$^{-1}$ so 10-meter class ground-based telescopes can perform redshift determinations as efficiently as space-based 2-meter.

8.5 JDEM candidate supernova missions

JDEM is one of the three Einstein Probes that make up NASA's Beyond Einstein Program, and is an interagency partnership with the Department of Energy. It is a mission that will specifically study dark energy and determine how it evolves with time. The mission is being designed by the JDEM Project Office with input from a Science Coordination Group. Whatever the ultimate mission turns out to be, it is interesting to see the range of possible space missions proffered by the three specific experiments that were awarded NASA funding to develop advanced mission concept studies. Two of these candidate JDEM experiments, SNAP and Destiny, use SN Ia distance indicators as one of their primary cosmological probes. The third experiment, ADEPT, is primarily a baryon oscillation survey with a secondary supernova survey.

The differences between the missions can be drawn along several lines. SNAP performs a photometric survey with triggered spectroscopic followup whereas the other two missions run spectroscopic surveys. All missions plan to combine their supernovae with a nearby sample collected from ground observatories. As SNAP can observe supernovae to much lower redshift than its competitors, its associated ground sample can be comparatively shallow. ADEPT does not claim the redshift depth of the other two missions. The restricted redshift range of Destiny and ADEPT allow them to be pure NIR missions. Perhaps most significantly, the three missions adopt different signal-to-noise requirements for the data necessary to constrain supernova heterogeneity and host-galaxy dust, for both the ground (low-redshift) and space (high-redshift) samples. As the development of the three proposals was dynamic and competition sensitive, we give a brief snapshot of how they look based on publicly available information.

Destiny (Benford and Lauer, 2006) proposes to observe and characterize $\sim$3000 SN Ia events over the redshift interval $0.4 < z < 1.7$ within a 3 square degree survey area in two years. Destiny's science instrumentation consists of a 1.65-meter space telescope featuring a grism-fed 0.85–1.7 μm slitless $R = 75$ spectrometer with a 0.12 square degree field of view. Destiny collects a spectroscopic supernova time series with 5-day cadence with a 4-hour exposure at each visit. These data provide synthetic broad-band photometry and supernova diagnostics. Destiny only obtains SN-frame B-band for $z > 1$ supernovae, so they must either be used jointly with ground data up to $z = 1$, or restrict their analysis to redder bands.

SNAP (Aldering *et al.*, 2004) proposes to discover and image over 3000 SNe Ia over the redshift interval $0.1 < z < 1.7$ in a 7.5 square degree area over the course of 22 months. A subset of 2000 SNe have targeted spectroscopic observations. SNAP has a nine-band imager with a 0.7 square degree instrumented field of view and an integral field unit spectrometer, both covering $0.35 < \lambda < 1.7$ μm. The supernova survey runs with a 4-day cadence obtaining optical exposure times of 1200 seconds and double that in NIR bands at each visit. Spectroscopic followup is triggered using imaging discoveries, with exposure times scaled to the object brightness. Imaging and spectroscopic data can be obtained concurrently. SNAP supernovae are supplemented with supernovae discovered from the ground at $0 < z < 0.3$. The expected performance of SNAP is shown in Fig. 8.4, which shows how the data resolve different dark energy models in the Hubble diagram, and in Fig. 8.7 that shows constraints in terms of w_0 and w_a for the combined supernova and weak-lensing surveys. The uncertainties in these plots are dominated by projected systematic uncertainties (Kim *et al.*, 2004).

Very little information on ADEPT is available to the public. The team proposes to obtain 1000 SNe in the redshift range $0.8 \leq z \leq 1.3$. The hardware is specifically designed for the baryon acoustic oscillation survey, consisting of a 1.3-meter telescope with a NIR low-dispersion multi-object spectrograph.

JEDI (Crotts *et al.*, 2005) is noteworthy as an alternative JDEM concept that proposed to go deeper into the infrared to observe restframe J and H bands. Dust is less of an issue and there is speculation that SNe may be better standardized at these longer wavelengths than in the optical.

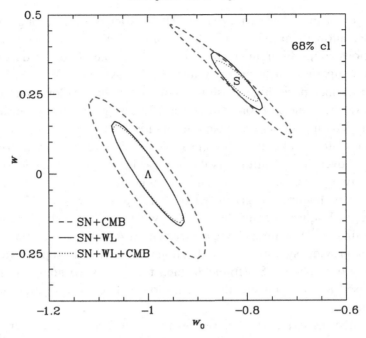

Fig. 8.7. The 68% confidence region in the dark energy parameters for fiducial cosmological constant (labeled with Λ) and supergravity (labeled with S) universes. The dashed curves show the results of the supernova simulation and associated systematic-error model, with a Planck prior and an assumed flat universe. In solid is the combined constraints with the SNAP weak-lensing surveys that combine statistical shear power-spectrum and bispectrum measurements and CCC with a 3% shear-calibration uncertainty. Adding CMB data (dotted curves) provides additional constraints for the supergravity universe, but not for the fiducial cosmological model. (From Aldering et al., 2004.)

8.6 Conclusions

Type Ia supernovae were used as distance indicators to discover the accelerating expansion of the universe and continue to provide the best constraints on the properties of dark energy. Improved discrimination of whether the acceleration is due to a cosmological constant or different esoteric physics requires extending the supernova Hubble diagram to higher redshifts and controlling sources of systematic uncertainty with precision measurements. The restframe optical and/or NIR observations needed to robustly determine distances are redshifted into observer near-infrared wavelengths. A space platform for observations, avoiding the Earth's atmosphere and with a stable optical chain, provides the ideal environment to obtain the required measurements. The scientific importance of the study of dark energy from space is recognized by NASA and the US Department of Energy:

JDEM is a key component of the Beyond Einstein Program. Assessment of a supernova component to JDEM is explicitly called for in the Science Coordination Group Charter.

The JWST is an upcoming space telescope that can and will observe high-redshift supernovae. Although it cannot itself generate the dataset necessary to probe dark energy, JWST can provide supplemental data for a subset of JDEM discoveries at wavelengths or faint magnitudes that would not be otherwise observed.

Although this note has focused on high-redshift supernovae, the low-redshift set is equally important for cosmological measurements. Certainly, the low-redshift discoveries ($z < 0.1$) at CTIO (Hamuy *et al.*, 1996a) and subsequent discoveries and observations (Riess *et al.*, 1999; Filippenko *et al.*, 2001; Jha, 2002) were instrumental in the discovery of the accelerated expansion of the universe. Searches for nearby supernovae continue (Filippenko *et al.*, 2001; Copin *et al.*, 2006; Wood-Vasey *et al.*, 2007b) and new experiments (Keller *et al.*, 2007) are planned for the future. A next generation ground-based program that provides low-redshift supernova data equal to or better than what is required for JDEM (Linder, 2006), together with JDEM, provides new discovery opportunities in our understanding of the dark energy that causes the accelerated expansion of our universe.

References

Aldering, G., *et al.* (2004). [astro-ph/0405232].

Aldering, G., *et al.* (2006). *Astrophys. J.* **650**, 510.

Aldering, G., Kim, A. G., Kowalski, M., Linder, E.V., and Perlmutter, S. (2007). *Astropart. Phys.* **27**, 213.

Arnett, W. D. (1969). *Astrophys. Space Sci.* **5**, 180.

Arnett, W. D. (1982). *Astrophys. J.* **253**, 785.

Astier, P., *et al.* (2006). *Astron. Astrophys.* **447**, 31.

Bebek, C. J., *et al.* (2004). *SPIE Conf. Ser.* **5167**, 50.

Benetti, S., *et al.* (2005). *Astrophys. J.* **623**, 1011.

Benford, D. J., and Lauer, T. R. (2006). *SPIE Conf. Ser.* **6265**.

Blondin, S., and Tonry, J. L. (2007). *Astrophys. J.* **666**, 1024.

Bohlin, R. C. (2007). *Astron. Soc. Pacif. Conf. Ser.* **364**, 315.

Bongard, S., Baron, E., Smajda, G., Branch, D., and Hauschildt, P. H. (2006). *Astrophys. J.* **647**, 513.

Brown, M. G., *et al.* (2006). *SPIE Conf. Ser.* **6265**.

Burstein, D., and Heiles, C. (1982). *Astron. J.* **87**, 1165.

Caldwell, R. R., and Linder, E. V. (2005). *Phys. Rev. Lett.* **95**, 141301.

Cardelli, J. A., Clayton, G. C., and Mathis, J. S. (1989). *Astrophys. J.* **345**, 245.

Clocchiatti, A., *et al.* (2006). *Astrophys. J.* **642**, 1.

Conley, A., *et al.* (2008). *Astrophys. J.* **681**, 482.

Copin, Y., *et al.* (2006). *New Astron. Rev.* **50**, 436.

Cousins, A. W. (1976). *Mon. Not. R. Astron. Soc.* **81**, 25.

Crotts, A., *et al.* (2005). [astro-ph/0507043].

Daly, R. A., and Djorgovski, S. G. (2004). *Astrophys. J.* **612**, 652.

Deustua, S. (2007). *Astron. Soc. Pacif. Conf. Ser.* **364**, 355.

Filippenko, A. V., Li, W. D., Treffers, R. R., and Modjaz, M. (2001). *Astron. Soc. Pacif. Conf. Ser.* **183**, 121.

Fitzpatrick, E. L., and Massa, D. (2007). *Astrophys. J.* **663**, 320.

Freedman, W. L., *et al.* (2001). *Astrophys. J.* **553**, 47.

Garnavich, P. M., *et al.* (1998). *Astrophys. J.* **493**, L53.

Goobar, A., and Perlmutter, S. (1995). *Astrophys. J.* **450**, 14.

Gunnarsson, C., Dahlén, T., Goobar, A., Jönsson, J., and Mörtsell, E. (2006). *Astrophys. J.* **640**, 417.

Guy, J., *et al.* (2007). *Astron. Astrophys.* **466**, 11.

Hamuy, M., *et al.* (1996a). *Astron. J.* **112**, 2408.

Hamuy, M., *et al.* (1996b). *Astron. J.* **112**, 2438.

Hamuy, M., *et al.* (2003). *Nature (London)* **424**, 651.

Hayes, D. S. (1985). In *Calibration of Fundamental Stellar Quantities*, D.S. Hayes, L.E. Pasinetti, and A.G.D. Philip, Eds., Dordrecht: D. Reidel Publishing Co., p. 225.

Höflich, P., Gerardy, C., Linder, E., *et al.* (2003). In *Stellar Candles for the Extragalactic Distance Scale*, D. Alloin and W. Gieren, Eds., Berlin: Springer–Verlag, p. 203.

Holland, S. E., Groom, D. E., Palaio, N. P., Stover, R. J., and Wei, M. (2003). *IEEE Trans. Elec. Dev.* **50**, 225.

Holz, D. E. (1998). *Astrophys. J.* **506**, L1.

Holz, D. E., and Liender, E. V. (2005). *Astrophys. J.* **631**, 678.

Howell, D. A., *et al.* (2005). *Astrophys. J.* **634**, 1190.

Hoyle, F., and Fowler, W. A. (1960). *Astrophys. J.* **132**, 565.

Hsiao, E. Y., *et al.* (2007). *Astrophys. J.* **663**, 1187.

James, J. B., Davis, T. M., Schmidt, B. P., and Kim, A. G. (2006). *Mon. Not. R. Astron. Soc.* **370**, 933.

Jha, S. (2002). Harvard University D. Phil. thesis.

Jha, S., Riess, A. G., and Kirshner, R. P. (2007). *Astrophys. J.* **659**, 122.

Johnson, H. L. and Morgan, W. W. (1953). *Astrophys. J.* **117**, 313.

Jönsson, J., Dahlén, T., Goobar, A., *et al.* (2006). *Astrophys. J.* **639**, 991.

Kaiser, M. E., Kruk, J. W., McCandliss, S. R., *et al.* (2007). *Astron. Soc. Pacif. Conf. Ser.* **364**, 361.

Kasen, D. (2006). *Astrophys. J.* **649**, 939.

Keller, S. C., *et al.* (2007). *Publ. Astron. Soc. Austral.* **24**, 1.

Khokhlov, A. M. (1991). *Astron. Astrophys.* **245**, 114.

Kim, A. G. and Miquel, R. (2006). *Astropart. Phys.* **24**, 451.

Kim, A. G. and Miquel, R. (2007). *Astropart. Phys.* **28**, 448.

Kim, A., Goobar, A., and Perlmutter, S. (1996). *Publ. Astron. Soc. Pacif.* **108**, 190.

Kim, A. G., Linder, E. V., Miquel, R., and Mostek, N. (2004). *Mon. Not. R. Astron. Soc.* **347**, 909.

Knop, R. A., *et al.* (2003). *Astrophys. J.* **598**, 102.

Krisciunas, K., Phillips, M. M., and Suntzeff, N. B. (2004a). *Astrophys. J.* **602**, L81.

Krisciunas, K., *et al.* (2004b). *Astron. J.* **128**, 3034.

Leonard, D. C., Li, W., Filippenko, A. V., Foley, R. J., and Chornock, R. (2005). *Astrophys. J.* **632**, 450.

Linder, E. V. (2006). *Phys. Rev. D* **74**, 103518.

Linder, E. V. (2007). *J. Phys. A* **40**, 6697.

Nobili, S., and Goobar, A. (2008). *Astron. Astrophys.* **487**, 19.

Nomoto, K., Sugimoto, D., and Neo, S. (1976). *Astrophys. Space Sci.* **39**, L37.

Norgaard-Nielsen, H. U., Hansen, L., Jorgensen, H. E., Aragon-Salamanca, A., and Ellis, R. S. (1989). *Nature (London)* **339**, 523.

Nugent, P., Phillips, M., Baron, E., Branch, D., and Hauschildt, P. (1995). *Astrophys. J.* **455**, L147.

Nugent, P., Kim, A., and Perlmutter, S. (2002). *Publ. Astron. Soc. Pacif.* **114**, 803.

Padmanabhan, N., *et al.* (2008). *Astrophys. J.* **674**, 1217.

Perlmutter, S., *et al.* (1995). *Astrophys. J.* **440**, L41.

Perlmutter, S., *et al.* (1997). *Astrophys. J.* **483**, 565.

Perlmutter, S., *et al.* (1998). *Nature (London)* **391**, 51.

Perlmutter, S., *et al.* (1999). *Astrophys. J.* **517**, 565.

Phillips, M. M., Lira, P., Suntzeff, N. B., Schommer, R. A., Hamuy, M., and Maza, J. (1999). *Astron. J.* **118**, 1766.

Plewa, T., Calder, A. C., and Lamb, D. Q. (2004). *Astrophys. J.* **612**, L37.

Prieto, J. L., *et al.* (2007). [arXiv:0706.4088].

Rapetti, D., Allen, S. W., Amin, M. A., and Blandford, R. D. (2007). *Mon. Not. R. Astron. Soc.* **375**, 1510.

Riess, A. G., Press, W. H., and Kirshner, R. P. (1996). *Astrophys. J.* **473**, 88.

Riess, A. G., *et al.* (1998). *Astron. J.* **116**, 1009.

Riess, A. G., *et al.* (1999). *Astrophys. J.* **117**, 707.

Riess, A. G., *et al.* (2001). *Astrophys. J.* **560**, 49.

Riess, A. G., *et al.* (2004). *Astrophys. J.* **607**, 665.

Riess, A. G., *et al.* (2007). *Astrophys. J.* **659**, 98.

Schlegel, D. J., Finkbeiner, D. P., and Davis, M. (1998). *Astrophys. J.* **500**, 525.

Smith, J. A., *et al.* (2002). *Astron. J.* **123**, 2121.

Stubbs, C. W., and Tonry, J. L. (2006). *Astrophys. J.* **646**, 1436.

Sweeney, D. W. (2006). *SPIE Conf. Ser.* **6267**.

Tripp, R., and Branch, D. (1999). *Astrophys. J.* **525**, 209.

Wang, L., Baade, D., and Patat, F. (2007). *Science* **315**, 212.

Wang, L., Goldhaber, G., Aldering, G., and Perlmutter, S. (2003). *Astrophys. J.* **590**, 944.

Wang, Y., and Tegmark, M. (2005). *Phys. Rev. D* **71**, 103513.

Weller, J., and Albrecht, A. (2002). *Phys. Rev. D* **65**, 103512.

Wood-Vasey, W. M. (2002). *IAU Circ.* **8019**, 2.

Wood-Vasey, W. M., *et al.* (2007a). *Astrophys. J.* **666**, 694.

Wood-Vasey, W. M., *et al.* (2007b). [arXiv:0711.2068].

9

Baryon acoustic oscillations

BRUCE BASSETT AND RENÉE HLOZEK

9.1 Introduction

Whilst often phrased in terms of discovering the nature of dark energy, cosmology in the twenty-first century might also aptly be described as 'the distance revolution'. With new knowledge of the extragalactic distance ladder we are, for the first time, beginning to accurately probe the expansion history of the cosmos beyond the local universe, at redshifts $z > 0.1$. While standard candles – most notably Type Ia supernovae (SN Ia) – kicked off the distance revolution, it is clear that standard rulers, and the baryon acoustic oscillations (BAO) in particular, will play an important role in the coming revolution.

Here we review the theoretical, observational and statistical aspects of the BAO as standard rulers and examine the impact BAO will have on our understanding of dark energy, the distance and expansion ladder.

9.1.1 A brief history of standard rulers and the BAO

Let us start by putting the BAO in context. The idea of a standard ruler is one familiar from everyday life. We judge the distance of an object of known length (such as a person) by its angular size. The further away it is, the smaller it appears. The same idea applies in cosmology, with one major complication: space can be curved. This is similar to trying to judge the distance of our known object through a smooth lens of unknown curvature. Now when it appears small, we are no longer sure it is because it is far away. It may be near and simply appear small because the lens is distorting the image. This degeneracy between the curvature of space and radial distance has not been the major practical complication in cosmology over the past century, however. That honour goes to a fact that has plagued us since the beginning of cosmology: we don't know how big extragalactic objects

Dark Energy: Observational and Theoretical Approaches, ed. Pilar Ruiz-Lapuente. Published by Cambridge University Press. © Cambridge University Press 2010.

are in general, in the same way that we don't know how bright they intrinsically are. This problem was at the heart of the great debate between Shapley and Curtis over the nature of galaxies. Shapley argued that they were small and inside our own galaxy while Curtis maintained that they were extragalactic and hence much larger.

To be useful for cosmology, we need a *standard ruler*: an object of a known size at a single redshift, z, or a population of objects at different redshifts whose size changes in a well-known way (or is actually constant) with redshift. Ideally the standard ruler falls into both classes, which, as we will argue below, is the case for the BAO, to good approximation. BAO are however a new addition to the family of putative standard rulers. A few that have been considered in the past include ultra-compact radio sources (Kellermann, 1993; Gurvits, 1994) which indeed led in 1996, prior to the SN Ia results, to claims that the density of dark matter was low, $\Omega_m < 0.3$, with non-zero cosmological constant of indeterminate sign (Jackson and Dodgson, 1997). Another radio standard ruler candidate is provided by double-lobed radio sources (Buchalter *et al.*, 1998). These Fanaroff–Riley Type IIb radio galaxies were suggested as cosmological probes as early as 1994 (Daly, 1994), and subsequent analyses have given results consistent with those from SN Ia (Daly and Guerra, 2002; Daly, Mory and O'Dea, 2007). An alternative approach uses galaxy clusters. Allen *et al.* relate the X-ray flux to the cluster gas mass and, in turn, its size, providing another standard ruler under the assumption that the gas fraction is constant in time. This too leads to results consistent with those from SN Ia (Allen, Schmidt and Fabian, 2002; Mantz *et al.*, 2008).

Beyond this we move into the realm of Statistical Standard Rulers (SSR), of which BAO are the archetype. SSR exploit the idea that the clustering of galaxies may have a preferred scale in it which, when observed at different redshifts, can be used to constrain the angular diameter distance. The idea of using a preferred clustering scale to learn about the expansion history of the cosmos has a fairly long history in cosmology, dating back at least to 1987 and perhaps earlier. In their conclusions Shanks *et al.* (1987) make the prescient comment:

There is one further important reason for searching for weak features in $\xi_{qq}(r)$, the quasar–quasar correlation function, at large separations. If a particular feature were found to appear in both the galaxy correlation function at low redshift and the QSO correlation function at high redshift, then a promising new cosmological test for q_0 might be possible.

A series of later analyses further built up the idea of SSR in cosmology using variously as motivation the turn-over in the power spectrum due to the transition from radiation to matter domination, the mysterious $128h^{-1}$ Mpc feature detected in early pencil-beam surveys (Broadhurst *et al.*, 1990) and the realisation that inflation could inject a preferred scale into the primordial power spectrum. These early studies often found tentative evidence for a low-density universe and/or non-zero cosmological constant e.g. Deng, Xia and Fang (1994), Broadhurst and Jaffe (2000). Since the preferred scale could not be accurately predicted *a priori* these studies only provided the relative size of the SSR at different redshifts. Nevertheless, an analysis of this type using quasars from the 2dF survey found $\Omega_m = 0.30 \pm 0.15$ assuming a flat ΛCDM universe (Roukema and Mamon, 2000; Roukema, Mamon and Bajtlik, 2002) shortly after the 1998 watershed discovery of acceleration by SN Ia.

BAO entered the fray initially as a putative explanation for the $\sim 100h^{-1}$ Mpc clustering but were found to be too weak to be the origin for the apparent excess (Eisenstein, Hu, and Tegmark, 1998; Meiksin, White and Peacock, 1999). The idea of using BAO themselves to learn about cosmological parameters seems to date first from Eisenstein, Hu and Tegmark (1998) who wrote:

Detection of acoustic oscillations in the matter power spectrum would be a triumph for cosmology, as it would confirm the standard thermal history and the gravitational instability paradigm. Moreover, because the matter power spectrum displays these oscillations in a different manner than does the CMB, we would gain new leverage on cosmological parameters.

The first photometric proposal for using the BAO as standard rulers appears to date from 2001 (Cooray *et al.*, 2001), while the foundations of the modern ideas on spectroscopic BAO surveys were laid by Eisenstein (2003), Blake and Glazebrook (2003) and a slew of later papers which developed hand-in-hand with the analysis of real data. Tantalising hints for the existence of the BAO were already visible in the Abell cluster catalogue (Miller, Nichol and Chen, 2002), but definitive detections had to wait for the increased survey volume and number of galaxies achieved in the SDSS and 2dF redshift surveys, yielding strong constraints on both curvature and dark energy at $z < 0.5$ (Eisenstein *et al.*, 2005; Cole *et al.*, 2005; Tegmark *et al.*, 2006; Percival *et al.*, 2007; Gaztanaga, Cabre and Hui, 2008). Figures 9.1 and 9.2 show the original evidence for the acoustic signature in the correlation function and power spectrum. Extracting the BAO scale from the matter power spectrum remains a thriving area of research in contemporary cosmology, as we discuss later in Section 9.5 on current and future BAO surveys.

9.1.2 Cosmological observables

We now discuss the relevant cosmological observables that are derived from standard rulers in general, and the BAO in particular. The BAO in the radial and tangential directions provide measurements of the Hubble parameter and angular diameter distance respectively. The Hubble parameter, $H \equiv \dot{a}/a$ – where a is

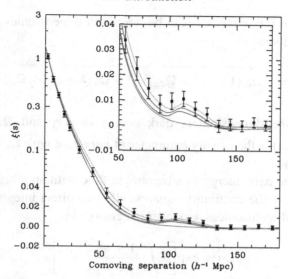

Fig. 9.1. Baryon acoustic peak (BAP) in the correlation function. The BAP is visible in the clustering of the SDSS LRG galaxy sample, and is sensitive to the matter density (shown are models with $\Omega_m h^2 = 0.12$ (*top*), 0.13 (*second*) and 0.14 (*third*), all with $\Omega_b h^2 = 0.024$). The bottom line without a BAP is the correlation function in the pure CDM model, with $\Omega_b = 0$. (From Eisenstein *et al.*, 2005.)

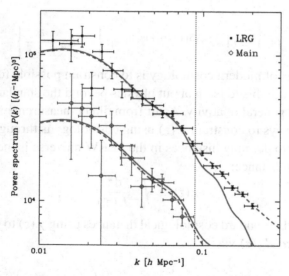

Fig. 9.2. Baryon acoustic oscillations (BAO) in the SDSS power spectra. The BAP of the previous figure now becomes a series of oscillations in the matter power spectrum of the SDSS sample. The power spectrum is computed for both the main SDSS sample (*bottom curve*) and the LRG sample (*top curve*), illustrating how LRGs are significantly more biased than average galaxies. The solid lines show the ΛCDM fits to the WMAP3 data (Spergel *et al.*, 2007), while the dashed lines include nonlinear corrections. (Figure from Tegmark *et al.*, 2006.)

the scale factor of the universe – can be written in dimensionless form using the Friedmann equation as

$$E(z) \equiv \frac{H(z)}{H_0} = \sqrt{\Omega_m(1+z)^3 + \Omega_{de}f(z) + \Omega_k(1+z)^2 + \Omega_{rad}(1+z)^4}, \quad (9.1)$$

where $f(z)$ is the dimensionless dark energy density and $\Omega_k = -\frac{k}{H_0^2 a^2} = 1 - (\Omega_m + \Omega_{de} + \Omega_{rad})$ is the density parameter of curvature with $\Omega_k = 0$ corresponding to a flat cosmos.

If one treats the dark energy as a barotropic fluid with an arbitrary equation of state, $w(z) \equiv p/\rho$, the continuity equation allows a direct integration to give the evolution of the dimensionless dark energy density via

$$f(z) = \exp\left[3\int_0^z \frac{1+w(z')}{1+z'}dz'\right]. \quad (9.2)$$

When we quote constraints on dark energy it will typically be in terms of the CPL parameterisation (Chevallier and Polarski, 2001; Linder, 2003):

$$w(z) = w_0 + w_a\frac{z}{1+z}, \quad (9.3)$$

which has

$$f(z) = (1+z)^{3(1+w_0+w_a)} \exp\left\{-3w_a\frac{z}{1+z}\right\}. \quad (9.4)$$

Much of the quest of modern cosmology is to constrain possible forms of $w(z)$ (or $f(z)$) and hence use this to learn about physics beyond the standard model of particle physics and general relativity. Apart from direct measurements of the Hubble rate, one of the ways to constrain $w(z)$ using cosmology is through distance measurements. Core to defining distances in the FLRW universe is the dimensionless, radial, comoving distance:

$$\chi(z) \equiv \int_0^z \frac{dz'}{E(z')}. \quad (9.5)$$

One then builds the standard cosmological distances using $\chi(z)$ to give the *angular diameter distance*, $d_A(z)$ via

$$d_A(z) = \frac{c}{H_0(1+z)\sqrt{-\Omega_k}} \sin\left(\sqrt{-\Omega_k}\chi(z)\right) \quad (9.6)$$

and the *luminosity distance*, $d_L(z)$, given via distance duality as

$$d_L(z) = (1+z)^2 d_A(z). \quad (9.7)$$

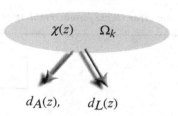

$$\chi(z) \qquad \Omega_k$$

$$d_A(z), \qquad d_L(z)$$

Fig. 9.3. Curvature and $\chi(z)$ define cosmological distances. In a flat universe, the cosmological distances are specified by $\chi(z) \propto \int_0^z dz'/E(z')$. In a general spacetime geometry, however, the curvature affects the geodesics, and hence the distance measured, in the universe.

The expression (Eq. (9.6)) for $d_A(z)$ holds for all values of the curvature Ω_k, since for $\Omega_k < 0$ the complex argument in Eq. (9.6) converts the sin function to a sinh function. Hence the two key quantities that determine distances in cosmology are the dimensionless distance $\chi(z)$ and Ω_k, shown schematically in Fig. 9.3. The link (Eq. (9.7)) between $d_A(z)$ and $d_L(z)$ holds in any metric theory of gravity as long as photon number is conserved. This distance duality can be tested and used to look for exotic physics (Bassett and Kunz, 2004).

Distances have a significant disadvantage over pure Hubble measurements: they require an integral over $f(z)$ that is itself an integral over $w(z)$. Hence, any interesting features in $w(z)$ tend to be washed out in distance measurements. There is also another problem: if we look at Eq. (9.6) for d_A we notice that if we make no assumptions about $f(z)$, then even perfect distance measurements cannot break the degeneracy between $f(z)$ and Ω_k (Weinberg, 1972). This is not a fundamental problem if one assumes that $w(z)$ has finite degrees of freedom, e.g. in Eq. (9.3), but one must remember that the degeneracy is being broken artificially by hand through one's choice of parameterisation and not by the data. An example is given in Fig. 9.4, which shows how a BAO survey at $z = 1$ suffers a perfect degeneracy with curvature, while the same survey split between $z = 1$ and $z = 3$ allows the curvature–dark energy degeneracy to be broken.

In principle this degeneracy can be broken even with arbitrary $f(z)$ by simultaneous measurements of both Hubble and distance; one can explicitly write Ω_k in any FLRW model (with no recourse to the Einstein field equations) as (Clarkson, Cortês and Bassett, 2007; Hlozek *et al.*, 2008)

$$\Omega_k = \frac{[H(z)D'(z)]^2 - H_0^2}{[H_0 D(z)]^2}, \tag{9.8}$$

where D is the dimensionless, transverse, comoving distance with $D(z) = (c/H_0)$ $(1+z)d_A(z)$. This relation gives the value of Ω_k *today*, as a function of

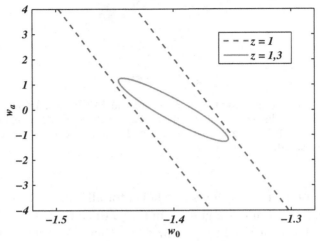

Fig. 9.4. Curvature–dark energy degeneracy. Error ellipses for the CPL parameters w_0, w_a, in two survey scenarios after marginalising over curvature: one with only one measurement at $z = 1$ of both the Hubble rate and the angular diameter distance, and another with two measurements, one at $z = 1$ and another at $z = 3$. The errors on the measurements in the two-bin case have all been increased by $\sqrt{2}$ to keep the total survey time roughly constant. Figure produced using Fisher4Cast. (Bassett *et al.*, 2008.)

measurements at *any* redshift z. Hence this can be turned into a powerful test of the Copernican principle (Clarkson, Bassett and Lu, 2008). Since Ω_k is a single number, the right hand side will have the same value when measured at any redshift if one is in a FLRW background. If it is found to vary with redshift, then we do not live in a FLRW universe.

The beauty of baryon acoustic oscillations is that they provide both $d_A(z)$ and $H(z)$ using almost completely linear physics.

In addition, they offer the as-yet unproven possibility of delivering constraints on growth though the change in the amplitude of the power spectrum. The time dependence of the matter density perturbations, $\delta\rho/\rho$ obeys the equation

$$\ddot{\delta} + 2H\dot{\delta} = \frac{3}{2}\Omega_m(t)H^2\delta. \tag{9.9}$$

The time dependence of the growing mode of this equation is given by the growth function, $G(z)$, which in a flat, ΛCDM model satisfies (Eisenstein, 1997)

$$G(z) = \frac{5\Omega_m E(z)}{2} \int_z^\infty \frac{(1 + z')dz'}{E(z')^3}, \tag{9.10}$$

and hence the growth contains a mixture of Hubble and 'distance' information. As a result, measurements of growth are potentially powerful probes of dark energy.

9.1.3 *Statistical standard rulers*

To illustrate the idea underlying statistical standard rulers (SSR), imagine that all galaxies are positioned at the intersections of a regular three-dimensional grid of known spacing L. Measuring angular diameter distances as a function of redshift would be trivial in this case (cf. Eq. (9.6)) and we would also have the expansion rate as a function of redshift, measured at a discrete set of redshifts corresponding to the mid-points between galaxies. Now imagine that we start to randomly insert galaxies into this regular grid. As the number of randomly distributed galaxies increases the regular grid pattern will rapidly become hard to see by eye. However, the underlying grid pattern would still be detectable statistically, for example in the Fourier transform.

However, a regular grid distributed throughout space would provide an absolute reference frame and would break the continuous homogeneity of space down to a discrete subgroup (formed by those translations that are multiples of the grid spacing L). To get to the core of SSR consider the following prescription for building up a galaxy distribution. Throw down a galaxy from a random distribution. Now with probability p put another galaxy at a distance L (in any direction) from it. Now randomly throw down another galaxy and repeat until you have the desired number of galaxies. Now there is no regular grid of galaxies but L is still a preferred length scale in the clustering of the galaxies and forms an SSR. To reconstruct this preferred scale is a statistical problem. This is illustrated schematically in Fig. 9.5,

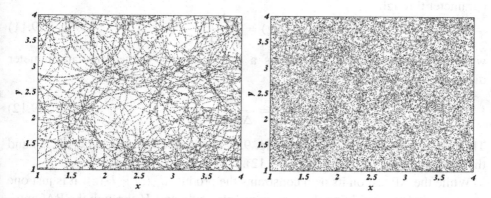

Fig. 9.5. Rings of power superposed. Schematic plot of many rings of the same characteristic scale L superimposed on the plane. The preferred radial scale is visible in the left hand panel where a few rings have been plotted with many points per ring. The right hand panel shows a scenario more like what we find in the universe – many centres of the particle rings are distributed over the $x-y$ plane and each ring contains relatively few points. This superposition in space of the rings smears out the visible imprint of the preferred scale, and it must be recovered statistically.

which shows many rings of the same characteristic radius L superimposed on one another. This superposition of rings on the plane visually 'hides' the characteristic scale, as the number of rings increases, and the sampling of each individual ring is reduced.

BAO, as we discuss below, provide an elegant SSR hidden between the rest of the galaxy clustering, but they are not the only possible SSR. Any preferred scale in the clustering provides an opportunity to apply a relative (absolute) Alcock–Paczynski test (Alcock and Paczynski, 1979) in the case that we don't (do) know *a priori* the size of the preferred scale. Other preferred scales include the Hubble scale at matter–radiation equality (which controls the scale of the turnover in the matter power spectrum) and the Silk damping scale. There may well be other preferred scales imprinted into the primordial clustering of matter. These can be naturally achieved if one inserts a short period of fast-rolling into the otherwise slow-roll of inflation. The sudden change of inflaton velocity creates a bump in the matter power spectrum that can serve the same purpose as the BAO. The required fast-roll can be achieved in multi-field models of inflation.

The SSR provided by the BAO has an additional advantage: it is primarily a linear physics phenomenon, which means we can ignore nonlinear effects[1] to good approximation (we will discuss them later however). This also means we can turn the BAO into a calibrated or absolute Alcock–Paczynski test since the characteristic scale of the BAO is set by the sound horizon at decoupling. As a result the angular diameter distance and Hubble rate can be obtained separately. The characteristic scale, $s_{||}(z)$, along the line-of-sight provides a measurement of the Hubble parameter through

$$H(z) = \frac{c\Delta z}{s_{||}(z)}, \tag{9.11}$$

while the tangential mode provides a measurement of the angular diameter distance,

$$d_A(z) = \frac{s_\perp}{\Delta\theta(1+z)}. \tag{9.12}$$

This is illustrated in the schematic Fig. 9.6. The horizontal axis is Eq. (9.11) and the vertical axis is $c\Delta z/H(z)$, Eq. (9.12).

While the AP test on its own constrains the product $d_A(z) \times H(z)$, it is just one function, and so combining the measurements of d_A and H through the BAO provides tighter constraints on cosmological parameters. This is illustrated in Fig. 9.7, which shows the error ellipse in the dark energy parameters from a hypothetical galaxy redshift survey with constraints from the angular diameter distance

---

[1] These nonlinear effects include redshift space distortions and nonlinear gravitational clustering, which will in general change the spherical nature of the oscillation scale.

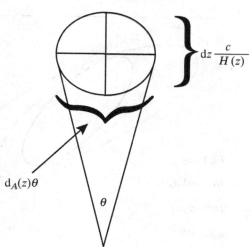

Fig. 9.6. The Alcock–Paczynski (AP) test: assuming that the source is spherical, the ratio of the vertical to the horizontal axis gives constraints on dark energy through the product $d_A(z) \times H(z)$. In the case of the baryon acoustic oscillations, the physical scales are known from theory, and so they provide an absolute AP test.

and Hubble parameter and the product $H \times d_A$ from the AP test. Constraints on the dark energy parameters from the AP test are similar to those from d_A alone, while the constraints are significantly improved when combining measurements of both H and d_A.

One method of extracting a statistical scale from the clustering of galaxies is via the two-point correlation function, $\xi(r)$, which quantifies the excess clustering on a given scale relative to a uniform distribution with the same mean density. The correlation function of galaxies is approximately described by a power law (Totsuji and Kihara, 1969),

$$\xi(r) \propto \left(\frac{r_0}{r}\right)^{\gamma}, \tag{9.13}$$

with $r_0 \sim 5h^{-1}\,\mathrm{Mpc}^{-1}$.

A characteristic scale in the clustering of galaxies will appear as a peak or dip in the correlation function, depending on whether there is an excess or deficiency of clustering at that scale. Any characteristic features will also be present in the power spectrum (we consider, for simplicity, the simple one-dimensional spherically averaged power specturm), since the correlation function and power spectrum form a Fourier pair:

$$P(k) = \int_{-\infty}^{\infty} \xi(r) \exp(-ikr)\, r^2 dr. \tag{9.14}$$

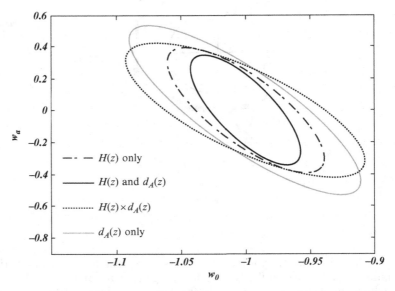

Fig. 9.7. Constraints from the Alcock–Paczynski (AP) test. Comparing the constraints from the AP test on the dark energy parameters w_0, w_a in the CPL parameterisation. Constraints from a hypothetical galaxy redshift survey for $0.1 < z < 2$, in redshift bins of width $\delta z = 0.1$ with 1% measurements on the angular diameter distance and Hubble rate. The parameters of interest are the Hubble constant, the densities of curvature and matter, and the dark energy parameters w_0, w_a in the CPL. The fiducial model is given as $(H_0, \Omega_m, \Omega_k, w_0, w_a) = (70, 0.3, 0, -1, 0)$, and the priors on the model are given as $\mathrm{Prior}(H_0, \Omega_m, \Omega_k, w_0, w_a) = (10^4, 10^4, 10^4, 0, 0)$. In addition we show constraints from the AP test, which constrains the product $H(z) \times d_A(z)$, whereas the BAO constrain d_A, H separately, since the radial and transverse oscillation scales are known separately. The AP test provides comparable constraints to using only d_A, while BAO provide the tightest constraints, as they provide d_A and H simultaneously. (Figure produced using Fisher4Cast, Bassett *et al.*, 2008.)

We will now see how features in the two functions are related.

A δ function at a characteristic scale, say r_*, in $\xi(r)$ will result in power spectrum oscillations, $P(k) \propto e^{-ikr_*}$, as can be seen in Fig. 9.8. These are the baryon acoustic oscillations.

9.1.4 Physics of the BAO

Before recombination and decoupling the universe consisted of a hot plasma of photons and baryons that were tightly coupled via Thomson scattering. The competing forces of radiation pressure and gravity set up oscillations in the photon fluid. If we consider a single, spherical density perturbation in the tightly coupled baryon–photon plasma it will propagate outwards as an acoustic wave with

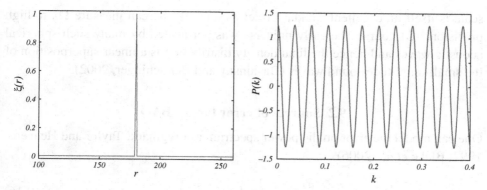

Fig. 9.8. Schematic illustration of the Fourier pairs $\xi(r)$, $P(k)$. A sharp peak in the correlation function (left panel) corresponds to a series of oscillations in $P(k)$ (right panel). The baryon acoustic peak in the correlation function will induce characteristic baryon acoustic oscillations in the power spectrum.

a speed $c_s = c/\sqrt{3(1 + R)}$, where $R \equiv 3\rho_b/4\rho_\gamma \propto \omega_b/(1 + z)$ (Eisenstein, Seo and White, 2007). At recombination the cosmos becomes neutral and the pressure on the baryons is removed. The baryon wave stalls while the photons freely propagate away forming what we now observe as the Cosmic Microwave Background (CMB). The characteristic radius of the spherical shell formed when the baryon wave stalled is imprinted on the distribution of the baryons as a density excess. The baryons and dark matter interact though gravity, and so the dark matter also preferentially clumps on this scale. There is thus an increased probability that a galaxy will form somewhere in the higher density remains of the stalled baryon wave.

Had a galaxy formed at the centre of our initial density perturbation there would be a bump in the two-point correlation function at the radius s of our spherical shell, reflecting the higher probability of finding two galaxies separated by s. The scale s is usually referred to as the sound horizon,[2] the comoving distance a sound wave could have travelled in the photon–baryon fluid by the time of decoupling, and depends on the baryon and matter densities via

$$s = \int_{z_{rec}}^{\infty} \frac{c_s \, dz}{H(z)} = \frac{1}{\sqrt{\Omega_m H_0^2}} \frac{2c_s}{\sqrt{3z_{eq}R_{eq}}} \ln \left[\frac{\sqrt{1 + R_{rec}} + \sqrt{R_{rec} + R_{eq}}}{1 + \sqrt{R_{eq}}} \right]. \quad (9.15)$$

The CMB strongly constrains the matter and baryon densities at decoupling and hence the sound horizon, 146.8 ± 1.8 Mpc (Komatsu *et al.*, 2008). Hence this

---

[2] This is perhaps a slight misnomer since s is not the maximum distance that could have been travelled by any fluid. A simple scalar field has speed of sound equal to c and hence travels faster than the baryon–photon sound wave we are discussing here.

scale is itself an excellent standard ruler as long as one can measure Ω_b to high precision.[3] Of course, the early universe was permeated by many such spherical acoustic waves and hence the final density distribution is a linear superposition of the small-amplitude sound waves (Bashinsky and Bertschinger, 2002).

9.2 Sources of error for the BAO

One can model the error on the power spectrum as (Tegmark, Taylor and Heavens, 1997; Blake *et al.*, 2006)

$$\frac{\delta P}{P} = \frac{1}{\sqrt{m}}\left(1 + \frac{1}{nP}\right),\qquad(9.16)$$

where m is the total number of independent Fourier modes contributing to the measurement of the oscillation scale, and $P \equiv P(k^*),\, k^* \simeq 0.2h\,\mathrm{Mpc}^{-1}$ is the value of the power spectrum amplitude at an average scale k^*, characteristic of the acoustic oscillations. From Eq. (9.16) we note that the two competing sources of error in reconstructing the baryon acoustic oscillation scale are cosmic variance and shot noise, represented by the first and second terms, respectively.

9.2.1 Shot noise

Our goal is to reconstruct the underlying dark matter distribution from discrete tracers such as galaxies. As is illustrated schematically in Fig. 9.9, this reconstruction is very difficult if we have few objects, an effect known as Poisson shot noise. Shot noise error decreases as the density of targets in a given volume increases: the complex underlying pattern in Fig. 9.9 becomes clearer as the number of points increases and the pattern is sufficiently sampled. Increasing the number of targets requires longer integration times on the same patch of sky to go deeper, leading to a reduction in the area surveyed in a fixed observing time.

9.2.2 Cosmic variance

Cosmic variance is the error arising when we can't see the big picture: we cannot estimate the clustering on scales larger than our survey size. This is illustrated schematically in Fig. 9.10, where points are distributed according to a pattern consisting of Fourier modes with a variety of wavelengths and directions. The survey size increases in a clockwise direction, starting from top left, allowing the large-wavelength modes to become visible. To reconstruct these large-scale patterns requires an increased survey volume. In cosmology, it is impossible to keep increasing the size of the survey indefinitely as we are limited to the observable

---

[3] Eisenstein and White (2004) argue that the key parameter in determining the sound horizon is the ratio of the matter and radiation densities and the redshift of the matter–radiation equality.

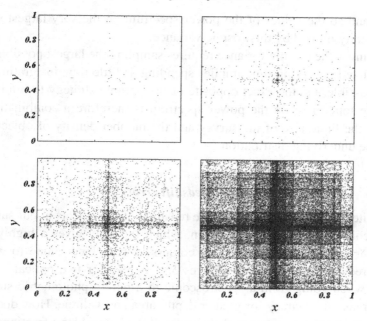

Fig. 9.9. The effect of *shot noise*. As the number of points (galaxies) in a survey increases, one is able to reconstruct the pattern in the distribution of those points more reliably. This is illustrated in the progression from the top left hand panel (100 points) to the bottom right hand panel (100 000 points). As the number of points increases, the substructure of the pattern becomes visible.

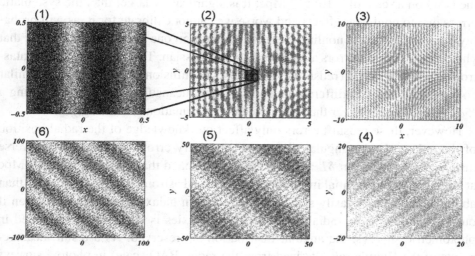

Fig. 9.10. Cosmic variance: as the window expands (clockwise from top left), so the large-scale modes in the distribution of points become visible, due to the reduction in cosmic variance. The total number of points is the same in each panel.

universe and so the values of the power spectrum on the very largest scales are fundamentally limited by this cosmic variance.

To minimise the cosmic variance requires sampling the largest possible volume, which at fixed observing time implies spending as little time integrating on each field as possible, i.e. the exact opposite of the optimal strategy to minimise shot noise. The total error on the power spectrum is therefore a combination of the effects of the finite size of the survey and the number density of objects used to sample the underlying distribution.

9.2.3 Redshift errors

To compute $P(k)$, or $\xi(r)$, we need the redshifts of the objects in our survey. This can be achieved by taking a spectrum, which typically gives a highly accurate redshift, $\delta z < 0.1\%$, or by using the colours of the object alone to get a photometric redshift, which typically gives $\delta z \sim 3\text{--}5\%$, for five optical bands. The trade-off when going from a spectroscopic to a purely photometric survey is to sacrifice redshift accuracy for greater depth, area and volume. How does degrading redshift affect the accuracy of the survey? Well, consider a fractional error σ_z on the redshift z. From our discussion of the radial comoving distance and the Alcock–Paczynski effect, we know that an error δz will result in an uncertainty in the radial position of $\delta L = c\delta z / H(z)$. If we consider $z \sim 1$ we see that even if $\delta z / (1+z) = 0.01$, the radial uncertainty is significant at around 1% of the Hubble scale or about $40h^{-1}$ Mpc. This is a large error when we are trying to measure the BAO on a scale of $\sim 105h^{-1}$ Mpc. It is significantly larger than the systematic errors due to nonlinear effects and worse we cannot calibrate for it, as will probably be possible for the nonlinear effects. In addition one must include the fact that photometric redshift errors are typically non-Gaussian. They usually exhibit catastrophic wings at certain redshifts, which reflect confusion with objects with similar colours at completely different redshifts, which means the likelihood of getting a '5σ' error is much larger than a Gaussian approximation would suggest.

However, these redshift errors only affect our knowledge of the radial position of the galaxy. In the angular direction, astrometry errors ($\delta\theta$) induce transverse distance errors of order $\delta L \sim \delta\theta d_A(z)$. If we demand that this be less than 1 Mpc at all redshifts of potential interest ($z < 4$) we need astrometry accuracy better than about $0.5'$, which is easily satisfied in all modern galaxy surveys. Hence even if our knowledge of the radial distribution of galaxies is significantly degraded in photometric surveys, the angular distribution is preserved. As a result measurements of the Hubble rate (derived from the radial BAO scale) in photo-z surveys are more strongly affected than measurements of d_A (derived from the transverse BAO scale).

We can model the radial degradation as a suppression of the radial power spectrum by a factor $\exp[-(k_{\parallel}\sigma_r)^2]$, where σ_r converts the redshift uncertainty into a physical distance as per our earlier discussion $\sigma_r = \sigma_z(1+z)/H(z)$. In this sense the effect of photo-z errors is similar to the nonlinear effects discussed later, which wipe out information about the higher order oscillations.

A simulation-calibrated fitting formula for the accuracy with which photo-z surveys can measure the power spectrum is given by Blake *et al.* (2006), which is included in the Fisher4Cast code (Bassett *et al.*, 2008). The key scaling is given by

$$\frac{\delta P}{P} \propto \left(\frac{\sigma_r}{V}\right), \tag{9.17}$$

where V is the survey volume and σ_r is, as before, the radial positional uncertainty due to the photo-z error. We see that sacrificing redshift accuracy has a similar effect to decreasing the survey volume. In particular, Blake and Bridle (2005) estimate that with $\delta z/(1+z) = 0.03$ a photometric galaxy survey requires approximately an order of magnitude greater area (and hence volume) to match the BAO accuracy achieved by a spectroscopic survey with the same magnitude limit (in the magnitude range $r = 21$–23). Recent claims suggest that there may be useful information in the small-scale power spectrum, which reduces this to a factor five (Cai, Angulo and Baugh, 2008) and which would have important implications for the attractiveness of future photometric BAO surveys.

In the limit of large photo-z errors, it is standard to bin the galaxies up into common redshift bins and project the galaxies in each bin onto a sphere at the central redshift of the bin. As a result one considers instead the angular power spectrum in each bin. This essentially throws away all of the radial information and is therefore maximally conservative (Blake *et al.*, 2007; Padmanabhan *et al.*, 2007).

9.3 Nonlinear theory

Our analysis so far has been predicated on the belief that the BAO are reliable standard rulers, which is based on the fact that the BAO scale is $\sim 105 h^{-1}$ Mpc which, within the context of FLRW models, is squarely in the linear regime since the quasi-linear regime only extends to about $30 h^{-1}$ Mpc, even at $z = 0$ (and is significantly smaller at higher redshift). However, every candidate standard ruler or candle has a limit beyond which it cannot be trusted, for either theoretical or observational reasons. In the case of BAO the factors that contribute to the breakdown of confidence are nonlinear clustering and scale-dependent bias. But as we will discuss below, there are reasons to be optimistic even about these potential problems.

A key advantage of the BAO as a cosmological probe is that nonlinearities such as those induced from nonlinear gravitational clustering induce predictable shifts in the oscillation scale and hence can be modelled both analytically and through numerical simulations. The effect of the nonlinearities can then be calibrated for, something that is not possible for many other standard rulers and candles. Here we briefly outline the effects of nonlinearity and techniques to correct for such nonlinearities.

Different prescriptions exist for the method of using the BAO in the power spectrum as cosmological tools. The full Fourier-space galaxy correlation method uses the entire power spectrum (including the shape) (Seo and Eisenstein, 2003), but is sensitive to nonlinearities such as scale-dependent bias and nonlinear redshift space distortions. The effect of these systematics is reduced if one removes the overall shape of the power spectrum by dividing by some reference cosmology (Blake *et al.*, 2006; Seo and Eisenstein, 2007), however you also lose any information contained in the overall shape and amplitude and so constraints on cosmological parameters will be weaker.

9.3.1 Nonlinear bias

Measurements of galaxy clustering from redshift surveys yield the galaxy power spectrum, which is traditionally related to the power spectrum of dark matter $P_{DM}(k)$, (which we are interested in) through the bias $b(k, z)$,

$$P_{gal}(k, z) = b^2(k, z) P_{DM}(k, z), \tag{9.18}$$

which in principle can be both redshift and scale dependent (Coles, 1993; Weinberg, 1995; Fry, 1996; Hui and Parfrey, 2007; Smith, Scoccimarro and Sheth, 2007; Cresswell and Percival, 2008; Sánchez and Cole, 2008).

Even a moderate scale-dependent bias will shift the peaks of the BAO and cause a systematic error in the standard ruler. In the extreme case one could even imprint oscillations not present in the underlying dark matter distribution. Fortunately there is a way out of this degeneracy. Clustering in redshift space is anisotropic due to redshift distortions. The radial component of the galaxy peculiar velocity contaminates the cosmological redshift in a characteristic, scale-dependent manner, which means that the power spectrum in redshift space is not isotropic, $P(\mathbf{k}) \neq P(k)$. On large scales galaxies falling into overdensities (such as clusters) are 'squashed' along the line of sight (the Kaiser effect), while on scales smaller than clusters the velocity dispersion of the galaxies within the cluster leads to the 'finger of god' effect – clusters appear elongated along the line of sight (Kaiser, 1987; Hamilton, 1998).

We can expand the anisotropic power spectrum as

$$P(\mathbf{k}, z) = \sum_{l=0,2,4,\ldots} P_l(k, z) \, \mathcal{L}_l(\mu), \tag{9.19}$$

where $\mathcal{L}_l(\mu)$ is the Legendre polynomial, $\mu = \cos(\theta)$, $k = |\mathbf{k}|$ and the monopole $P_0(k, z)$ is the spherically averaged power spectrum. The odd moments vanish by symmetry. Studies have shown that the extra information in the higher order moments P_l allows the recovery of essentially all the standard ruler information, even marginalising over a reasonable redshift and scale-dependent bias (e.g. a four-parameter model), with future experiments (Yamamoto, Basett and Nishioka, 2005). To understand why the different multipoles would break the bias degeneracy, remember that the amplitude of the redshift distortions is controlled by the parameter $\beta = \Omega_m^\gamma/b$, where $\gamma \sim 0.6$. Imagine an observed monopole galaxy power spectrum. If $b \to 0$ then the dark matter power spectrum must increase to leave the galaxy clustering unchanged. The larger dark matter clustering will lead to larger velocities and hence larger redshift distortions, which will be visible in the dipole and quadrupole power spectra. Including information from the full power spectrum allows one to calibrate for such a scale-dependent galaxy bias (Zhang, 2008; Percival and White, 2008).

Before moving on we note that the dark matter power spectrum is the product of the initial power spectrum of the universe, the growth function $G(z)$ (defined in Eq. (9.10)) and the transfer function $T(k)$:

$$P_{\mathrm{DM}}(k, z) = G^2(z) \, T^2(k) \, P_{\mathrm{I}}(k). \tag{9.20}$$

It is clear from Eq. (9.20) that at the level of the monopole power spectrum the growth function is completely degenerate with a general bias. Redshift distortions and non-Gaussian clustering (measured e.g. through the bispectrum) offer the opportunity of determining the growth independent of the bias using the same principle we have discussed above.

9.3.2 Movement and broadening of the peak

The main systematic error due to nonlinearity is the shift in the peak of the correlation function due to mode–mode coupling, as has been studied extensively (Crocce and Scoccimarro, 2008; Smith, Scoccimarro and Sheth, 2007, 2008). There are a couple of effects at play here. Firstly, if the broadband correlation function (i.e. without a BAP) changes with time, the acoustic peak will shift, as can be seen with elementary calculus. Secondly, consider the simple physical model introduced in Section 9.1.4, where we thought of the correlation function peak as arising from

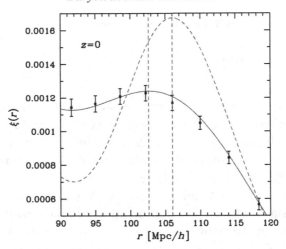

Fig. 9.11. Movement of the baryon acoustic peak (BAP). The correlation function at $z = 0$ (Figure 5 from Crocce and Scoccimarro, 2008) illustrates how the transfer of power to smaller scales due to nonlinearities leads to a shift of the BAP in the correlation function, $\xi(r)$. The linear peak, indicated by the *dashed* line is both broadened and shifted towards smaller scales. The *solid* line shows the prediction for the shift from renormalised perturbation theory (RPT) (Crocce and Scoccimarro, 2006a,b). The vertical lines denote the corresponding maxima of the linear and nonlinear correlation functions.

an acoustic wave that moves outwards before stalling at recombination. If we consider the nonlinear evolution of such a sharp density shell, at rest shortly after recombination, we would expect it to undergo some collapse over the history of the cosmos due to its own self-gravity, thereby shrinking the radius and hence the standard ruler length, by a small but systematic amount.

As can be seen in Fig. 9.11, nonlinearities not only shift the peak, but also smooth out and broaden the peak of the correlation function, which equivalently damps the oscillations in the power spectrum on small scales.[4] Broadening the peak obviously makes reconstruction of the position of the peak – and hence the standard ruler length – less accurate, hence degrading dark energy constraints. Broadening the peak in the correlation function washes out the oscillations in $P(k)$ at large wavenumbers or small scales.

We can illustrate this analytically as follows. Let us model the correlation function as a Gaussian bump shifted so it is centred at a scale r_*, or

$$\xi(r) = \exp\left(-\frac{(r - r_*)^2}{2\sigma^2}\right). \tag{9.21}$$

---

[4] The correlation function and power spectrum are introduced in Section 9.1.3.

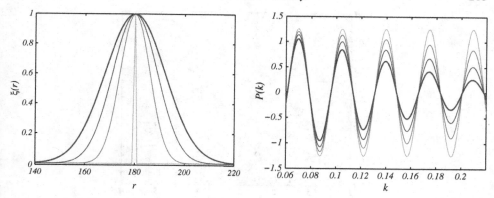

Fig. 9.12. Smoothing out the baryon acoustic signal in the Fourier pair $\xi(r)$, $P(k)$. Increasing the width of the correlation function bump corresponds to the damping of the acoustic oscillations in the power spectrum, particularly severely at large k. Both effects make reconstruction of the standard ruler length more noisy.

Hence the power spectrum is given by

$$P(k) = \int_{-\infty}^{\infty} \exp\left(-\frac{(r-r_*)^2}{2\sigma^2}\right) \exp\left(-ikr\right) dr$$

$$= \sqrt{\frac{\pi}{2}} \exp\left(-ikr_*\right) \exp\left(-k^2\sigma^2\right). \tag{9.22}$$

Figure 9.12 illustrates this toy-model correlation function consisting of a Gaussian shifted to some preferred scale r_*, and the corresponding power spectrum $P(k)$. The oscillations are given for a range of widths of the Gaussian bump, $10 < \sigma < 35$. Clearly, as the Gaussian broadens, the oscillations in the power spectrum are washed out, making their detection harder. Recalling the illustration of rings of power in Figure 9.5, we can examine the effect of successively broadening the rings from which the points are drawn. This is shown in Fig. 9.13, which shows the smearing of the characteristic radius implying an increased error in the standard ruler measurement.

9.3.3 Reconstruction

While nonlinear gravitational collapse broadens and shifts the peak of the correlation function, Eisenstein *et al.* (2007) point out that the map of galaxies used to extract the power spectrum in redshift space can also be used to map the velocity field. Since the galaxies are essentially test particles in the standard ΛCDM

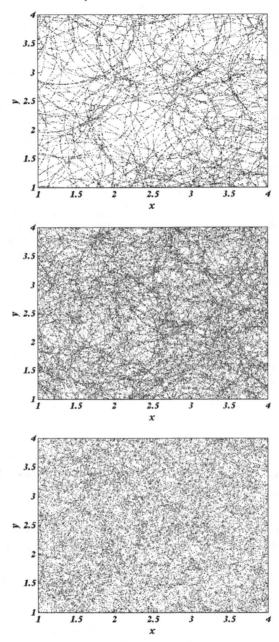

Fig. 9.13. Hiding the characteristic scale. As the peak is broadened from top to bottom as shown schematically in Fig. 9.12, the underlying rings of power are lost, and must be recovered statistically; cf. Fig. 9.5. The number of points is kept the same in each panel.

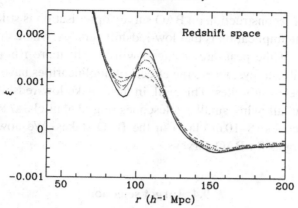

Fig. 9.14. Reconstructing the baryon acoustic peak (BAP). The nonlinear effects generate bulk flows that can be reconstructed from the galaxy distribution itself, hence allowing the nonlinear movements of densities to be undone to good accuracy, thereby both sharpening the acoustic peak and moving it to the correct linear position. (From Eisenstein *et al.*, 2007.)

paradigm, this velocity field can then be used to undo the effects of the nonlinear clustering or equivalently to reconstruct the position and sharpness of the linear acoustic oscillation peak by moving densities to where they would have been had linear theory held at all times. By considering a pair of galaxies separated by the characteristic BAO scale, Eisenstein, Seo and White (2007) show that the majority of the corrupting signal comes from wavenumbers $k \sim 0.02$–$0.2h\,\mathrm{Mpc}^{-1}$. Larger wavelengths coherently move both galaxies while smaller scales are weak because the power spectrum has little power there. The typical distances induced by nonlinear corrections are around 10 Mpc.

Various methods can be followed to reconstruct the velocity and density field as summarised in Eisenstein *et al.* (2007). They move the measured densities back to their linear locations using the following prescription. Firstly they smooth the density field on about 10 Mpc scales. Then they compute the Lagrangian displacement field, $\vec{q}$, which is assumed to be irrotational (no vector perturbations) and which obeys the linear prediction $\Delta \cdot \vec{q} = -\delta$. All particles are then shifted by $-\vec{q}$. In redshift space the densities are then boosted by $1 + \mathrm{d}(\ln G)/\mathrm{d}(\ln a)$, where G is the growth factor, to account for the linear redshift distortions.

Figure 9.14 shows the reconstruction of the acoustic peak in the real-space correlation function using these techniques with improvement in the accuracy of the peak by a factor of 2 to 3, with similar results for the redshift-space correlation function.

The impact of reconstruction on BAO survey optimisation is still to be explored but an immediate implication is that low-redshift surveys, where the nonlinearities and broadening of the peak are stronger, will benefit more from reconstruction than high-redshift surveys, where the effects of nonlinearities have not had time to imprint on the relevant scales. This will, in turn, make low-redshift BAO surveys more interesting, allowing smaller telescopes (e.g. 2–4 m class) to compete with the larger telescopes (8–10 m class) in the BAO stakes. We now move on to a discussion of targets for BAO surveys.

9.4 Target selection

A key decision in undertaking any BAO survey is the choice of target, since the bias, b, of different potential targets differs considerably as a function of morphology, colour etc. relative to the underlying dark matter distribution, in addition to any redshift or scale dependence; $P_{\text{target}}(k) = b^2 P_{\text{DM}}(k)$. This translates into different optimal target densities, since if one requires $n P_{\text{target}} \sim 1$ (the criterion translates to $n \sim 1/(b^2 P_{\text{DM}})$), one needs fewer more highly biased tracers of clustering than weakly biased targets. For example, luminous red galaxies (LRGs) are highly biased tracers, $b_{\text{LRG}} \sim 1.9$ (Tegmark *et al.*, 2006) since they are typically found in clusters, while blue spirals are typically field galaxies and hence are not strongly biased. Mixing blue and red targets would then lead to a complex biasing that would need to be understood and compensated for. Instead, all modern studies of BAO use dedicated targets, the choice of which typically trades off bias versus integration time. Integration times for various types of possible targets for a large 5000 fibre 10 m class survey (such as WFMOS) are shown in Fig. 9.15.

9.4.1 Luminous red galaxies (LRGs)

Luminous red galaxies rose to prominence with the SDSS LRG survey (Eisenstein *et al.*, 2005). They are typically 'red and dead', passive elliptical galaxies with featureless spectra. A high S/N composite LRG spectrum is shown in Fig. 9.16. The redshift is derived from the continuum part of the spectrum and in particular the position of the 4000 Å break, which therefore requires long integration times (>1 hour) even on a 10 m class telescope. This is counteracted by the large bias of LRGs, which means that the required target density is significantly lower. The latter advantage, plus the ability to efficiently find LRGs in optical photometric surveys like the SDSS survey, has led to LRGs being chosen as the targets for the BOSS.[5]

---

[5] www.sdss3.org http://sdss3.org/collaboration/description.pdf

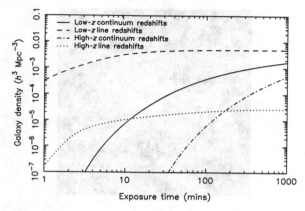

Fig. 9.15. Number densities as a function of target. Figure 2 in Parkinson *et al.* (2007) shows the galaxy number density for four different classes of targets: pre-selected red galaxies at $z \sim 1$ (*solid line*) where multi-colour photometric data are used to select targets, and strong continuum features (such as the 4000 Å break) are used to obtain redshifts; blue galaxies at $z \sim 1$ (*dashed line*) where spectroscopic redshifts are obtained using galaxy emission lines; red galaxies at high redshift ($z \sim 3$; *dot-dashed line*) and blue galaxies again at high redshift (*dotted line*). The plateau in the galaxy density occurs when the surface density reaches the spectroscopic fibre density.

9.4.2 *Blue galaxies*

While redshifts for LRGs are obtained from the continuum spectrum, blue, star-forming galaxies have strong emission lines that provide good redshifts (see Fig. 9.17) from for example the OII doublet at 3727 Å, which is within the optical band at redshifts $z < 1.4$ – expected integration times for this line using a 10 m class telescope are around 15 minutes. Despite this, the low bias means that a much higher target density is required compared to LRGs. Selection of star-forming targets is achieved with a combination of optical and UV imaging and forms the basis for the WiggleZ survey, which uses a combination of SDSS and GALEX (UV) imaging for selection (Blake *et al.*, 2009).

9.4.3 *Lyman break galaxies*

The standard emission lines go out of the optical band at $z \sim 1.4$ leading to the redshift desert for optical surveys because of a dearth of emission lines at wavelengths <3000 Å . This 'drought' is broken by Ly-α at the wavelength of 1216 Å, which comes into the optical passbands around a redshift of $z \sim 2.3$ and remains there until $z \sim 6.4$, making it an ideal target at high redshift. For galaxies at higher redshifts, the Ly-α break moves into different bands and the galaxy will have negligible flux in (for example) the U band, but strong flux in the V band – hence the UV 'drop-out'. Multi-colour imaging of the galaxy is hence used to photometrically determine the redshift of the galaxy. Lyman break galaxies (LBG) take long

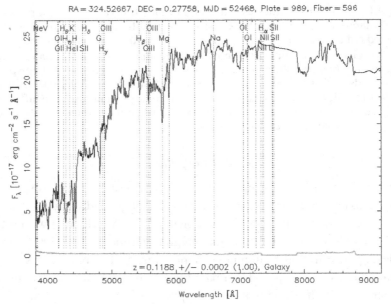

Fig. 9.16. Example of a luminous red galaxy spectrum. The Sloan Digital Sky Survey DR6 image *(top panel)* and spectrum *(bottom panel)* of a typical LRG with (ra, dec) = (324.5266635, 0.27757882). (Downloaded from http://skyserver. sdss.org/)

integration times (see the high-z continuum curve in Fig. 9.15) but there are large numbers of them.

9.4.4 Lyman emitting galaxies

A small set of LBGs also have strong Ly-α emission lines. When they exist they provide ideal targets for redshifts due to the strong emission, however their number density is somewhat unknown. They are the target of preference for the HETDEX[6]

---

[6] http://hetdex.org

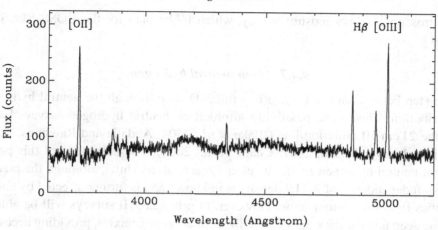

Fig. 9.17. A star-forming galaxy spectrum showing the characteristic emission lines used for redshift determination taken as part of the WiggleZ survey.

(Hill *et al.*, 2004) instrument planned for the Hobby–Eberly Telescope in the redshift interval $1.9 < z < 3.5$, which is expected to detect $\sim 8 \times 10^5$ of these Ly-α emitting galaxies.

9.4.5 Supernovae

LSST will potentially detect millions of photometric Type Ia supernovae (SN Ia). Zhan *et al.* have proposed that these supernovae could be used to measure BAO (Zhan *et al.*, 2008) at $z < 1$. Hence the same data could provide both d_L and d_A, providing constraints that are insensitive to cosmic microwave background priors. The advantage of SN Ia as photometric BAO targets over galaxies is that the photometric redshift error is typically significantly smaller ($\sigma_z \sim 0.02$) due to the well-sampled, multi-epoch spectral templates that will be available from current and future low-z supernova surveys.

9.4.6 Lyman-alpha forest

Sampling the underlying dark matter distribution at a discrete set of N points makes it difficult to uncover subtle underlying patterns due to shot noise. Instead, a potentially superior method would be to take one-dimensional slices through the density distribution. This is the idea behind using the Lyman-alpha forest to probe the BAO. McDonald and Eisenstein (2007) discuss such a survey at redshift $2.2 < z < 3.3$, and project constraints on the radial and tangential oscillation scales of order 1.4%. Such a survey could be performed at the same time as a spectroscopic galaxy redshift survey, which is the plan for the BOSS SDSS-III survey.

9.4.7 21 cm neutral hydrogen

One step better than a 1-D slice is a full 3-D slice through the neutral hydrogen distribution. This is the possibility afforded by neutral hydrogen surveys based on the 21 cm HI emission line (Blake *et al.*, 2004; Abdalla and Rawlings, 2005; Wyithe, Loeb and Geil, 2007; Chang *et al.*, 2008). The advantage of this probe is that neutral hydrogen should be ubiquitous at all redshifts, although the precise redshift dependence of the HI density is unknown and is further affected by uncertainties in cosmic reionisation. However, in principle, HI surveys will be able to probe deep into the dark ages before the formation of galaxies, providing access to the cosmic density field uncontaminated by nonlinearities.

9.5 Current and future BAO surveys

The key elements for a BAO survey are redshift accuracy, redshift coverage, area and volume (of course the latter three are not independent at fixed total survey time). The ideal instrument therefore has large field of view (the area it can see at any one time), large mirror size allowing short integration times and, if it is taking spectra, the ability to harvest large numbers of spectra simultaneously. For future surveys being considered now, this means fields of view in excess of $1\,\square°$, mirrors greater than 4 m in size and, for spectroscopic surveys, the ability to take at least 1000 spectra simultaneously, using a multi-fibre or other technology.

9.5.1 Spectroscopic

We now discuss in rough chronological order the current and future spectroscopic BAO surveys.[7] First up are the final Sloan Digital Sky Survey (SDSS-II) LRG and main galaxy surveys at $z \simeq 0.35$ and $z \simeq 0.1$ respectively and which will cover about $10\,000\,\square°$ in the northern hemisphere. Next is the WiggleZ survey using the 400 fibres on the AAT (Blake *et al.*, 2009) and covering $1000\,\square°$ over the redshift range $0.2 < z < 1.0$, with a median redshift of $z \sim 0.6$, which will be completed in 2010 and will measure $H(z)$ and $d_A(z)$ to around 5%. Beyond that is the Baryon Oscillation Spectroscopic Survey (BOSS), part of the SDSS-III[8] survey and scheduled to operate over the period 2009–14. The relatively small diameter (2.5 m) of the SDSS telescope combined with the large field of view means that BOSS will focus on a wide-and-shallow survey measuring approximately 1.5 million LRGs at

---

[7] These would be called Stage II, III and IV surveys in the DETF report (Albrecht and Bernstein, 2006).
[8] www.sdss3.org

$z \leq 0.7$ and around 160 000 Ly-α forest lines at $2.2 < z < 3$ and giving projected absolute distance measurements of 1.0% at $z = 0.35$, 1.1% at $z = 0.6$, and 1.5% at $z = 2.5$.[9]

Projects on a similar, 2010–15, timescale include FMOS and LAMOST, although their BAO status is currently unclear. FMOS is an infrared spectrograph for Subaru[10] with 400 fibres, which would enable a moderate BAO survey in the redshift desert at $z \sim 1 - 1.7$ over $\sim 300 \,\square°$ while LAMOST[11] is the Chinese 4 m telescope with a planned 4000-fibre spectroscopic capability with a large field of view, which would enable a very effective BAO survey at $z \sim 1$ (Wang *et al.*, 2008). Another interesting BAO survey is the Hobby–Eberly Dark Energy eXperiment (HETDEX)[12] (Hill *et al.*, 2004), which will target the highly biased Ly-α emitting galaxies over the range $1.8 < z < 3.7$. Such a survey over $200 \,\square°$ would probe about $5h^{-3}$ Gpc^3 with approximately one million galaxies, allowing HETDEX to provide $\sim 1\%$ measurements of $d_A(z)$ and $H(z)$ at three redshifts over the survey range. An attractive feature of HETDEX is that it does not need any pre-selection imaging; targets are acquired purely by chance using integral field spectrographs (Hill *et al.*, 2008).

The next major advance in the spectroscopic BAO domain would be enabled by the Wide Field Multi-Object Spectrograph (WFMOS) on a 10 m class telescope, such as Subaru (Bassett, Nichol and Eisenstein, 2005; Glazebrook *et al.*, 2005b). While WFMOS has been cancelled as a Gemini project, it may still take place in a different form. The default plan for WFMOS called for a large field of view ($>1 \,\square°$) and a large number of fibres (at least 3000, although the optimal number is still being investigated: Parkinson *et al.*, 2006). Slated for a $c.2015$ start, the default WFMOS-like surveys would measure in excess of one million LRGs or blue galaxies at $z = 0.8$–1.3 over an area of 2000–$6000 \,\square°$ and of order one million Lyman Break Galaxies at $z = 2.5$–4 over a somewhat smaller area, providing percentage level measurements of both $d_A(z)$ and $H(z)$ at $z = 1$ and $z = 3$ (Eisenstein, 2003; Seo and Eisenstein, 2003; Parkinson *et al.*, 2007, 2009). The high-z component of the survey would provide a key leverage against uncertainties in curvature and $w(z)$ at $z > 1$. It would also be a powerful probe of modified gravity (Yamamoto, Basett and Nishioka, 2005) and allow high spectral resolution archaeology of the Milky Way to understand the origins of its stellar populations (Glazebrook *et al.*, 2005b).

Beyond the 2015 timescale there are a number of planned and proposed missions in various stages of planning and preparation. An interesting proposal is to

---

[9] http://sdss3.org/collaboration/description.pdf
[10] http://www.sstd.rl.ac.uk/fmos/
[11] http://www.lamost.org/en/
[12] http://hetdex.org

use slitless spectroscopy (Glazebrook *et al.*, 2005a; Robberto *et al.*, 2007). This is one possibility for the spectroscopy component of the proposed EUCLID survey, which is a combination of the SPACE (Cimatti *et al.*, 2009) and DUNE (Refregier *et al.*, 2008) missions. The spectroscopic component posits an all-sky near-IR survey down to $H = 22$ which would provide of order 150 million redshifts. EUCLID is slated for launch around 2018 if it is chosen as the winner of the ESA Cosmic Visions programme. Another space BAO programme is ADEPT, a component of the DOE-NASA JDEM mission. ADEPT would also gather around 100 million redshifts over the range $z < 2$.

Despite the inherent weakness of the 21 cm signal, it is likely that radio telescopes will play an important role in future cosmology. This is primarily driven by the fact that the sensitivity of radio telescopes for projects such as BAO scales as the square of their area, unlike optical telescopes whose sensitivity scales linearly with their diameter. This, together with technologies such as synthetic aperture arrays that allow very large fields of view, offers the appealing possibility of surveying huge volumes at very high target densities.

An exciting proposal in this direction is the 21 cm Hubble Sphere Hydrogen Survey (HSHS),[13] which would measure the BAO in neutral hydrogen over the whole sky out to $z = 1.5$. This highly ambitious proposal would provide essentially cosmic variance-limited measurements of the power spectrum in bins of width $\Delta z \sim 0.1$ and exquisite accuracy on d_A and $H(z)$ in the same bins. The key to the HSHS concept is simultaneously combining a huge collecting area with a very large field of view. This can be achieved, at what is hoped to be low cost, by using multiple fixed parabolic cylinders that provide drift scans of the entire sky every day. In this sense, one of the Fourier transforms needed to form an image is undertaken in software ('along the cylinder') while the other is done in hardware ('in the parabolic direction'). HSHS is unusual for a galaxy survey because of its low angular resolution of around $1'$, adapted for statistical analysis of the BAO rather than producing a galaxy catalogue as its primary output. In this sense HSHS resembles a CMB experiment for neutral hydrogen.

A more ambitious proposal is that of the full Square Kilometre Array (SKA),[14] which may be a fully software telescope at 21 cm frequencies, with both Fourier transforms being done in software and using completely flat reflectors. The great advantage of such purely synthetic apertures would be that detectors would essentially see all of the visible sky all of the time, providing the ultimate field of view (Carilli and Rawlings, 2004). This idea appears to have been rediscovered in the form of the Fourier Transform Telescope (Tegmark and Zaldarriaga, 2008). The

---

[13] http://h1survey.phys.cmu.edu/
[14] http://www.skatelescope.org/

SKA would provide essentially cosmic variance limited BAO measurements out to $z = 1.4$ and beyond with 10^9 redshifts, but with sub-arcsecond angular resolution allowing excellent weak lensing measurements as well (Blake *et al.*, 2004). While SKA will be an exceptional BAO machine, pathfinders leading to the full SKA will also provide the first detections of the BAO in the radio (Abdalla, Blake and Rawlings, 2009).

Beyond that, one can imagine using radio surveys to probe the BAO at very high redshifts, $z > 10$, where many more modes are in the linear regime. Since there are essentially no galaxies above this redshift, neutral hydrogen will likely be the only way to test dark energy in the dark ages.

9.5.2 Photometric

Spectra are slow and expensive to obtain and it is tempting to try to study the BAO with only multi-band imaging. A large number of photometrically harvested galaxies might provide a useful probe of the BAO, provided the photometric redshift error is small enough, as discussed in Section 9.2.3.

The current state of the art of photometric redshift surveys is provided by the MegaZ and related catalogues (Blake *et al.*, 2007; Collister *et al.*, 2007; Padmanabhan *et al.*, 2007) based on the SDSS photometry. These catalogues target LRGs and typically achieve $\delta z \simeq 0.03(1 + z)$ redshift accuracy with approximately 1% contamination from M-star interlopers, after suitable cuts. Although they include more than 1 million LRGs out to $z \sim 0.7$ and cover $10\,000\,\square°$ they do not detect the BAO with any significance due to projection effects arising from the photometric redshift errors.

Beyond SDSS there are a number of exciting photometric surveys. SkyMapper will essentially provide Sloan in the southern hemisphere (Cai *et al.*, 2008), while the Dark Energy Survey (DES)[15] (Annis *et al.*, 2005) will use 30% of the 4 m CTIO telescope time to cover around $5000\,\square°$ and detect of order 300 million galaxies in the five Sloan photometric bands, u, g, r, i, z, over the redshift range $0.2 < z < 1.3$. The PS1 phase of the Pan-STARRS project could cover 3π steradians of the sky and detect of order 100 million LRGs, again in five, slightly redder, passbands (Cai *et al.*, 2008). Both surveys should provide compelling BAO detections in addition to the wealth of other science including lensing and a rich SNIa dataset.

Beyond these surveys the Large Synoptic Survey Telescope (LSST) will likely provide the definitive photometric survey for the next two decades. Covering $20\,000\,\square°$ of the sky visible from Chile, LSST should detect every galaxy visible in the optical down to a co-added limiting magnitude of $r = 27.5$, or about 10 billion galaxies. With science operations slated to begin in 2015 or soon thereafter, LSST

[15] https://www.darkenergysurvey.org/

will yield exquisite detections of the angular BAO as a function of redshift, albeit without the radial information provided by spectroscopic or ultra-photometric surveys (Tyson, 2002).

9.6 Conclusions

In the era of precision cosmology, standard rulers of ever-increasing accuracy provide powerful constraints on dark energy and other cosmic parameters. The baryon acoustic oscillations are rooted primarily in linear physics with nonlinearities that can be well modelled and corrected for. As a result the characteristic scale of these 'frozen relics' imprinted into the cosmic plasma before decoupling will likely remain as the most reliable of statistical standard rulers in the coming decade.

Acknowledgements

We would like to thank Chris Blake for detailed comments and Daniel Eisenstein, Varun Sahni, Alexei Starobinsky, Will Percival, Jacques Kotze and Yabebal Fantaye for insightful discussions. We thank Martin Crocce, Daniel Eisenstein, David Parkinson, Will Percival, Kevin Pimbblet, Roman Scoccimarro and Max Tegmark for permission to reproduce figures in this review. BB thanks his WFMOS Team A colleagues and in particular Chris Blake, Martin Kunz, Bob Nichol and David Parkinson for their collaborations and discussions over the years, and the Perimeter Institute for hospitality on his visit during which part of this work was completed. RH would like to thank Princeton University and acknowledges support from the NSF PIRE grant OISE/0530095BB during her visit there. BB and RH acknowledge funding from the NRF and SA SKA, respectively, and RH acknowledges funding from the Rhodes Trust.

References

Abdalla, F. B., and Rawlings, S. (2005). *Mon. Not. R. Astron. Soc.* **360**, 27.
Abdalla, F., Blake, C., and Rawlings, S. (2009). [arXiv:0905.4311].
Albrecht, A., and Bernstein, G. (2006). [astro-ph/0608269].
Alcock, C., and Paczynski, B. (1979). *Nature (London)* **281**, 358.
Allen, S. W., Schmidt, R. W., and Fabian, A. C. (2002). *Mon. Not. R. Astron. Soc.* **334**, L11.
Annis, J., Bridle, S., Castander, F. J., *et al.* (2005). [astro-ph/0510195].
Bashinsky, S., and Bertschinger, E. (2002). *Phys. Rev. D* **65**, 123008.
Bassett, B. A., and Kunz, M. (2004). *Phys. Rev. D* **69** 101305.
Bassett, B. A., Nichol, B., and Eisenstein, D. J. (2005). *Astron. Geophys.* **46**, 26.
Bassett, B. A., Fantaye, Y. T., Hlozek, R. A., and Kotze, J. (2008). Code available at www.cosmology.org.za.
Blake, C., and Bridle, S. (2005). *Mon. Not. R. Astron. Soc.* **363**, 1329.
Blake, C., and Glazebrook, K. (2003). *Astrophys. J.* **594**, 665.
Blake, C. A., Abdalla, F. B., Bridle, S. L., and Rawlings, S. (2004). *New Astron. Rev.* **48**, 1063.
Blake, C., Parkinson, D., Bassett, B. A. *et al.* (2006). *Mon. Not. R. Astron. Soc.* **365**, 255.
Blake, C., Collister, A., Bridle, S., and Lahav, O. (2007). *Mon. Not. R. Astron. Soc.* **374**, 1527.

Blake, C., Jurek, R. J., Brough, S., *et al.* (2009). *Mon. Not. R. Astron. Soc.* **395**, 240.

Broadhurst, T., and Jaffe, A. (2000). In *Clustering at High Redshift*, A. Mazure, O. Le Fèvre, and V. Le Brun, Eds., *Astron. Soc. Pacif. Conf. Ser.* **200**, 241.

Broadhurst, T. J., Ellis, R. S., Koo, D. C., and Szalay, A. S. (1990). *Nature (London)* **343**, 726.

Buchalter, A., Helfand, D. J., Becker, R. H., and White, R. L. (1998). *Astrophys. J.* **494**, 503.

Cai, Y.-C., Angulo, R. E., Baugh, C. M., *et al.* (2008). [arXiv:0810.2300].

Carilli, C., and Rawlings, S. (2004). [astro-ph/0409274].

Chang, T.-C., Pen, U.-L., Peterson, J. B., and McDonald, P. (2008). *Phys. Rev. Lett.* **100**, 091303.

Chevallier, M., and Polarski, D. (2001). *Int. J. Mod. Phys. D* **10**, 213.

Cimatti, A., Robberto, M., Baugh, C., *et al.* (2009). *Exp. Astron.* **23**, 39.

Clarkson, C., Cortês, M., and Bassett, B. (2007). *JCAP* **8**, 11.

Clarkson, C., Bassett, B., and Lu, T. H. C. (2008). *Phys. Rev. Lett.* **101**, 011301.

Cole, S., Percival, W. J., Peacock, J. A., *et al.* (2005). *Mon. Not. R. Astron. Soc.* **362**, 505.

Coles, P. (1993). *Mon. Not. R. Astron. Soc.* **262**, 1065.

Collister, A., Lahav, O., Blake, C., *et al.* (2007). *Mon. Not. R. Astron. Soc.* **375**, 68.

Cooray, A., Hu, W., Huterer, D., and Joffre, M. (2001). *Astrophys. J.* **557**, L7.

Cresswell, J. G., and Percival, W. J. (2008). [arXiv:0808.1101].

Crocce, M., and Scoccimarro, R. (2006a). *Phys. Rev. D* **73**, 063519.

Crocce, M., and Scoccimarro, R. (2006b). *Phys. Rev. D* **73**, 063520.

Crocce, M., and Scoccimarro, R. (2008). *Phys. Rev. D* **77**, 023533.

Daly, R. A. (1994). *Astrophys. J.* **426**, 38.

Daly, R. A., and Guerra, E. J. (2002). *Astron. J.* **124**, 1831.

Daly, R. A., Mory, M. P., O'Dea, C. P., *et al.* (2007). [arXiv:0710.5112].

Deng, Z., Xia, X., and Fang, L.-Z. (1994). *Astrophys. J.* **431**, 506.

Eisenstein, D. J. (1997). [astro-ph/9709054].

Eisenstein, D. (2003). [astro-ph/0301623]

Eisenstein, D., and White, M. (2004). *Phys. Rev. D* **70**, 103523.

Eisenstein, D. J., Hu, W., Silk, J., and Szalay, A. S. (1998). *Astrophys. J.* **494**, L1.

Eisenstein, D. J., Hu, W., and Tegmark, M. (1998). *Astrophys. J.* **504**, L57.

Eisenstein, D. J., Zehavi, I., Hogg, D. W., *et al.* (2005). *Astrophys. J.* **633**, 560.

Eisenstein, D. J., Seo, H.-J., and White, M. (2007). *Astrophys. J.* **664**, 660.

Eisenstein, D. J., Seo, H.-J., Sirko, E., and Spergel, D. N. (2007). *Astrophys. J.* **664**, 675.

Fry, J. N. (1996). *Astrophys. J.* **461**, L65.

Gaztanaga, E., Cabre, A., and Hui, L. (2008). [arXiv:0807.3551].

Glazebrook, K., Baldry, I., Moos, W., Kruk, J., and McCandliss, S. (2005a). *New Astron. Rev.* **49**, 374.

Glazebrook, K., Eisenstein, D., Dey, A., Nichol, B., and The WFMOS Feasibility Study Dark Energy Team (2005b). [astro-ph/0507457].

Gurvits, L. I. (1994). *Astrophys. J.* **425**, 442.

Hamilton, A. J. S. (1998). In *The Evolving Universe*, D. Hamilton, Ed., *Astrophys. Space Sci. Libr.* **231**, 185.

Hill, G. J., Gebhardt, K., Komatsu, E., and MacQueen, P. J. (2004). In *The New Cosmology: Conference on Strings and Cosmology*, R. E. Allen, D. V. Nanopoulos, and C. C. Pope, Eds., *Amer. Inst. Phys. Conf. Ser.* **743**, 224.

Hill, G. J., Gebhardt, K., Komatsu, E., *et al.* (2008). [arXiv:0806.0183].

Hlozek, R., Cortês, M., Clarkson, C., and Bassett, B. (2008). *Gen. Relat. Grav.* **40**, 285.

Hui, L., and Parfrey, K. P. (2007). [arXiv:0712.1162].

Jackson, J. C., and Dodgson, M. (1997). *Mon. Not. R. Astron. Soc.* **285**, 806.

Kaiser, N. (1987). *Mon. Not. R. Astron. Soc.* **227**, 1.

Kellermann, K. I. (1993). *Nature (London)* **361**, 134.

Komatsu, E., Dunkley, J., Nolta, M. R., *et al.* (2008). [arXiv:0803.0547].

Linder, E. V. (2003). *Phys. Rev. Lett.* **90**, 091301.

Mantz, A., Allen, S. W., Ebeling, H., and Rapetti, D. (2008). *Mon. Not. R. Astron. Soc.* **387**, 1179.

McDonald, P., and Eisenstein, D. J. (2007). *Phys. Rev. D* **76**, 063009.

Meiksin, A., White, M. and Peacock, J. A. (1999). *Mon. Not. R. Astron. Soc.* **304**, 851.

Miller, C. J., Nichol, R. C. and Chen, X. (2002). *Astrophys. J.* **579**, 483.

Padmanabhan, N., Schlegel, D. J., Seljak, U. *et al.* (2007). *Mon. Not. R. Astron. Soc.* **378**, 852.

Parkinson, D., Blake, C., Kunz, M., *et al.* (2007). *Mon. Not. R. Astron. Soc.* **377**, 185.

Parkinson, D., Kunz, M., Liddle, A. R., *et al.* (2009). [arXiv:0905.3410].

Percival, W. J., and White, M. (2008). [arXiv:0808.0003].

Percival, W. J., Nichol, R. C., Eisenstein, D. J., *et al.* (2007). *Astrophys. J.* **657**, 51.

Refregier, A., Douspis, M., and the DUNE collaboration (2008). [arXiv:0807.4036].

Robberto, M., Cimatti, A., and The Space Science Team (2007). *Nuovo Cimento B* **122**, 1467.

Roukema, B. F., and Mamon, G. A. (2000). *Astron. Astrophys.* **358**, 395–408.

Roukema, B. F., Mamon, G. A., and Bajtlik, S. (2002). *Astron. Astrophys.* **382**, 397.

Sánchez, A. G., and Cole, S. (2008). *Mon. Not. R. Astron. Soc.* **385**, 830.

Seo, H.-J., and Eisenstein, D. J. (2003). *Astrophys. J.* **598**, 720.

Seo, H.-J., and Eisenstein, D. J. (2007). *Astrophys. J.* **665**, 14.

Shanks, T., Fong, T., Boyle, B. J., and Peterson, B. A. (1987). *Mon. Not. R. Astron. Soc.* **227**, 739.

Smith, R. E., Scoccimarro, R., and Sheth, R. K. (2007). *Phys. Rev. D* **75**, 063512.

Smith, R. E., Scoccimarro, R. and Sheth, R. K. (2008). *Phys. Rev. D* **77**, 043525.

Spergel, D. N., Bean, R., Doré, O., *et al.* (2007). *Astrophys. J. Suppl.* **170**, 377.

Tegmark, M., and Zaldarriaga, M. (2008). [arXiv:0805.4414].

Tegmark, M., Taylor, A. N., and Heavens, A. F. (1997). *Astrophys. J.* **480**, 22.

Tegmark, M., Eisenstein, D. J., Strauss, M. A., *et al.* (2006). *Phys. Rev. D* **74**, 123507.

Totsuji, H., and Kihara, T. (1969). *Publ. Astron. Soc. Japan* **21**, 221.

Tyson, J. A. (2002). In *Survey and Other Telescope Technologies and Discoveries*, J. Tyson and S. Wolff, Eds., *Proc. SPIE* **4836**, 10.

Wang, X., Chen, X., Zheng, Z., *et al.* (2008). [arXiv:0809.3002].

Weinberg, D. H. (1995). *Bull. Amer. Astron. Soc.* **27**, 853.

Weinberg, S. (1972). *Gravitation and Cosmology: Principles and Applications of the General Theory of Relativity*, New York: Wiley.

Wyithe, S., Loeb, A., and Geil, P. (2007). [arXiv:0709.2955].

Yamamoto, K., Bassett, K., and Nishioka, H. (2005). *Phys. Rev. Lett.* **94**, 051301.

Zhan, H., Wang, L., Pinto, P., and Tyson, J. A. (2008). *Astrophys. J.* **675**, L1.

Zhang, P. (2008). [arXiv:0802.2416].

10

Weak gravitational lensing, dark energy and modified gravity

10.1 Introduction

Our state of knowledge of the universe has improved markedly in the last twenty years. Observations of the microwave background radiation, large-scale structure and distant supernovae provide strong evidence for a universe composed of about 4% baryonic matter, 20% non-baryonic dark matter, and 76% dark energy (Spergel *et al.*, 2007). Of the last two components, we know rather little in detail, so the obvious question for the next decades is what is the nature of the dark matter and dark energy? The last component is the most unexpected, and might correspond to Einstein's cosmological constant, or to a dynamical substance with an equation of state parameter $w \equiv p/(\rho c^2)$ currently close to -1, but which may evolve with time. More intriguingly still, it might not be dark energy at all, but rather a manifestation of a law of gravity that is not Einstein's, and might be a result of a higher-dimensional universe. Dark energy and modified gravity manifest themselves in two general ways that open up possibilities for confronting theories with observation. Firstly, they affect the expansion history of the universe, and secondly they influence the growth rate of perturbations. The expansion history can be probed by observations of distant supernovae or baryon oscillations, and these are covered elsewhere in this volume. Gravitational lensing can probe both, by distorting the images of distant galaxies as the light passes through the non-uniform universe. This can be an advantage in that if one considers only the expansion history, then there is a degeneracy between modified gravity models and dark energy plus general relativity. Gravitational lensing lifts that degeneracy, and with a sufficiently large survey can distinguish between modified gravity and dark energy. A major advantage of gravitational lensing is that it is sensitive to the distribution

Dark Energy: Observational and Theoretical Approaches, ed. Pilar Ruiz-Lapuente. Published by Cambridge University Press. © Cambridge University Press 2010.

of matter within the universe, whether it is baryonic or not, and this is generally more easily predicted *ab initio* from theory than, for example, the distribution of galaxies. The robust connection between theory and observation is an advantage that has been exploited very successfully in microwave background radiation studies, and one should not underestimate the power of using well-understood physics in attempting high-precision cosmological measurements. There is no doubt, however, that the study of weak gravitational lensing is challenging from an observational point of view, and excellent image quality is a strong requirement. Furthermore, there are some complex physical processes that complicate the simple picture, and these issues need to be addressed in order to probe the rather subtle effects of changing the dark energy properties. In this article, we will explore the relevant physics, as well as the prospects and problems that need to be dealt with.

10.2 Sensitivity to dark energy

The degree of distortion of distant images depends on the distance–redshift relation $r(z)$, and the growth rate of perturbations. Without addressing subtleties here, this makes intuitive sense: the larger the perturbations, the greater the distortion, and if r is greater for a source at given redshift z, the greater the path length for the distortions to operate over. Assuming general relativity, we summarise here the effects of dark energy on the $r(z)$ relation and the growth rate. For background, see any cosmology textbook, such as Peacock (1999).

We characterise the unperturbed universe with the Friedmann–Robertson–Walker metric:

$$ds^2 = c^2 \, dt^2 - R^2(t) \left[dr^2 + S_k^2(r) \left(d\theta^2 + \sin^2 \theta \, d\varphi^2 \right) \right], \qquad (10.1)$$

where t is cosmic time, $R(t)$ is the cosmic scale factor, and $S_k(r) = r, \sin r, \sinh r$, depending on whether the curvature $k = 0$, 1 or -1. θ and φ are the usual spherical polar angular coordinates. The radial distance is related to the Hubble expansion parameter $H(t) \equiv R^{-1} dR/dt$ by

$$r = \int_0^z dz' \, \frac{c}{H(z')}, \qquad (10.2)$$

where the argument of H is the redshift z, where $1 + z = a^{-1} \equiv R(t_0)/R(t)$, and t_0 is the current cosmic time. For an equation of state parameter for dark energy $w(a)$, the Hubble parameter is given by

$$H^2(a) = H_0^2 \left[\Omega_m a^{-3} + \Omega_k a^{-2} + \Omega_{DE} \exp\left(3 \int_1^a \frac{da'}{a'} \left[1 + w(a') \right] \right) \right]. \qquad (10.3)$$

Ω_m, Ω_k and H_0 are the present matter density parameter, curvature density parameter and Hubble parameter. We see therefore that the dark energy sensitivity is via the Hubble parameter, or equivalently through its effect on the expansion history of the universe.

The dark energy also affects the growth rate via the Hubble parameter, since in general relativity, the fractional overdensity $\delta \equiv \delta\rho/\bar{\rho} -1$ (where $\bar{\rho}$ is the mean density) grows to linear order according to

$$\ddot{\delta} + 2H\dot{\delta} - 4\pi G\rho_m\delta = 0, \tag{10.4}$$

where ρ_m is the matter density and we assume the dark energy density is not perturbed.

10.3 Gravitational lensing

Gravitational lensing arises when the light from distant objects is deflected by mass concentrations close to the line-of-sight. These deflections distort the images, changing shapes, sizes and brightnesses. Passage close to a large mass concentration can cause large distortions, leading to what is referred to as strong lensing. In this article we will focus largely on weak lensing on a cosmological scale, being small changes in image properties, at the percentage level, arising from the passage of the light through the non-uniform universe. For the most part we will concentrate on statistical analysis of these distortions, but we will also make some remarks on potential and mass reconstruction. Recent reviews of this area include Bartelmann and Schneider (2001), Schneider (2003), van Waerbeke and Mellier (2003), and Munshi *et al.* (2006).

The basic effect of weak lensing is for the clumpy matter distribution to perturb slightly the trajectories of photon paths. By considering how nearby light paths are perturbed, one finds that the shapes of distant objects are changed slightly. Associated with the size change is a change in the brightness of the source. The size and magnitude changes can, in principle, be used to constrain the properties of the matter distribution along the line-of-sight (and cosmological parameters as well), but it is the change in shape of background sources that has almost exclusively been used in cosmological weak lensing studies. The reason for this is that the signal-to-noise is usually much better. These notes will concentrate on shear (= shape changes), but the magnification and amplification of sources can also be used and will probably be used in future when the surveys are larger. The great promise of lensing is that it acts as a direct probe of the matter distribution (whether dark or not), and avoids the use of objects that are assumed to trace the mass distribution in some way, such as galaxies in large-scale structure studies. Theoretically, lensing is very appealing, as the physics is very simple, and very robust, direct connections can be made between weak lensing observables and the statistical properties of the matter distribution. These statistical properties are dependent on cosmological parameters in a known way, so weak lensing can be employed as a cosmological tool. The main uncertainties in lensing are observational – it is very challenging to make images of the necessary quality. Lensing can, of course, be used to investigate the mass distribution of a discrete lens lying along the line-of-sight, but I will concentrate here on the weak effects on a cosmological scale of the non-uniform distribution of matter all the way between the source and observer, an effect often referred to as *cosmic shear*.

10.3.1 Distortion of light bundles

The distortion of a light bundle has to be treated with general relativity, but if one is prepared to accept one modification to Newtonian physics, one can do without GR.

In an expanding universe, it is usual to define a *comoving coordinate* **x**, such that 'fundamental observers' retain the same comoving coordinate. Fundamental observers are characterised by the property of seeing the universe as isotropic; the Earth is not (quite) a fundamental observer, as from here the cosmic microwave background looks slightly anisotropic. The equation of motion for the transverse coordinates $(i = 1, 2)$, about some fiducial direction, of a photon in a flat universe is

$$\frac{d^2 x_i}{d\eta^2} = -\frac{2}{c^2}\frac{\partial \Phi}{\partial x_i}, \qquad i = 1, 2. \tag{10.5}$$

η is the conformal time, related to the coordinate t by $d\eta = c\, dt / R(t)$. $\Phi(x_i, r)$ is the peculiar gravitational potential. We assume a flat universe throughout, but the generalisation to non-flat universes is straightforward (there is an extra term in the equation above, and some r symbols need to be changed to an angular diameter distance). Φ is related to the matter overdensity field $\delta \equiv \delta\rho/\rho$ by Poisson's equation:

$$\nabla^2_{3D}\Phi = \frac{3H_0^2 \Omega_m}{2a(t)}\delta, \tag{10.6}$$

where H_0 is the Hubble constant, Ω_m is the present matter density parameter, and $a(t) = R(t)/R_0 = (1 + z)^{-1}$, where z is redshift.

The equation of motion can be derived using general relativity, using the (nearly-flat) Newtonian metric:

$$ds^2 = (1 + 2\Phi/c^2)c^2\, dt^2 - (1 - 2\Phi/c^2)R^2(t)\left[dr^2 + r^2\, d\theta^2 + r^2 \sin^2\theta\, d\varphi^2\right]. \tag{10.7}$$

From a Newtonian point-of-view, Eq. (10.5) is understandable if we note that time is replaced by η (which arises because we are using comoving coordinates), and there is a factor 2, which does not appear in Newtonian physics. This same factor of 2 gives rise to the famous result that in GR the angle of light bending round the Sun is double that of Newtonian theory.

The coordinates x_i are related to the (small) angles of the photon to the fiducial direction $\theta = (\theta_x, \theta_y)$ by $x_i = r\theta_i$.

10.3.2 Lensing potential

The solution to (10.5) is obtained by first noting that the zero-order ray has $ds^2 = 0 \Rightarrow dr = -d\eta$, where we take the negative root because the light ray is incoming. Integrating twice, and reversing the order of integration gives

$$x_i = r\theta_i - \frac{2}{c^2} \int_0^r dr' \frac{\partial \Phi}{\partial x_i'} (r - r'). \tag{10.8}$$

We now perform a Taylor expansion of $\partial \Phi / \partial x_i'$, and find the deviation of two nearby light rays is

$$\Delta x_i = r\Delta\theta_i - \frac{2}{c^2} \Delta\theta_j \int_0^r dr' r' (r - r') \frac{\partial^2 \Phi}{\partial x_i' \partial x_j'}, \tag{10.9}$$

which we may write as

$$\Delta x_i = r\Delta\theta_j (\delta_{ij} - \phi_{ij}), \tag{10.10}$$

where δ_{ij} is the Kronecker delta ($i = 1, 2$) and we define

$$\phi_{ij}(\mathbf{r}) \equiv \frac{2}{c^2} \int_0^r dr' \frac{(r - r')}{rr'} \frac{\partial^2 \Phi(\mathbf{r}')}{\partial \theta_i \partial \theta_j}. \tag{10.11}$$

The integral is understood to be along a radial line (i.e. $\mathbf{r} \parallel \mathbf{r}'$); this is the *Born approximation*, which is a very good approximation for weak lensing (Bernardeau, van Waerbeke and Mellier, 1997; Schneider *et al.*, 1998; van Waerbeke *et al.*, 2002). In reality the light path is not quite radial.

It is convenient to introduce the *(cosmological) lensing potential*, which controls the distortion of the ray bundle:

$$\phi(\mathbf{r}) \equiv \frac{2}{c^2} \int_0^r dr' \frac{(r - r')}{rr'} \Phi(\mathbf{r}'). \tag{10.12}$$

Note that $\phi_{ij}(\mathbf{r}) = \partial^2 \phi(\mathbf{r}) / \partial \theta_i \partial \theta_j$. So, remarkably, we can describe the distortion of an image as it passes through a clumpy universe in a rather simple way. The shear and convergence of a source image is obtained from the potential in the same way as the thin lens case (Eq. (10.17)), although the relationship between the convergence and the foreground density is more complicated, as we see next.

10.3.2.1 Convergence, magnification and shear for general thin lenses

The magnification and distortion of an infinitesimal source are given by the transformation matrix from the source position x_i to the apparent image position $r\Delta\theta$. The (inverse) amplification matrix is

$$A_{ij} \equiv \frac{\partial x_i}{\partial (r\Delta\theta_j)} = \delta_{ij} - \phi_{ij}, \tag{10.13}$$

then we see that A is symmetric, and it can be decomposed into an isotropic expansion term, and a shear. A general amplification matrix also includes a rotation term (the final degree of freedom being the rotation angle), but we see that weak lensing doesn't introduce rotation of the image, and has only three degrees of freedom, rather than the four possible in a 2×2 matrix. We decompose the amplification matrix as follows:

$$A_{ij} = \begin{pmatrix} 1 - \kappa & 0 \\ 0 & 1 - \kappa \end{pmatrix} + \begin{pmatrix} -\gamma_1 & -\gamma_2 \\ -\gamma_2 & \gamma_1 \end{pmatrix}, \tag{10.14}$$

where κ is called the *convergence* and

$$\gamma = \gamma_1 + i\gamma_2 \tag{10.15}$$

is the *complex shear*. For weak lensing, both $|\kappa|$ and $|\gamma_i|$ are $\ll 1$. A non-zero κ represents an isotropic expansion or contraction of a source; $\gamma_1 > 0$ represents an elongation of the image along the x-axis and contraction along y. $\gamma_1 < 0$ stretches along y and contracts along x. $\gamma_2 \neq 0$ represents stretching along $x = \pm y$ directions. The effects are shown in Fig. 10.1.

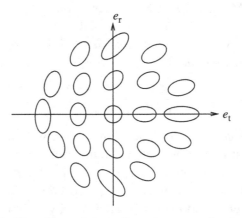

Fig. 10.1. The effect of shear distortions on a circular source.

Making the decomposition, we find that

$$\kappa = \frac{1}{2}(\phi_{11} + \phi_{22}),$$

$$\gamma_1 = \frac{1}{2}(\phi_{11} - \phi_{22}) \equiv D_1\phi, \qquad (10.16)$$

$$\gamma_2 = \phi_{12} \equiv D_2\phi.$$

Note that $\kappa > 0$ corresponds to *magnification* of the image. Lensing preserves surface brightness, so this also amounts to *amplification* of the source flux. The magnification is

$$A = \frac{1}{\det A_{ij}} = \frac{1}{(1-\kappa)^2 - |\gamma|^2}. \qquad (10.17)$$

We see that we may have infinite amplifications if $\kappa \geq 1$. Such effects apply only for infinitesimal sources, and places in the source plane that lead to infinite magnifications are referred to as *caustics*, and lead to highly distorted images along lines called *critical lines* in the lens plane. The giant arcs visible in images of some rich clusters lie on or close to critical lines. For cosmic shear, κ and $|\gamma|$ are typically 0.01, and the lensing is weak.

It is worth noting that the amplification matrix may be written

$$A_{ij} = (1 - \kappa)\begin{pmatrix} 1 - g_1 & -g_2 \\ -g_2 & 1 + g_1 \end{pmatrix}, \qquad (10.18)$$

where $g \equiv \gamma/(1-\kappa)$ is called the *reduced shear*. Since the $1 - \kappa$ multiplier affects only the overall size (and hence brightness) of the source, but not its shape, we see that shape measurements can determine only the reduced shear, and not the shear itself. For weak lensing, $\kappa \ll 1$, so the two coincide to linear order.

Note that a rotation of $\pi/4$ in the coordinate system changes a real shear into an imaginary shear – i.e. the complex shear rotates by $\pi/2$, twice the angle of rotation of the coordinate system. This behaviour is characteristic of a *spin-weight 2* field, and is encountered also in microwave background polarisation and gravitational wave fields.

10.3.2.2 Relationship to matter density field

Since the gravitational potential Φ is related to the matter overdensity field $\delta \equiv \delta\rho/\rho$ by Poisson's equation (10.6), the convergence is

$$\kappa(\mathbf{r}) = \frac{3H_0^2\Omega_m}{2c^2} \int_0^r dr' \frac{r'(r - r')}{r} \frac{\delta(\mathbf{r}')}{a(r')}. \qquad (10.19)$$

Note that there is an extra term $\partial^2 \Phi / \partial r'^2$ in ∇^2_{3D} which integrates to zero to the order to which we are working.

10.3.2.3 Averaging over a distribution of sources

If we consider the distortion averaged over a distribution of sources with a radial distribution $p(r)$ (normalised such that $\int dr\, p(r) = 1$), the average distortion is again obtained by reversing the order of integration:

$$\Delta x_i = r \Delta \theta_j \left(\delta_{ij} - \frac{2}{c^2} \int_0^r \frac{dr'}{r'} G(r') \frac{\partial^2 \Phi(\mathbf{r}')}{\partial \theta_i\, \partial \theta_j} \right), \tag{10.20}$$

where

$$G(r) \equiv \int_r^\infty dr'\, p(r') \frac{r' - r}{r'}. \tag{10.21}$$

In order to estimate $p(r)$, surveys began to estimate distances to source galaxies using photometric redshifts. This has opened up the prospect of a full 3D analysis of the shear field, which we will discuss briefly later in this article.

10.3.3 Three-dimensional potential and mass density reconstruction

As we have already seen, it is possible to reconstruct the surface density of a lens system by analysing the shear pattern of galaxies in the background. An interesting question is then whether the reconstruction can be done in three dimensions, when distance information is available for the sources. It is probably self-evident that mass distributions can be *constrained* by the shear pattern, but the more interesting possibility is that one may be able to *determine* the 3D mass density in an essentially non-parametric way from the shear data.

The idea (Taylor, 2001) is that the shear pattern is derivable from the lensing potential $\phi(\mathbf{r})$, which is dependent on the gravitational potential $\Phi(\mathbf{r})$ through the integral equation

$$\phi(\mathbf{r}) = \frac{2}{c^2} \int_0^r dr' \left(\frac{1}{r'} - \frac{1}{r} \right) \Phi(\mathbf{r}'), \tag{10.22}$$

where the integral is understood to be along a radial path (the Born approximation), and a flat universe is assumed in Eq. (10.22). The gravitational potential is related to the density field via Poisson's equation (10.6). There are two problems to solve here: one is to construct ϕ from the lensing data, the second is to invert Eq. (10.22). The second problem is straightforward: the solution is

$$\Phi(\mathbf{r}) = \frac{c^2}{2} \frac{\partial}{\partial r} \left[r^2 \frac{\partial}{\partial r} \phi(\mathbf{r}) \right]. \tag{10.23}$$

From this and Poisson's equation $\nabla^2\Phi = (3/2)H_0^2\Omega_m\delta/a(t)$, we can reconstruct the mass overdensity field:

$$\delta(\mathbf{r}) = \frac{a(t)c^2}{3H_0^2\Omega_m}\nabla^2\left\{\frac{\partial}{\partial r}\left[r^2\frac{\partial}{\partial r}\phi(\mathbf{r})\right]\right\}. \tag{10.24}$$

The construction of ϕ is more tricky, as it is not directly observable, but must be estimated from the shear field. This reconstruction of the lensing potential suffers from a similar ambiguity to the mass-sheet degeneracy for simple lenses. To see how, we first note that the complex shear field γ is the second derivative of the lensing potential:

$$\gamma(\mathbf{r}) = \left[\frac{1}{2}\left(\frac{\partial^2}{\partial\theta_x^2} - \frac{\partial^2}{\partial\theta_y^2}\right) + i\frac{\partial^2}{\partial\theta_x\,\partial\theta_y}\right]\phi(\mathbf{r}). \tag{10.25}$$

As a consequence, since the lensing potential is real, its estimate is ambiguous up to the addition of any field $f(\mathbf{r})$ for which

$$\frac{\partial^2 f(\mathbf{r})}{\partial\theta_x^2} - \frac{\partial^2 f(\mathbf{r})}{\partial\theta_y^2} = \frac{\partial^2 f(\mathbf{r})}{\partial\theta_x\,\partial\theta_y} = 0. \tag{10.26}$$

Since ϕ must be real, the general solution to this is

$$f(\mathbf{r}) = F(r) + G(r)\theta_x + H(r)\theta_y + P(r)(\theta_x^2 + \theta_y^2), \tag{10.27}$$

where F, G, H and P are arbitrary functions of $r \equiv |\mathbf{r}|$. Assuming these functions vary smoothly with r, only the last of these survives at a significant level to the mass density, and corresponds to a sheet of overdensity

$$\delta = \frac{4a(t)c^2}{3H_0^2\Omega_m r^2}\frac{\partial}{\partial r}\left[r^2\frac{\partial}{\partial r}P(r)\right]. \tag{10.28}$$

There are a couple of ways to deal with this problem. For a reasonably large survey, one can assume that the potential and its derivatives are zero on average, at each r, or that the overdensity has average value zero. For further details, see Bacon and Taylor (2003). Note that the relationship between the overdensity field and the lensing potential is a linear one, so if one chooses a discrete binning of the quantities, one can use standard linear algebra methods to attempt an inversion, subject to some constraints such as minimising the expected reconstruction errors. With prior knowledge of the signal properties, this is the Wiener filter. See Hu and Keeton (2002) for further details of this approach.

This method was first applied to COMBO-17 data (Taylor *et al.*, 2004), and more recently to COSMOS HST data (Massey *et al.*, 2007) – see Fig. 10.2.

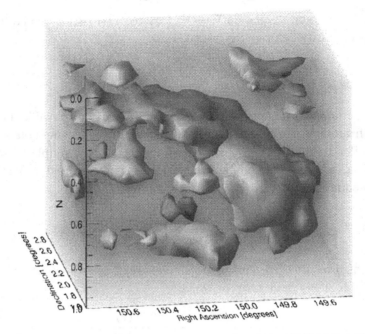

Fig. 10.2. 3D reconstruction of matter density from the COSMOS ACS data. (Massey *et al.*, 2007.)

10.3.3.1 Convergence power spectrum and shear correlation function

The average shear is zero, so the most common statistics to use for cosmology are two-point statistics, quadratic in the shear. These may be in 'configuration' ('real') space, or in transform space (using Fourier coefficients or similar). I will focus on two quadratic measures, the convergence power spectrum and the shear–shear correlation function.

To find the expectation value of a quadratic quantity, it is convenient to make use of the matter density power spectrum, $P(k)$, defined by the following relation between the overdensity Fourier coefficients:

$$\langle \delta_{\mathbf{k}} \delta_{\mathbf{k}'}^* \rangle = (2\pi)^3 \delta^D(\mathbf{k} - \mathbf{k}') P(k). \tag{10.29}$$

where δ^D is the Dirac delta function. $P(k)$ is evolving, so we write it as $P(k; r)$ in future, where r and t are related through the lookback time. (This r-dependence may look strange; there is a subtlety: (10.29) holds if the field is homogeneous and isotropic, which the field on the past light cone is not, since it evolves. In the radial integrals, one has to consider the homogeneous field at the same cosmic time as the time of emission of the source.) The trick is to get the desired quadratic quantity into a form that includes $P(k; r)$.

For the convergence power spectrum, we first transform the convergence in a 2D Fourier transform on the sky, where ℓ is a 2D dimensionless wavenumber:

$$\kappa_\ell = \int d^2\theta\, \kappa(\theta) e^{-i\ell.\theta} \tag{10.30}$$

$$= A \int_0^\infty dr\, r\, \frac{G(r)}{a(r)} \int d^2\theta\, \delta(r\theta, r) e^{-i\ell.\theta}, \tag{10.31}$$

where $A \equiv 3H_0^2 \Omega_m / 2c^2$. We expand the overdensity field in a Fourier transform,

$$\delta(r\theta, r) = \int \frac{d^3k}{(2\pi)^3} \delta_k e^{ik_\parallel r} e^{ik_\perp.r\theta}, \tag{10.32}$$

and substitute into (10.31). We form the quantity $\langle \kappa_\ell \kappa_{\ell'}^* \rangle$, which, by analogy with (10.29), is related to the (2D) *convergence power spectrum* by

$$\langle \kappa_\ell \kappa_{\ell'}^* \rangle = (2\pi)^2 \delta^D(\ell - \ell') P_\kappa(|\ell|). \tag{10.33}$$

Straightforwardly,

$$\langle \kappa_\ell \kappa_{\ell'}^* \rangle = A^2 \int_0^\infty dr\, G(r) \int_0^\infty dr'\, G(r') \int d^2\theta\, d^2\theta'\, \frac{d^3k}{(2\pi)^3} \frac{d^3k'}{(2\pi)^3} \tag{10.34}$$

$$\langle \delta_k \delta_{k'}^* \rangle \exp(ik_\parallel r - ik_\parallel' r') \exp(ik_\perp.\theta - ik_\perp'.\theta') \exp(-i\ell.\theta + i\ell'.\theta'),$$

where $G(r) \equiv rg(r)/a(r)$. Using (10.29) we remove the k' integration, replacing k' by k and introducing the power spectrum $P(k) = P\left(\sqrt{k_\parallel^2 + |k_\perp|^2}\right)$. For small-angle surveys, most of the signal comes from short wavelengths, and the $k_\parallel$ is negligible, so $P(k) \simeq P(|k_\perp|)$. The only $k_\parallel$ term remaining is the exponential, which integrates to $(2\pi)\delta^D(r - r')$. The integrals over θ and θ' give $(2\pi)^2 \delta^D(\ell - rk_\perp)$ and $(2\pi)^2 \delta^D(\ell' - rk_\perp')$ respectively, so the whole lot simplifies to give the convergence power spectrum as

$$P_\kappa(\ell) = \left(\frac{3H_0^2 \Omega_m}{2c^2}\right)^2 \int_0^\infty dr \left[\frac{g(r)}{a(r)}\right]^2 P(\ell/r; r). \tag{10.35}$$

An exercise for the reader is to show that the power spectrum for γ is the same: $P_\gamma = P_\kappa$. The shear correlation function, for points separated by an angle θ is

$$
\begin{aligned}
\langle \gamma \gamma^* \rangle_\theta &= \int \frac{d^2\ell}{(2\pi)^2} P_\gamma(\ell) \, e^{i\ell \cdot \theta} \\
&= \int \frac{\ell \, d\ell}{(2\pi)^2} P_\kappa(\ell) e^{i\ell\theta \cos\varphi} d\varphi \\
&= \int \frac{d\ell}{2\pi} \ell P_\kappa(\ell) J_0(\ell\theta),
\end{aligned} \tag{10.36}
$$

where we have used polar coordinates, with φ the angle between ℓ and θ, and we have exploited the isotropy (P_κ depends only on the modulus of ℓ). J_0 is a Bessel function.

Alternatively, the shears may be referred to axes oriented tangentially (t) and at 45 degrees to the radius ($\times$), defined with respect to each pair of galaxies used in the averaging. The rotations $\gamma \to \gamma' = \gamma \, e^{-2i\psi}$, where ψ is the position angle of the pair, give tangential and cross components of the rotated shear as $\gamma' = -\gamma_t - i\gamma_\times$, where the components have correlation functions ξ_{tt} and $\xi_{\times\times}$ respectively. It is common to define a pair of correlations

$$
\xi_\pm(\theta) = \xi_{tt} \pm \xi_{\times\times}, \tag{10.37}
$$

which can be related to the convergence power spectrum by (Kaiser, 2002)

$$
\begin{aligned}
\xi_+(\theta) &= \int_0^\infty \frac{d\ell}{2\pi} \ell P_\kappa(\ell) J_0(\ell\theta), \\
\xi_-(\theta) &= \int_0^\infty \frac{d\ell}{2\pi} \ell P_\kappa(\ell) J_4(\ell\theta).
\end{aligned} \tag{10.38}
$$

As we will see later, these statistics are useful for assessing systematic errors in the lensing analysis.

Finally, there is a general class of statistics referred to as *aperture masses* associated with *compensated filters*, which are defined as the convergence field smoothed with a compensated filter. These can be computed by averaging the tangential shear (Schneider, 1996):

$$
M_{ap}(\theta) \equiv \int d\theta' U_{M_{ap}}(|\theta' - \theta|)\kappa(\theta') = \int d\theta' Q_{M_{ap}}(|\theta' - \theta|)\gamma_t(\theta'), \tag{10.39}
$$

where

$$
Q_{M_{ap}}(\theta) = -U_{M_{ap}}(\theta) + \frac{2}{\theta^2} \int_0^\theta d\theta' \, \theta' \, U_{M_{ap}}(\theta'). \tag{10.40}
$$

Several forms of $U_{M_{ap}}$ have been suggested (e.g. Crittenden *et al.*, 2002), which trade locality in real space with locality in ℓ space, or have $U = $ constant within a certain radius (where $Q = 0$), potentially avoiding the strong lensing regime, or which are matched to the expected mass profile, to improve signal-to-noise (Dietrich *et al.*, 2007). Other forms have been suggested (see e.g. Crittenden *et al.*, 2002), which are broader in real space, but pick up a narrower range of ℓ power for a given θ. All of these two-point statistics can be written as integrals over ℓ of the convergence power spectrum $P_\kappa(\ell)$ multiplied by some kernel function.

10.3.4 Matter power spectrum

As we have seen, the two-point statistics of the shear and convergence fields depend on the power spectrum of the matter, $P(k; t)$. The power spectrum grows in a simple way when the perturbations in the overdensity are small, $|\delta| \ll 1$, when the power spectrum grows in amplitude whilst keeping the same shape as a function of k. However, gravitational lensing can still be weak, even if the overdensity field is nonlinear. Poisson's equation still holds provided we are in the weak-field limit as far as general relativity is concerned, and this essentially always holds for cases of practical interest. In order to get as much statistical power out of lensing, one must probe the nonlinear regime, so it is necessary for parameter estimation to know how the power spectrum grows. Through the extensive use of numerical simulations, the growth of dark matter clustering is well understood down to quite small scales, where uncertainties in modelling, or uncertain physics, such as the influence of baryons on the dark matter (White, 2004), make the predictions unreliable. Accurate fits for the nonlinear power spectrum have been found (Smith *et al.*, 2003) up to $k > 10h\,\mathrm{Mpc}^{-1}$, which is far beyond the linear/nonlinear transition $k \sim 0.2h\,\mathrm{Mpc}^{-1}$. Figure 10.3 shows fits for a number of CDM models. For precision use, one must make sure that the statistics do not have substantial contributions from the high-k end where the nonlinear power spectrum is uncertain. This can be explored by looking at the kernel functions implicit in the quantities such as the shear correlation function (10.36).

10.4 Observations, status and prospects

10.4.1 Estimating the shear field

We need to have some estimator for the shear field. The most obvious is to use the shape of a galaxy somehow. We need some measure of shape, and we require to know how it is altered by cosmic shear. I will describe the most common shape measurement used, but note that there are new techniques (shapelets, polar

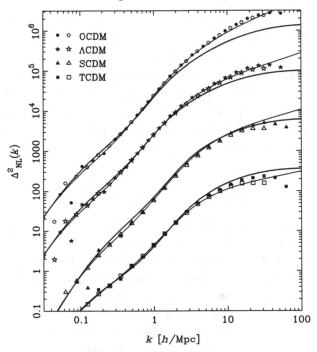

Fig. 10.3. The nonlinear power spectrum from numerical simulations, along with fitting functions (from Smith *et al.*, 2003).

shapelets, etc.) (Bernstein and Jarvis, 2002; Chang and Refregier, 2002) coming into use.

The method I describe is commonly referred to as KSB (Kaiser, Squires and Broadhurst, 1995). The shape measurement part of KSB uses moments of the surface brightness distribution of the source galaxy (here assumed to be centred on the origin):

$$Q_{ij} \equiv \int d^2\theta \, \theta_i \theta_j \, I(\boldsymbol{\theta}) \, W(\boldsymbol{\theta}), \qquad (10.41)$$

where I is the surface brightness of the galaxy on the sky, and W is an optional weight function. The shape and orientation of the galaxy are encapsulated in the two *ellipticities*:

$$e_1' \equiv \left(\frac{Q_{11} - Q_{22}}{Q_{11} + Q_{22}} \right), \qquad e_2' = \frac{2Q_{12}}{Q_{11} + Q_{22}} \qquad (10.42)$$

and it is sometimes convenient to define a *complex ellipticity* $e' \equiv e_1' + ie_2'$.

It is straightforward to show that the source moments Q_{ij}^s are related to the observed moments by

$$Q_{ij}^s = A_{il} Q_{lm} A_{jm}. \tag{10.43}$$

A straightforward but tedious calculation relates the complex source and image ellipticities:

$$e' = \frac{e'^s + 2\gamma + g^2 e'^{s*}}{1 + |g|^2 + 2\mathrm{Re}(g e'^{s*})}, \tag{10.44}$$

where $g \equiv \gamma/(1-\kappa)$. To estimate the shear correlation function $\langle \gamma \gamma^* \rangle$, one can use $\langle e' e'^* \rangle$, provided the terms $\langle e'^s e'^{s*} \rangle$ and $\langle e'^s g^* \rangle$ are zero. The first term corresponds to intrinsic alignments of galaxies, which may be non-zero due to tidal effects. We will look at this later. Even the cross-term may be non-zero, but this term has received little attention to date (Hirata and Seljak, 2004).

The estimate of the shear will be noisy, as galaxies are not round and have an intrinsic ellipticity dispersion of around $\sigma_e \simeq 0.3$. With many (N) sources, the error on the average ellipticity can be reduced to $\sigma_e'/\sqrt{N}$, and progress can be made.

In some ways, a more convenient definition of ellipticity is (Seitz and Schneider, 1995)

$$e = \frac{Q_{11} - Q_{22} + 2iQ_{12}}{Q_{11} + Q_{22} + 2(Q_{11}Q_{22} - Q_{12}^2)^{1/2}}, \tag{10.45}$$

which, although a more complicated definition, has simpler transformation properties:

$$e_S = \frac{e - g}{1 - g^* e}. \tag{10.46}$$

If we average over some galaxies, $\langle e \rangle = g$, which is simpler than the average of the alternative ellipticity definition. Note that e is dominated by the intrinsic ellipticity, and many source galaxies are needed to get a robust measurement of cosmic shear. This results in estimators of averaged quantities, such as the average shear in an aperture, or a weighted average in the case of M_{ap}. Any analysis of these quantities needs to take account of their noise properties, and more generally of their covariance properties. We will look only at a couple of examples here; a more detailed discussion of covariance of estimators, including nonlinear cumulants, appears in Munshi and Valageas (2005).

10.4.1.1 Linear estimators

Perhaps the simplest average statistic to use is the average (of N) galaxy ellipticities in a 2D aperture on the sky:

$$\bar{\gamma} \equiv \frac{1}{N} \sum_{i=1}^{N} \frac{e_i}{2}. \tag{10.47}$$

The covariance of two of these estimators $\bar{\gamma}_\alpha$ and $\bar{\gamma}_\beta^*$ is

$$\langle \bar{\gamma}_\alpha \bar{\gamma}_\beta^* \rangle = \frac{1}{4NM} \sum_{i=1}^{N} \sum_{j=1}^{M} \langle (e_i^s + 2\gamma_i)(e_j^{s*} + 2\gamma_j^*) \rangle, \tag{10.48}$$

where the apertures have N and M galaxies respectively. If we assume (almost certainly incorrectly; see later) that the source ellipticities are uncorrelated with each other, and with the shear, then for distinct apertures the estimator is an unbiased estimator of the shear correlation function averaged over the pair separations. If the apertures overlap, then this is not the case. For example, in the shear variance, the apertures are the same, and

$$\langle |\bar{\gamma}^2| \rangle = \frac{1}{N^2} \sum_{i=1}^{N} \sum_{j=1}^{N} \langle (|e_{Si}|^2 \delta_{ij} + \gamma_i \gamma_j^*) \rangle, \tag{10.49}$$

which is dominated by the presence of the intrinsic ellipticity variance, $\sigma_e^2 \equiv \langle |e^S|^2 \rangle \simeq 0.3^2 - 0.4^2$. The average shear therefore has a variance of $\sigma_e^2/4N$. If we use the (quadratic) shear variance itself as a statistic, then it is estimated by omitting the diagonal terms:

$$|\bar{\gamma}^2| = \frac{1}{4N(N-1)} \sum_{i=1}^{N} \sum_{j \neq i} e_i e_j^*. \tag{10.50}$$

For aperture masses (Eq. (10.39)), the intrinsic ellipticity distribution leads to a shot noise term from the finite number of galaxies. Again we simplify the discussion here by neglecting correlations of source ellipticities. The shot noise can be calculated by the standard method (see e.g. Peebles, 1980) of dividing the integration solid angle into cells i of size $\Delta^2 \theta_i$ containing $n_i = 0$ or 1 galaxy:

$$M_{\rm ap} \simeq \sum_i \Delta^2 \theta_i \, n_i \, Q(|\theta_i|) \, (e_i^s/2 + \gamma_i)_{\rm t}. \tag{10.51}$$

Squaring and taking the ensemble average, noting that $\langle e_{i,t}^s e_{j,t}^s \rangle = \sigma_e^2 \delta_{ij}/2$, $n_i^2 = n_i$, and rewriting as a continuous integral gives

$$\langle M_{\rm ap}^2 \rangle_{\rm SN} = \frac{\sigma_e^2}{8} \int d^2\theta \, Q^2(|\theta|). \tag{10.52}$$

Shot noise terms for other statistics are calculated in similar fashion. In addition to the covariance from shot noise, there can be signal covariance, for example from samples of different depths in the same area of sky. Both samples are affected by the lensing by the common low-redshift foreground structure. See Munshi and Valageas (2005) for more detail.

10.4.1.2 Quadratic estimators

We have already seen how to estimate in an unbiased way the shear variance. The shear correlation functions can similarly be estimated:

$$\hat{\xi}_{\pm}(\theta) = \frac{\sum_{ij} w_i w_j (e_{it} e_{jt} \pm e_{i\times} e_{j\times})}{4 \sum_{ij} w_i w_j}, \tag{10.53}$$

where the w_i are arbitrary weights, and the sum extends over all pairs of source galaxies with separations close to θ. Only in the absence of intrinsic correlations, $\langle e_{it} e_{jt} \pm e_{i\times} e_{j\times} \rangle = \sigma_e^2 \delta_{ij} + 4\xi_{\pm}(|\theta_i - \theta_j|)$, are these estimators unbiased.

As with any quadratic quantity, the covariance of these estimators depends on the 4-point function of the source ellipticities and the shear. These expressions can be evaluated if the shear field is assumed to be Gaussian, but the expressions for this (and the squared aperture mass covariance) are too cumbersome to be given here, so the reader is directed to Schneider *et al.* (2002b).

In harmonic space, the convergence power spectrum may be estimated from either ξ_+ or ξ_- (or both), using Eq. (10.38). From the orthonormality of the Bessel functions,

$$P_\kappa(\ell) = \int_0^\infty d\theta \, \theta \, \xi_{\pm}(\theta) \, J_{0,4}(\ell\theta), \tag{10.54}$$

where the 0, 4 correspond to the $+/-$ cases. In practice, $\xi_{\pm}(\theta)$ is not known for all θ, and the integral is truncated on both small and large scales. This can lead to inaccuracies in the estimation of $P_\kappa(\ell)$ (see Schneider *et al.*, 2002b). An alternative method is to parametrise $P_\kappa(\ell)$ in band-powers, and to use parameter estimation techniques to estimate it from the shear correlation functions (Hu and White, 2001; Brown *et al.*, 2003).

10.4.2 Systematics

10.4.2.1 Point spread function

The main practical difficulty of lensing experiments is that the atmosphere and telescope affect the shape of the images. These modifications to the shape may arise due to such things as the point spread function, or poor tracking of the telescope. The former needs to be treated with great care. Stars (whose images should be round) can be used to remove image distortions to very high accuracy, although a possibly fundamental limitation may arise because of the finite number of stars in an image. Interpolation of the anisotropy of the PSF needs to be done carefully, and examples of how this can be done in an optimal way are given in van Waerbeke, Mellier and Hoekstra (2003).

10.4.2.2 Photometric redshift errors

E/B decomposition Weak gravitational lensing does not produce the full range of locally linear distortions possible. These are characterised by translation, rotation, dilation and shear, with six free parameters. Translation is not readily observable, but weak lensing is specified by three parameters rather than the four remaining degrees of freedom permitted by local affine transformations. This restriction is manifested in a number of ways: for example, the transformation of angles involves a 2×2 matrix which is symmetric, so not completely general, see Eq. (10.13). Alternatively, a general spin-weight 2 field can be written in terms of second derivatives of a *complex* potential, whereas the lensing potential is real. As noticed below Eq. (10.6) and in Eq. (10.54), this also implies that there are many other consistency relations that have to hold if lensing is responsible for the observed shear field. In practice the observed ellipticity field may not satisfy the expected relations, if it is contaminated by distortions not associated with weak lensing, so the failure to satisfy the expected relations may indicate the presence of some systematic error. The most obvious of these is optical distortions of the telescope system, but could also involve physical effects such as intrinsic alignment of galaxy ellipticities, which we will consider later.

A convenient way to characterise the distortions is via E/B decomposition, where the shear field is described in terms of an 'E-mode', which is allowed by weak lensing, and a 'B-mode', which is not. These terms are borrowed from similar decompositions in polarisation fields. In fact weak lensing can generate B-modes, but they are expected to be very small (e.g. Schneider, van Waerbeke and Mellier, 2002a), so the existence of a significant B-mode in the observed shear pattern is indicative of some non-lensing contamination. The easiest way to introduce a B-mode mathematically is to make the lensing potential complex:

$$\phi = \phi_E + i\phi_B. \tag{10.55}$$

There are various ways to determine whether a B-mode is present. A neat way is to generalise the aperture mass to a complex $M = M_{ap} + iM_{\perp}$, where the real part picks up the E-modes, and the imaginary part the B-modes. Alternatively, the $\xi_{\pm}$ can be used (Crittenden *et al.*, 2002; Schneider *et al.*, 2002b):

$$P_{\kappa\pm}(\ell) = \pi \int_0^\infty d\theta \, \theta \, [J_0(\ell\theta)\xi_+(\theta) \pm J_4(\ell\theta)\xi_-(\theta)], \tag{10.56}$$

where the $\pm$ power spectra refer to E- and B-mode powers. In principle this requires the correlation functions to be known over all scales from 0 to ∞. Variants of this (Crittenden *et al.*, 2002) allow the E/B-mode correlation functions to be written in terms of integrals of $\xi_\pm$ over a finite range:

$$\xi_E(\theta) = \frac{1}{2}\left[\xi_-(\theta) + \xi'_+(\theta)\right],$$

$$\xi_B(\theta) = -\frac{1}{2}\left[\xi_-(\theta) - \xi'_+(\theta)\right], \tag{10.57}$$

where

$$\xi'_+(\theta) = \xi_+(\theta) + 4\int_0^\theta \frac{d\theta'}{\theta'}\xi_+(\theta') - 12\theta^2 \int_0^\theta \frac{d\theta'}{\theta'^3}\xi_+(\theta'). \tag{10.58}$$

This avoids the need to know the correlation functions on large scales, but needs the observed correlation functions to be extrapolated to small scales; this was one of the approaches taken in the analysis of the CFHTLS data (Hoekstra *et al.*, 2006). Difficulties with estimating the correlation functions on small scales have led others to prefer to extrapolate to large scales, such as in the analysis of the GEMS (Heymans *et al.*, 2005) and William Herschel data (Massey *et al.*, 2005). Note that without full-sky coverage, the decomposition into E- and B-modes is ambiguous, although for scales much smaller than the survey it is not an issue.

Intrinsic alignments The main signature of weak lensing is a small alignment of the images, at the level of a correlation of ellipticities of $\sim10^{-4}$. One might be concerned that physical processes might also induce an alignment of the galaxies themselves. In the traditional lensing observations, the distances of individual galaxies are ignored, and one simply uses the alignment on the sky of galaxies, and hopes that the galaxies will typically be at such large separations along the line of sight that any physical interactions are rare and can be ignored. However, the lensing signal is very small, so the assumption that intrinsic alignment effects are sufficiently small needs to be tested. This was first done in a series of papers by a number of groups in 2000–1 (e.g. Croft and Metzler, 2000; Heavens, Refregier and Heymans, 2000; Catelan, Kamionkowski, and Blanford, 2001; Crittenden *et al.*, 2001), and the answer is that the effects may not be negligible. The contamination by intrinsic alignments is highly depth-dependent. This is easy to see, since at fixed angular separation, galaxies in a shallow survey will be physically closer together in space, and hence more likely to experience tidal interactions that might align the galaxies. In addition to this, the shallower the survey, the smaller the lensing signal. In a pioneering study, the alignments of nearby galaxies in the SuperCOSMOS

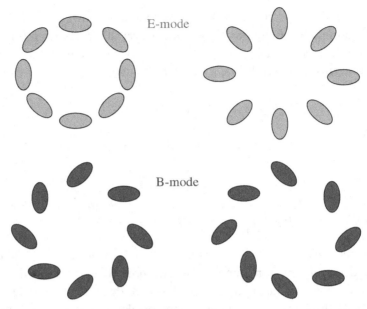

Fig. 10.4. Example patterns from E-mode and B-mode fields. (From van Waerbeke and Mellier, 2003.) Weak lensing only produces E-modes at any significant level, so the presence of B-modes can indicate systematic errors.

survey were investigated (Brown *et al.*, 2002). This survey is so shallow (median redshift ∼0.1) that the expected lensing signal is tiny. A non-zero alignment was found, which agrees with at least some of the theoretical estimates of the effect. The main exception is the numerical study of Jing (2002), which predicts a contamination so high that it could dominate even deep surveys. For deep surveys, the effect is expected to be rather small, but if one wants to use weak lensing as a probe of subtle effects such as the effects of altering the equation of state of dark energy, then one has to do something. There are essentially two options – either one tries to calculate the intrinsic alignment signal and subtract it, or one tries to remove it altogether. The former approach is not practical, as, although there is some agreement as to the general level of the contamination, the details are not accurately enough known. The latter approach is becoming possible, as lensing surveys are now obtaining estimates of the distance to each galaxy, via photometric redshifts. (Spectroscopic redshifts are difficult to obtain, because one needs a rather deep sample, with median redshift at least 0.6 or so, and large numbers, to reduce shot noise due to the random orientations of ellipticities.) With photometric redshifts, one can remove physically close galaxies from the pair statistics (such as the shear correlation function) (Heymans and Heavens, 2003; King and Schneider, 2003).

Thus one removes a systematic error in favour of a slightly increased statistical error. The analysis in Heymans *et al.* (2004) is the only study that has explicitly removed close pairs.

10.4.3 Results

The first results from cosmic shear were published in 2000 (Bacon, Refregier and Ellis, 2000; Kaiser, Wilson and Luppino, 2000; van Waerbeke *et al.*, 2000; Wittman *et al.*, 2000), so as an observational science, cosmological weak lensing is very young. To date, the surveys have been able to show clear detections of the effect, and reasonably accurate determination of some cosmological parameters, usually the amplitude of the dark matter perturbations (measured by the rms fractional fluctuations in an $8h^{-1}$Mpc sphere and denoted σ_8), and the matter density parameter Ω_m. Current surveys cannot lift a near-degeneracy between these two, and usually a combination (typically $\sigma_8\Omega_m^{0.5}$) is quoted. This makes sense – it is difficult, but certainly not impossible, to distinguish between a highly clumped low-density universe, and a modestly clumped high-density universe. There is no question that the surveys do not yet have the size, or the careful control of systematics required to compete with the microwave background and other techniques used for cosmological parameter estimation. However, this situation is changing fast, particularly with the CFHT Legacy Survey, which is underway, and the upcoming VST weak lensing survey. There are other more ambitious surveys that are in some stage of planning or funding, such as Pan-STARRS, the Dark Energy Survey and VISTA. A summary of recent results is in the review by van Waerbeke and Mellier (2003), and I will not list the results here. I mention only one recent study, a reanalysis of the VIRMOS-DESCART survey (van Waerbeke, Mellier and Hoekstra, 2005), as this one uses a neat trick to deal with varying point spread function and quotes a small error ($\sigma_8 = (0.83 \pm 0.07)(\Omega_m/0.3)^{-0.49}$). The technique employed was to make use of the fact that lensing does not induce a significant B-mode. As discussed earlier, one effectively needs to interpolate the shape of the point spread function between the positions of the stars in the image, and this can be done in a number of ways. Van Waerbeke *et al.* (2005) chose the interpolation scheme that minimised the B-mode. Although not foolproof in principle, the results are impressive, and are shown in Fig. 10.5. Finally, a number of recent surveys have been reanalysed, using photometric redshifts from the surveys to give better measurements of the redshift distribution, rather than, for example, relying on redshift distributions from the Hubble Deep Field, which is small and subject to large sample variance. Interestingly, this study (Benjamin *et al.*, 2007) resulted in a lower value of the perturbation amplitude $\sigma_8 = 0.84 \pm 0.07$ for $\Omega_m = 0.24$, thus reducing the previous tension with the results from microwave background analysis of WMAP (Spergel *et al.*, 2007); see Fig. 10.7; Fig. 10.6 shows the earlier results.

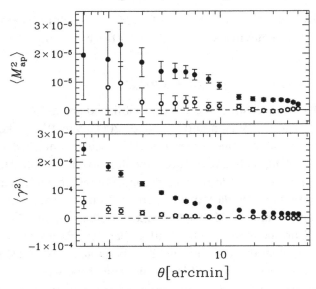

Fig. 10.5. E- and B-modes from a reanalysis of the VIRMOS-DESCART survey (van Waerbeke, Mellier and Hoekstra, 2005). Top panel shows the aperture mass statistic (a weighted quadratic measure) and the lower panel the shear variance on different scales. Top points are the E-modes and bottom the B-modes.

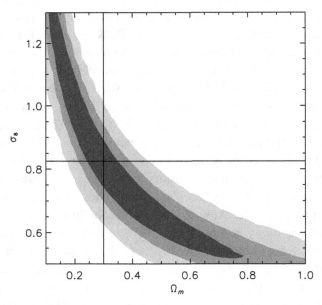

Fig. 10.6. Cosmological parameters from a reanalysis of the VIRMOS-DESCART survey (van Waerbeke, Mellier and Hoekstra, 2005; data from Smith *et al.*, 2003). Contours are 68%, 95% and 99.9% joint confidence levels. The vertical line is at $\Omega_m = 0.3$, a little higher than what is favoured by microwave background and large-scale structure studies.

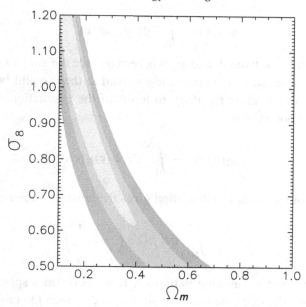

Fig. 10.7. Cosmological parameters from a reanalysis of 100 square degrees of lensing surveys (Benjamin *et al.*, 2007). Contours are 68%, 95% and 99.9% joint confidence levels. Better estimation of the redshift distribution has reduced the estimated amplitude of clustering.

10.5 Dark energy: lensing in 3D

10.5.1 Tomography

In the case where one has distance information for individual sources, it makes sense to employ the information for statistical studies. A natural course of action is to divide the survey into slices at different distances, and perform a study of the shear pattern on each slice. In order to use the information effectively, it is necessary to look at cross-correlations of the shear fields in the slices, as well as correlations within each slice (Hu, 1999). This procedure is usually referred to as tomography, although the term does not seem entirely appropriate.

We start by considering the average shear in a shell, which is characterised by a probability distribution for the source redshifts $z = z(r)$, $p(z)$. The shear field is the second edth derivative of the lensing potential (e.g. Castro, Heavens and Kitching, 2006)

$$\gamma(\mathbf{r}) = \frac{1}{2}\,\tilde{\partial}\,\tilde{\partial}\,\phi(\mathbf{r}) \simeq \frac{1}{2}(\partial_x + i\partial_y)^2\phi(\mathbf{r}), \qquad (10.59)$$

where the derivatives are in the angular direction, and the last equality holds in the flat-sky limit. If we average the shear in a shell, giving equal weight to each galaxy, then the average shear can be written in terms of an effective lensing potential

$$\phi_{\text{eff}}(\boldsymbol{\theta}) = \int_0^\infty dz \, p(z)\phi(\mathbf{r}), \tag{10.60}$$

where the integral is at fixed $\boldsymbol{\theta}$, and $p(z)$ is zero outside the slice (we ignore errors in distance estimates such as photometric redshifts; these could be incorporated with a suitable modification to $p(z)$). In terms of the gravitational potential, the effective lensing potential is

$$\phi_{\text{eff}}(\boldsymbol{\theta}) = \frac{2}{c^2} \int_0^\infty dr \, \Phi(\mathbf{r})g(r), \tag{10.61}$$

where reversal of the order of integration gives the lensing efficiency to be

$$g(r) = \int_{z(r)}^\infty dz' \, p(z') \left(\frac{1}{r} - \frac{1}{r'} \right), \tag{10.62}$$

where $z' = z'(r')$ and we assume flat space. If we perform a spherical harmonic transform of the effective potentials for slices i and j, then the cross power spectrum can be related to the power spectrum of the gravitational potential $P_\Phi(k)$ via a version of Limber's equation:

$$\left\langle \phi_{\ell m}^{(i)} \phi_{\ell' m'}^{*(j)} \right\rangle = C_{\ell,ij}^{\phi\phi} \, \delta_{\ell'\ell} \delta_{m'm}, \tag{10.63}$$

where

$$C_{\ell,ij}^{\phi\phi} = \left(\frac{2}{c^2} \right)^2 \int_0^\infty dr \, \frac{g^{(i)}(r) \, g^{(j)}(r)}{r^2} \, P_\Phi(\ell/r; r) \tag{10.64}$$

is the cross power spectrum of the lensing potentials. The last argument in P_Φ allows for evolution of the power spectrum with time, or equivalently distance. The power spectra of the convergence and shear are related to $C_{\ell,ij}^{\phi\phi}$ by (Hu, 2000)

$$C_{\ell,ij}^{\kappa\kappa} = \frac{\ell^2(\ell+1)^2}{4} C_{\ell,ij}^{\phi\phi},$$

$$C_{\ell,ij}^{\gamma\gamma} = \frac{1}{4} \frac{(\ell+2)!}{(\ell-2)!} C_{\ell,ij}^{\phi\phi}. \tag{10.65}$$

The sensitivity of the cross power spectra to cosmological parameters is through various effects, as in 2D lensing: the shape of the linear gravitational potential power spectrum is dependent on some parameters, as is its nonlinear evolution; in addition the $z(r)$ relation probes cosmology. The reader is referred to standard cosmological texts for more details of the dependence of the distance–redshift relation on cosmological parameters.

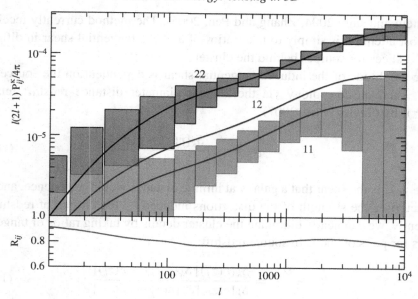

Fig. 10.8. The power spectra of two slices, their cross power spectrum, and their correlation coefficient. (From Hu, 1999.)

Hu (1999) illustrates the power and limitation of tomography, with two shells (Fig. 10.8). As expected, the deeper shell (2) has a larger lensing power spectrum than the nearby shell (1), but it is no surprise to find that the power spectra from shells are correlated, since the light from both passes through some common material. Thus one does gain from tomography, but, depending on what one wants to measure, the gains may or may not be very great. For example, tomography adds rather little to the accuracy of the amplitude of the power spectrum, but far more to studies of dark energy properties. One also needs to worry about systematic effects, as leakage of galaxies from one shell to another, through noisy or biased photometric redshifts, can degrade the accuracy of parameter estimation (Huterer *et al.*, 2006; Ma, Hu and Huterer, 2006).

10.5.2 The shear ratio test

The shear contributed by the general large-scale structure is typically about 1%, but the shear behind a cluster of galaxies can far exceed this. As always, the shear of a background source is dependent on its redshift, and on cosmology, but also on the mass distribution in the cluster. This can be difficult to model, so it is attractive to consider methods that are decoupled from the details of the mass distribution of the cluster. Various methods have been proposed (e.g. Jain and Taylor, 2003;

Bernstein and Jain, 2004; Zhang and Pen, 2005). The method currently receiving the most attention is simply to take ratios of average tangential shear in different redshift slices for sources behind the cluster.

The amplitude of the induced tangential shear is dependent on the source redshift z, and on cosmology via the angular diameter distance–redshift relation $S_k[r(z)]$ by (Taylor *et al.*, 2007)

$$\gamma_t(z) = \gamma_t(z = \infty) \frac{S_k[r(z) - r(z_l)]}{S_k[r(z)]}, \qquad (10.66)$$

where $\gamma_{t,\infty}$ is the shear that a galaxy at infinite distance would experience, and that characterises the strength of the distortions induced by the cluster, at redshift z_l. Evidently, we can neatly eliminate the cluster details by taking ratios of tangential shears for pairs of shells in source redshift:

$$R_{ij} \equiv \frac{\gamma_{t,i}}{\gamma_{t,j}} = \frac{S_k[r(z_j)]\, S_k[r(z_i) - r(z_l)]}{S_k[r(z_i)]\, S_k[r(z_j) - r(z_l)]}. \qquad (10.67)$$

In reality, the light from the more distant shell passes through an extra pathlength of clumpy matter, so suffers an additional source of shear. This can be treated as a noise term (Taylor *et al.*, 2007). This approach is attractive in that it probes cosmology through the distance–redshift relation alone, being (at least to good approximation) independent of the growth rate of the fluctuations. Its dependence on cosmological parameters is therefore rather simpler, as many parameters (such as the amplitude of matter fluctuations) do not affect the ratio except through minor side-effects. More significantly, it can be used in conjunction with lensing methods that probe both the distance–redshift relation and the growth rate of structure. Such a dual approach can in principle distinguish between quintessence-type dark energy models and modifications of Einstein gravity. This possibility arises because the effect on global properties (e.g. $z(r)$) is different from the effect on perturbed quantities (e.g. the growth rate of the power spectrum) in the two cases. The method has a signal-to-noise that is limited by the finite number of clusters massive enough to have measurable tangential shear. In an all-sky survey, the bulk of the signal would come from the 10^5–10^6 clusters above a mass limit of $10^{14}\, M_\odot$.

10.5.3 Full 3D analysis of the shear field

An alternative approach to take is to recognise that, with photometric redshift estimates for individual sources, the data one is working with are a very noisy 3D shear field, which is sampled at a number of discrete locations, and for which the locations are somewhat imprecisely known. It makes some sense, therefore, to deal with the data one has, and to compare the statistics of the discrete 3D field with theoretical predictions. This was the approach of Heavens (2003), Castro, Heavens and Kitching (2006), and Heavens, Kitching and Taylor (2006). It should yield smaller statistical errors than tomography, as it avoids the binning process which loses information.

In common with many other methods, one has to make a decision whether to analyse the data in configuration space or in the spectral domain. The former, usually studied via correlation functions, is advantageous for complex survey geometries, where the convolution with a complex window function implicit in spectral methods is avoided. However, the more readily computed correlation properties of a spectral analysis are a definite advantage for Bayesian parameter estimation, and we follow that approach here.

The natural expansion of a 3D scalar field (r, θ, ϕ) derived from a potential is in terms of products of spherical harmonics and spherical Bessel functions, $j_\ell(kr)Y_\ell^m(\theta)$. Such products, characterised by three spectral parameters (k, ℓ, m), are eigenfunctions of the Laplace operator, thus making it very easy to relate the expansion coefficients of the density field to that of the potential (essentially via $-k^2$ from the ∇^2 operator). Similarly, the 3D expansion of the lensing potential,

$$\phi_{\ell m}(k) \equiv \sqrt{\frac{2}{\pi}} \int d^3\mathbf{r}\, \phi(\mathbf{r})k j_\ell(kr)Y_\ell^m(\theta), \tag{10.68}$$

where the prefactor and the factor of k are introduced for convenience. The expansion of the complex shear field is most naturally made in terms of spin-weight 2 spherical harmonics $_2Y_\ell^m$ and spherical Bessel functions, since $\gamma = \frac{1}{2}\tilde{\partial}\,\tilde{\partial}\,\phi$, and $\tilde{\partial}\,\tilde{\partial}\,Y_\ell^m \propto_2 Y_\ell^m$:

$$\gamma(\mathbf{r}) = \sqrt{2\pi} \sum_{\ell m} \int dk\, \gamma_{\ell m}\, k\, j_\ell(kr)\, _2Y_\ell^m(\theta). \tag{10.69}$$

The choice of the expansion becomes clear when we see that the coefficients of the shear field are related very simply to those of the lensing potential:

$$\gamma_{\ell m}(k) = \frac{1}{2}\sqrt{\frac{(\ell + 2)!}{(\ell - 2)!}}\, \phi_{\ell m}(k). \tag{10.70}$$

The relation of the $\phi_{\ell m}(k)$ coefficients to the expansion of the density field is readily computed, but more complicated as the lensing potential is a weighted integral of the gravitational potential. The details will not be given here, but relevant effects such as photometric redshift errors, nonlinear evolution of the power spectrum, and the discreteness of the sampling are easily included. The reader is referred to the original papers for details (Heavens, 2003; Castro, Heavens and Kitching, 2006; Heavens, Kitching and Taylor, 2006).

In this way the correlation properties of the $\gamma_{\ell m}(k)$ coefficients can be related to an integral over the power spectrum, involving the $z(r)$ relation, so cosmological parameters can be estimated via standard Bayesian methods from the coefficients. Clearly, this method probes the dark energy effect on both the growth rate and the $z(r)$ relation.

10.5.4 Parameter forecasts from 3D lensing methods

In this section we summarise some of the forecasts for cosmological parameter estimation from 3D weak lensing. We concentrate on the statistical errors that should be achievable with the shear ratio test and with the 3D power spectrum techniques. Tomography should be similar to the latter. We show results from 3D weak lensing alone, as well as in combination with other experiments. These include CMB, supernova and baryon oscillation studies. The methods generally differ in the parameters that they constrain well, but also in terms of the degeneracies inherent in the techniques. Using more than one technique can be very effective at lifting the degeneracies, and very accurate determinations of cosmological parameters, in particular dark energy properties, may be achievable with 3D cosmic shear surveys covering thousands of square degrees of sky to median source redshifts of order unity.

The figures show the accuracy that might be achieved with a number of surveys designed to measure cosmological parameters. We concentrate here on the capabilities of each method, and the methods in combination, to constrain the dark energy equation of state, and its evolution, parametrised by (Chevallier and Polarski, 2001)

$$w(a) = \frac{p}{\rho c^2} = w_0 + w_a(1 - a), \tag{10.71}$$

where the behaviour as a function of scale factor a is, in the absence of a compelling theory, assumed to have this simple form. $w = -1$ would arise if the dark energy behaviour was actually a cosmological constant.

The assumed experiments are: a five-band 3D weak lensing survey, analysed either with the shear ratio test, or with the spectral method, covering 10 000 square degrees to a median redshift of 0.7, similar to the capabilities of a ground-based 4 m-class survey with a several square degrees field; the Planck CMB experiment (14-month mission); a spectroscopic survey to measure baryon oscillations (BAO) in the galaxy matter power spectrum, assuming constant bias, and covering 2000 square degrees to a median depth of unity, and a smaller $z = 3$ survey of 300 square degrees, similar to WFMOS capabilities on Subaru; a survey of 2000 Type Ia supernovae to $z = 1.5$, similar to SNAP's design capabilities.

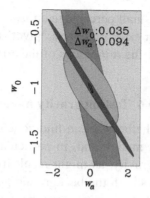

Fig. 10.9. The accuracy expected from the combination of experiments dedicated to studying dark energy properties. The equation of state of dark energy is assumed to vary with scale factor a as $w(a) = w_0 + w_a(1 - a)$, and the figures show the 1σ, two-parameter regions for the experiments individually and in combination. The supernova study fills the plot, the thin diagonal band is Planck, the near-vertical band is BAO, and the ellipse is the 3D lensing power spectrum method. The small ellipse is the expected accuracy from the combined experiments. (From Heavens *et al.*, 2006.)

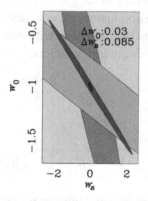

Fig. 10.10. As in Fig. 10.9, but with the shear ratio test as the lensing experiment. Supernovae fill the plot, Planck is the thin diagonal band, BAO is the near-vertical band, and the shear ratio is the remaining 45 degree band. The combination of all experiments is in the centre. (From Taylor *et al.*, 2006.)

We see that the experiments in combination are much more powerful than individually, as some of the degeneracies are lifted (see Figs. 10.9 and 10.10). Note that the combined experiments appear to have rather smaller error bars than is suggested by the single-experiment constraints. This is because the combined ellipse is the projection of the product of several multidimensional likelihood surfaces, which intersect in a small volume. (The projection of the intersection of two surfaces is not the same as the intersection of the projection of two surfaces.) The figures show that errors of a small percentage on w_0 are potentially achievable, or, with this parametrisation, an error of w at a 'pivot' redshift of $z \simeq 0.4$ of under 0.02. This error is essentially the minor axis of the error ellipses.

10.6 Testing gravity models

Here we ask the question whether we can find conclusive evidence that GR plus dark energy is not the right model, and, in particular, if there is evidence for a need for a modification to gravity, as might result from braneworld models. We have seen that lensing probes both the background geometry and the growth rate of perturbations. It is possible to mimic any global geometry relation $r(z)$ for any dark energy equation of state by altering the gravity law, so in principle any dark energy experiment (such as supernovae) that probes only $r(z)$ will not be able to lift the degeneracy. However, although we may be able to reproduce the *global* properties of a non-GR universe by choosing dark energy properties appropriately, it is harder (although not impossible – see Kunz and Sapone, 2007) to reproduce the properties of perturbation growth as well.

To see that there is a degeneracy in the global properties, we note that any gravity law will produce some expansion history or, equivalently, some $H(a)$. Can we find a dark energy equation of state that, together with general relativity, gives the same $H(a)$? We do this from the Friedmann equation:

$$H^2 + \frac{k}{a^2} = \frac{8\pi G\rho}{3} \tag{10.72}$$

and the energy equation:

$$\frac{d}{da}\left(\rho_q a^3\right) = -p_q a^2 = -w(a)\rho_q a^2. \tag{10.73}$$

It is a straightforward exercise to solve this for $w(a)$:

$$w(a) = -\frac{1}{3}\frac{d}{d\ln a}\ln\left[\frac{1}{\Omega_m(a)} - 1\right]. \tag{10.74}$$

Note that this does still depend on the Hubble parameter via

$$\Omega_m(a) = \frac{\rho(a)}{\rho_{\text{crit}}(a)} = \frac{\Omega_m}{a^3}\frac{H_0^2}{H^2(a)} \tag{10.75}$$

and Ω_m is the present matter density parameter.

However, in general, modified gravity models, such as those inspired by braneworlds, predict a different growth rate from GR, so could in principle be distinguished from GR by weak lensing observations. The question is, can this be done in practice? Evidently the issue here will be one of statistics – what size of survey would be needed to distinguish the standard cosmological model, including GR and dark energy, from braneworld gravity models? A comprehensive study of braneworld models is yet to be done, so we use the convenient minimal modified gravity parametrisation introduced by Linder (2005) and expanded by Huterer and Linder (2006) and Linder and Cahn (2007) to characterise beyond-Einstein models of gravity.

In this parametrisation, perturbations are described by a growth factor γ. In GR, the growth rate of perturbations in the matter density ρ_m, $\delta \equiv \delta \rho_m / \rho_m$, is accurately parametrised as a function of scale factor $a(t)$ by

$$\frac{\delta}{a} \equiv g(a) = \exp\left\{\int_0^a \frac{da'}{a'} \left[\Omega_m(a')^\gamma - 1\right]\right\}, \tag{10.76}$$

where $\Omega_m(a)$ is the density parameter of the matter, and the growth factor $\gamma \simeq 0.55$ for the standard general relativistic cosmological model. For modified gravity theories it deviates from this value – for example, for the DGP braneworld model (Dvali, Gabadze and Porrati, 2000), $\gamma \simeq 0.68$ (Linder and Cahn, 2007), on scales much smaller than those where cosmological acceleration is apparent.

Asking whether we need an extra parameter is a problem of *model selection*, which has gained attention recently in the cosmological literature (Hobson, Bridle and Lahav, 2002; Saini, Weller and Bridle, 2004; Mukherjee *et al.*, 2006; Liddle *et al.*, 2006; Pahud *et al.*, 2006, 2007; Szydlowski and Godlowski, 2006a,b; Serra, Heavens and Melchiorri, 2007). For this paper, we denote the beyond-Einstein gravity models as Model M, and GR plus dark energy as Model M'. γ is an additional parameter in Model M. We use DGP as a straw-man modified gravity model, despite its difficulties for other reasons as a viable theory.

10.6.1 Bayesian evidence

The Bayesian method to select between models (GR and modified gravity) is to consider the Bayesian evidence ratio. Essentially we want to know if, given the data, there is evidence that we need to expand the space of gravity models beyond GR. Assuming non-committal priors for the models (i.e. the same a-priori probability), the probability of the models given the data is simply proportional to the evidence.

We denote two competing models by M and M'. We assume that M' is a simpler model, which has fewer ($n' < n$) parameters in it. We further assume that it is *nested* in model M', i.e. the n' parameters of model M' are common to M, which has $p \equiv n - n'$ extra parameters in it. These parameters are fixed to fiducial values in M'.

We denote by D the data vector, and by θ and θ' the parameter vectors (of length n and n').

The posterior probability of each model comes from Bayes' theorem:

$$p(M|D) = \frac{p(D|M)\,p(M)}{p(D)} \tag{10.77}$$

and similarly for M'. By marginalisation $p(D|M)$, known as the *evidence*, is

$$p(D|M) = \int d\theta \; p(D|\theta, M)\,p(\theta|M), \tag{10.78}$$

which should be interpreted as a multidimensional integration. Hence the posterior relative probabilities of the two models, regardless of what their parameters are, is

$$\frac{p(M'|D)}{p(M|D)} = \frac{p(M')}{p(M)} \frac{\int d\theta' \; p(D|\theta', M')\,p(\theta'|M')}{\int d\theta \; p(D|\theta, M)\,p(\theta|M)}. \tag{10.79}$$

With non-committal priors on the models, $p(M') = p(M)$, this ratio simplifies to the ratio of evidences, called the *Bayes factor*:

$$B \equiv \frac{\int d\theta' \; p(D|\theta', M')\,p(\theta'|M')}{\int d\theta \; p(D|\theta, M)\,p(\theta|M)}. \tag{10.80}$$

Note that the more complicated model M will inevitably lead to a higher likelihood (or at least as high), but the evidence will favour the simpler model if the fit is nearly as good, through the smaller prior volume.

We assume uniform (and hence separable) priors in each parameter, over ranges $\Delta\theta$ (or $\Delta\theta'$). Hence $p(\theta|M) = (\Delta\theta_1 \ldots \Delta\theta_n)^{-1}$ and

$$B = \frac{\int d\theta' \; p(D|\theta', M')}{\int d\theta \; p(D|\theta, M)} \frac{\Delta\theta_1 \ldots \Delta\theta_n}{\Delta\theta'_1 \ldots \Delta\theta'_{n'}}. \tag{10.81}$$

Note that if the prior ranges are not large enough to contain essentially all the likelihood, then the position of the boundaries would influence the Bayes factor. In what follows, we will assume the prior range is large enough to encompass all the likelihood.

In the nested case, the ratio of prior hypervolumes simplifies to

$$\frac{\Delta\theta_1 \ldots \Delta\theta_n}{\Delta\theta'_1 \ldots \Delta\theta'_{n'}} = \Delta\theta_{n'+1} \ldots \Delta\theta_{n'+p}, \tag{10.82}$$

where $p \equiv n - n'$ is the number of extra parameters in the more complicated model.

The Bayes factor in Eq. (10.81) still depends on the specific dataset D. For future experiments, we do not yet have the data, so we compute the expectation value of the Bayes factor, given the statistical properties of D. The expectation is computed over the distribution of D for the correct model (assumed here to be M). To do this, we make two further approximations: first we note that B is a ratio, and we approximate $\langle B \rangle$ by the ratio of the expected values, rather than the expectation value of the ratio. This should be a good approximation if the evidences are sharply peaked.

We also make the Laplace approximation, that the expected likelihoods are given by multivariate Gaussians. For example,

$$\langle p(D|\theta, M) \rangle = L_0 \exp \left[-\frac{1}{2} (\theta - \theta^0)_\alpha F_{\alpha\beta} (\theta - \theta^0)_\beta \right] \tag{10.83}$$

and similarly for $\langle p(D|\theta', M') \rangle$. This assumes that a Taylor expansion of the likelihood around the peak value to second order can be extended throughout the parameter space. $F_{\alpha\beta}$ is the Fisher matrix, given for Gaussian-distributed data by (see e.g. Tegmark, Taylor and Heavens, 1997)

$$F_{\alpha\beta} = \frac{1}{2} \mathrm{Tr} \left[C^{-1} C_{,\alpha} C^{-1} C_{,\beta} + C^{-1} (\mu_{,\beta} \mu_{,\alpha}^t + \mu_{,\alpha} \mu_{,\beta}^t) \right]. \tag{10.84}$$

C is the covariance matrix of the data, and μ its mean (no noise). Commas indicate partial derivatives w.r.t. the parameters. For the correct model M, the peak of the expected likelihood is located at the true parameters θ^0. Note, however, that for the incorrect model M', the peak of the expected likelihood is not in general at the true parameters (see Fig. 10.11 for an illustration of this). This arises because the likelihood in the numerator of Eq. (10.81) is the probability of the dataset D given incorrect model assumptions.

The Laplace approximation is routinely used in forecasting marginal errors in parameters, using the Fisher matrix. Clearly the approximation may break down in some cases, but for Planck, the Fisher matrix errors are reasonably close to (within 30% of) those computed with Monte Carlo Markov chains.

If we assume that the posterior probability densities are small at the boundaries of the prior volume, then we can extend the integrations to infinity, and the integration over the multivariate Gaussians can be easily done. This gives, for M, $(2\pi)^{n/2} (\det F)^{-1/2}$, so for nested models,

$$\langle B \rangle = (2\pi)^{-p/2} \frac{\sqrt{\det F}}{\sqrt{\det F'}} \frac{L_0'}{L_0} \Delta\theta_{n'+1} \ldots \Delta\theta_{n'+p}. \tag{10.85}$$

An equivalent expression was obtained, using again the Laplace approximation by Lazarides, Ruiz de Austri and Trotta (2004). The point here is that with the

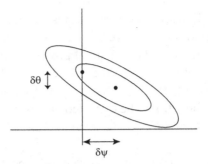

Fig. 10.11. Illustrating how assumption of a wrong parameter value can influence
the best-fitting value of other model parameters. Ellipses represent iso-likelihood
surfaces, and here in the simpler model the parameter on the horizontal axis
is assumed to take the value given by the vertical line. Filled circles show the
true parameters in the more complicated model and the best-fit parameters in the
simpler model.

Laplace approximation, one can compute the L_0'/L_0 ratio from the Fisher matrix.
To compute this ratio of likelihoods, we need to take into account the fact that, if
the true underlying model is M, in M' (the incorrect model) the maximum of the
expected likelihood will not in general be at the correct values of the parameters
(see Fig. 10.11). The n' parameters shift from their true values to compensate for
the fact that, effectively, the p additional parameters are being kept fixed at incorrect fiducial values. If in M' the additional p parameters are assumed to be fixed
at fiducial values that differ by $\delta\psi_\alpha$ from their true values, the others are shifted
on average by an amount that is readily computed under the assumption of the
multivariate Gaussian likelihood (see e.g. Taylor *et al.*, 2007):

$$\delta\theta_\alpha' = -(F'^{-1})_{\alpha\beta} G_{\beta\zeta} \delta\psi_\zeta, \qquad \alpha, \beta = 1\dots n', \quad \zeta = 1\dots p, \qquad (10.86)$$

where

$$G_{\beta\zeta} = \frac{1}{2}\mathrm{Tr}\left[C^{-1} C_{,\beta} C^{-1} C_{,\zeta} + C^{-1}(\mu_{,\zeta}\mu_{,\beta}^t + \mu_{,\beta}\mu_{,\zeta}^t) \right], \qquad (10.87)$$

which we recognise as a subset of the Fisher matrix. For clarity, we have given
the additional parameters the symbol ψ_ζ; $\zeta = 1\dots p$ to distinguish them from the
parameters in M'.

With these offsets in the maximum likelihood parameters in model M', the ratio
of likelihoods is given by

$$L_0' = L_0 \exp\left(-\frac{1}{2}\delta\theta_\alpha F_{\alpha\beta} \delta\theta_\beta \right), \qquad (10.88)$$

where the offsets are given by $\delta\theta_\alpha = \delta\theta'_\alpha$ for $\alpha \le n'$ (Eq. (10.86)), and $\delta\theta_\alpha = \delta\psi_{\alpha-n'}$ for $\alpha > n'$.

The final expression for the expected Bayes factor is then

$$\langle B \rangle = (2\pi)^{-p/2} \frac{\sqrt{\det F}}{\sqrt{\det F'}} \exp\left(-\frac{1}{2}\delta\theta_\alpha F_{\alpha\beta}\delta\theta_\beta\right) \prod_{q=1}^{p} \Delta\theta_{n'+q}. \tag{10.89}$$

Note that F and F^{-1} are $n \times n$ matrices, F' is $n' \times n'$, and G is an $n' \times p$ block of the full $n \times n$ Fisher matrix F. The expression we find is a specific example of the Savage–Dickey ratio (Trotta, 2007a); here we explicitly use the Laplace approximation to compute the offsets in the parameter estimates that accompany the wrong choice of model.

Note that the 'Occam's razor' term, common in evidence calculations, is encapsulated in the term $(2\pi)^{-p/2} \frac{\sqrt{\det F}}{\sqrt{\det F'}}$, multiplied by the prior product: models with more parameters are penalised in favour of simpler models, unless the data demand otherwise. Such terms should be treated with caution; as pointed out by Linder and Miquel (2007), simpler models do not always result in the most physically realistic conclusions (but see Liddle *et al.*, 2007 for a thorough discussion of the issues). In cases where the Laplace approximation is not a good one, other techniques must be used, at more computational expense (e.g. Skilling, 2004; Beltran *et al.*, 2005; Mukherjee, Parkinson and Liddle, 2006; Mukherjee *et al.*, 2006; Parkinson, Mukherjee and Liddle, 2006; Trotta, 2007a,b). Alternatively, a reparametrisation of the parameter space can make the likelihood closer to Gaussian (see e.g. Kosowsky, Milosavljevic and Jimenez (2002) for CMB).

From these Fisher matrices (the Fisher matrix of a combination of independent data sets is the sum of the individual Fisher matrices), we compute the expected evidence ratio assuming that the true model is a DGP braneworld, and take a prior range $\Delta\gamma = 1$. Table 10.1 shows the expected evidence for the 3D weak lensing surveys with and without *Planck*.

We find that $\ln\langle B \rangle$ obtained for the standard general relativity model is only ~ 1 for the Dark Energy Survey (DES) + *Planck*, whereas for Pan-STARRS + *Planck* we find that $\ln\langle B \rangle \sim 2$, for Pan-STARRS + *Planck* + SN + BAO, $\ln\langle B \rangle \sim 3.61$ and for WL$_{NG}$ + *Planck*, $\ln\langle B \rangle$ is a decisive 52.2. Furthermore a WL$_{NG}$ experiment could still decisively distinguish dark energy from modified gravity without *Planck*. The expected evidence in this case scales proportionally as the total number of galaxies in the survey. Pan-STARRS and *Planck* should be able to determine the expansion history, parametrised by $w(a)$ to very high accuracy in the context of the standard general relativity cosmological model,

Table 10.1. *The evidence ratio for the three weak lensing experiments considered with and without* Planck, *supernova and BAO priors*

Survey	ν	$\ln\langle B \rangle$	
DES + *Planck* + BAO + SN	3.5	1.28	substantial
DES + *Planck*	2.2	0.56	inconclusive
DES	0.7	0.54	inconclusive
PS1 + *Planck* + BAO + SN	2.9	3.78	strong
PS1 + *Planck*	2.6	2.04	substantial
PS1	1.0	0.62	inconclusive
WL_{NG} + *Planck* + BAO + SN	10.6	63.0	decisive
WL_{NG} + *Planck*	10.2	52.2	decisive
WL_{NG}	5.4	11.8	decisive

W L_{NG} is a next-generation space-based imaging survey such as proposed for DUNE or SNAP. $\nu\sigma$ is the frequentist significance with which GR would be expected to be ruled out, if the DGP braneworld were the correct model.

with an accuracy of 0.03 on $w(z \simeq 0.4)$. It will be able to substantially distinguish between general relativity and the simplification of the DGP braneworld model considered here, although this does depend on there being a strong CMB prior.

The relatively low evidence from DES + *Planck* in comparison to Pan-STARRS + *Planck* is due to the degeneracy between the running of the spectral index, α, and γ. The larger effect volume of Pan-STARRS, in comparison to DES, places a tighter constraint on this degeneracy. This could be improved by on-going high-resolution CMB experiments.

Alternatively, we can ask the question of how different the growth rate of a modified-gravity model would have to be for these experiments to be able to distinguish the model from general relativity, assuming that the expansion history in the modified gravity model is still well described by the w_0, w_a parametrisation. This is shown in Fig. 10.12. It shows how the expected evidence ratio changes with progressively greater differences from the general relativistic growth rate. We see that a WL_{NG} survey could even distinguish 'strongly' $\delta\gamma = 0.048$, Pan-STARRS $\delta\gamma = 0.137$ and DES $\delta\gamma = 0.179$. Note that changing the prior range $\Delta\gamma$ by a factor 10 changes the WL_{NG} + *Planck* numbers by ~ 0.012, so the dependence on the prior range is rather small.

If one prefers to ask a frequentist question, then a combination of WL_{NG} + *Planck* + BAO + SN should be able to distinguish $\delta\gamma = 0.13$, at 10.6σ. Results for other experiments are shown in Fig. 10.12. Alternatively, one can calculate the expected error on γ (Amendola, Kunz and Sapone, 2007) within the extended

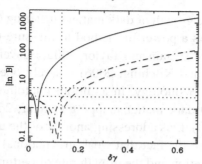

Fig. 10.12. The expected value of $|\ln\langle B\rangle|$ from WL$_{NG}$ (solid), Pan-STARRS (dot-dashed) and DES (dashed), in combination with CMB constraints from *Planck*, as a function of the difference in the growth rate between the modified-gravity model and general relativity. The crossover at small $\delta\gamma$ occurs because Occam's razor will favour the simpler (general relativity) model unless the data demand otherwise. To the left of the cusps, GR would be likely to be preferred by the data. The dotted vertical line shows the offset of the growth factor for the DGP model. The horizontal lines mark the boundaries between 'inconclusive', 'significant', 'strong', and 'decisive' in the terminology of Jeffreys (1961). We also include, for completeness, the anticipated frequentist significance of rejection of the simpler model by each experiment, assuming DGP braneworld is the correct model.

model M. In this chapter, we are asking a slightly different question of whether the data demand that a wider class of models needs to be considered at all, rather than estimating a parameter within that wider class.

The case for a large, space-based 3D weak lensing survey is strengthened, as it offers the possibility of conclusively distinguishing dark energy from at least some modified gravity models.

10.7 The future

The main promise of weak lensing in the future will come from larger surveys with optics designed for excellent image quality. Currently, the CFHTLS is underway, designed to cover 170 square degrees to a median redshift in excess of one. In future Pan-STARRS, VST and VISTA promise very small PSF distortions and large areal coverage, and in the far future LSST on the ground, and satellites such as SNAP or DUNE may deliver extremely potent lensing surveys. In parallel with these developments, the acquisition of photometric redshifts for the sources has opened up the exciting possibility of analysing weak lensing surveys in 3D. Each source represents a noisy estimate of the shear field at a location in 3D space, and this extra information turns out to be extremely valuable, increasing substantially the statistical power of lensing surveys. It can lift the degeneracy between σ_8 and

Ω_m, measure directly the growth of dark matter clustering (Bacon *et al.*, 2005) and in particular it represents a powerful method to measure the equation of state of dark energy (Heavens, 2003; Jain and Taylor, 2003; Heavens, Kitching and Taylor, 2006; Castro, Heavens and Kitching, 2006) – surely one of the most important remaining questions in cosmology. In addition, photometric redshifts allow the possibility of direct 3D dark matter mapping (Taylor, 2001; Bacon and Taylor, 2003; Taylor *et al.*, 2004), thus addressing another of the unsolved problems.

Finally, we can exploit the fact that weak gravitational lensing can probe both the distance–redshift relation and the growth rate of perturbations. This allows the lifting of a degeneracy between GR plus dark energy and beyond-Einstein gravity models. The distance–redshift relation of such models can always be modelled by dark energy within GR, if one has freedom in the evolution of the equation of state parameter $w(a)$. With a large and deep enough survey, weak lensing has the potential to decide whether GR and dark energy is preferred, or whether we need to expand the range of gravity models into braneworlds or other directions.

References

Amendola, L., Kunz M., and Sapone D. (2007). [astro-ph/0704.2421].

Bacon, D. J., and Taylor, A. N. (2003). *Mon. Not. R. Astron. Soc.* **344**, 1307.

Bacon, D. J., Refregier A., and Ellis, R. S. (2000). *Mon. Not. R. Astron. Soc.* **318**, 625.

Bacon, D. J., *et al.* (2005). *Mon. Not. R. Astron. Soc.* **363**, 723.

Bartelmann, M., and Schneider, P. (2001). *Phys. Rep.* **340**, 291 [astro-ph/9912508].

Beltran, M., Garcia-Bellido, J., Lesgourgues, J., Liddle, A. R., and Slosar, A. (2005). *Phys. Rev. D* **71**, 063532.

Benjamin, J., *et al.* (2007). [astro-ph/0703570].

Bernardeau, F., van Waerbeke, L., and Mellier, Y. (1997). *Astron. Astrophys.* **322**, 1.

Bernstein, G., and Jarvis, M. (2002). *Astron. J.* **123**, 583.

Bernstein, G., and Jain, B. (2004). *Astrophys. J.* **600**, 17.

Brown, M., *et al.* (2002). *Mon. Not. R. Astron. Soc.* **333**, 501.

Brown, M. L., Taylor, A. N., Bacon, D. J., *et al.* (2003). *Mon. Not. R. Astron. Soc.* **341**, 100.

Castro, P. G., Heavens, A. F., and Kitching, T. (2006). *Phys. Rev. D* **72**, 023516.

Catelan, P., Kamionkowski, M., Blandford, R. D. (2001). *Mon. Not. R. Astron. Soc.* **320**, L7.

Chang, T.-C., and Refregier, A. (2002). *Astrophys. J.* **570**, 447.

Chevallier, M., and Polarski , D. (2001). *Int. J. Mod. Phys. D* **10**, 213.

Crittenden, R., Natarajan, P., Pen, U.-L., and Theuns, T. (2001). *Astrophys. J.* **559**, 552.

Crittenden, R., Natarajan, P., Pen, U.-L., and Theuns, T. (2002). *Astrophys. J.* **568**, 20.

Croft, R. A. C., and Metzler, C. A. (2000). *Astrophys. J.* **545**, 561.

Dietrich, J. P., Erben, T., Lamer, G., *et al.* (2007). [astro-ph/0705.3455].

Dvali, G., Gabadaze, G., and Porrati, M. (2000). *Phys. Lett. B* **485**, 208.

Heavens, A. F. (2003). *Mon. Not. R. Astron. Soc.* **343**, 1327.

Heavens, A. F., Refregier, A., and Heymans, C. E. C. (2000). *Mon. Not. R. Astron. Soc.* **319**, 649.

Heavens, A. F., Kitching, T., and Taylor, A. N. (2006). *Mon. Not. R. Astron. Soc.* **373**, 105.

Heymans, C. E. C., and Heavens, A. F. (2003). *Mon. Not. R. Astron. Soc.* **337**, 711.

Heymans, C. E. C., *et al.* (2004). *Mon. Not. R. Astron. Soc.* **347**, 895.

Heymans, C. E. C., *et al.* (2005). *Mon. Not. R. Astron. Soc.* **361**, 160.

Hirata, C., and Seljak, U. (2004). *Phys. Rev. D* **70**, 063526.

Hobson, M. P., Bridle, S. L., and Lahav, O. (2002). *Mon. Not. R. Astron. Soc.* **335**, 377.

Hoekstra, H., Mellier, Y., van Waerbeke, L., *et al.* (2006). *Astrophys. J.* **647**, 116.

Hu, W. (1999). *Astrophys. J.* **522**, 21.

Hu, W. (2000). *Phys. Rev. D* **62**, 3007.

Hu, W., and Keeton, C. R. (2002). *Phys. Rev. D* **66**, 063506.

Hu, W., and White, M. (2001). *Astrophys. J.* **554**, 67.

Huterer, D., and Linder, E. V. (2006). *Phys. Rev. D* **75**, 023519.

Huterer, D., Takada, M., Bernstein, G., and Jain, B. (2006). *Mon. Not. R. Astron. Soc.* **366**, 101.

Jain, B., and Taylor, A. N. (2003). *Phys. Rev. Lett.* **91**, 141302.

Jeffreys, H. (1961). *Theory of Probability*, Oxford: Oxford University Press.

Jing, Y. P. (2002). *Mon. Not. R. Astron. Soc.* **335**, L89.

Kaiser, N. (2002). *Astrophys. J.* **388**, 272.

Kaiser, N., Squires, G., and Broadhurst, T. (1995). *Astrophys. J.* **449**, 460.

Kaiser, N., Wilson, G., and Luppino, G. (2000). [astro-ph/0003338].

King, L., and Schneider, P. (2003). *Astron. Astrophys.* **396**, 411.

Kosowsky, A., Milosavljevic, M., and Jimenez, R. (2002). *Phys. Rev. D* **66**, 063007.

Kunz, M., and Sapone, D. (2007). *Phys. Rev. Lett.* **98**, 121301.

Lazarides, G., Ruiz de Austri, R., and Trotta, R. (2004). *Phys. Rev. D* **70**, 123527.

Liddle, A., Mukherjee, P., Parkinson, D., and Wang, Y. (2006). [astro-ph/0610126].

Liddle, A., Corasaniti, P. S., Kunz, M., *et al.* (2007). [astro-ph/0703285].

Linder, E. V. (2005). *Phys. Rev. D* **72**, 043529.

Linder, E. V., and Cahn, R. N. (2007). [astro-ph/0701317].

Linder, E. V., and Miquel, R. (2007). [astro-ph/0702542].

Ma, Z., Hu, W., and Huterer, D. (2006). *Astrophys J.* **636**, 21.

Massey, R., Refregier, A., Bacon, D. J., Ellis, R., and Brown, M. L. (2005). *Mon. Not. R. Astron. Soc.* **359**, 1277.

Massey, R., *et al.* (2007). *Nature (London)* **445**, 286.

Mukherjee, P., Parkinson, D., and Liddle, A. R. (2006). *Astrophys. J.* **638**, L51.

Mukherjee, P., Parkinson, D., Corasaniti, P. S., Liddle, A. R., and Kunz, M. (2006). *Mon. Not. R. Astron. Soc.* **369**, 1725.

Munshi, D., and Valageas, P. (2005). *Mon. Not. R. Astron. Soc.* **360**, 1401.

Munshi, D., Valageas, P., van Waerbeke, L., and Heavens, A. F. (2006). *Phys. Rep.* [astro-ph/0612667].

Pahud, C., Liddle, A., Mukherjee, P., and Parkinson, D. (2006). *Phys. Rev. D* **73**, 123524

Pahud, C., Liddle, A., Mukherjee, P., and Parkinson, D. (2007). [astro-ph/0701481].

Parkinson, D., Mukherjee, P., and Liddle, A. R. (2006). *Phys. Rev. D* **73**, 123523.

Peacock, J. A. (1999). *Cosmological Physics*, Cambridge: Cambridge University Press.

Peebles, P. J. E. (1980). *The Large Scale Structure of the Universe*, Princeton: Princeton University Press.

Saini, T. D., Weller, J., and Bridle, S. L. (2004). *Mon. Not. R. Astron. Soc.* **348**, 603.

Schneider, P. (1996). *Mon. Not. R. Astron. Soc.* **283**, 837.

Schneider, P. (2003). In *Dark Matter and Dark Energy in the Universe.* [astro-ph/0306465].

Schneider, P., van Waerbeke, L., Jain, B., and Kruse, G. (1998). *Mon. Not. R. Astron. Soc.* **296**, 893.

Schneider, P., van Waerbeke, L., and Mellier, Y. (2002a). *Astron. Astrophys.* **389**, 729.

Schneider, P., van Waerbeke, L., Kilbinger. M., and Mellier, Y. (2002b). *Astron. Astrophys.* **396**, 1.

Seitz, S., and Schneider, P. (1994). *Astron. Astrophys.* **288**, 1.

Serra, P., Heavens, A. F., and Melchiorri A. (2007). *Mon. Not. R. Astron. Soc.* **379**, 169.

Skilling, J. (2004). Avaliable at www.inference.phy.cam.ac.uk/bayesys.

Smith, R. E., *et al.* (2003). *Mon. Not. R. Astron. Soc.* **341**, 1311.

Spergel, D., *et al.* (2007). *Astrophys. J. Suppl.* **170**, 377.

Szydlowski, M., and Godlowski, W. (2006a). *Phys. Lett. B* **633**, 427.

Szydlowski, M., and Godlowski, W. (2006b). *Phys. Lett. B* **639**, 5.

Taylor, A. N. (2001). [astro-ph/0111605].

Taylor, A. N., *et al.* (2004). *Mon. Not. R. Astron. Soc.* **353**, 1176.

Taylor, A. N., Kitching, T. D., Bacon, D., and Heavens, A. F. (2007). *Mon. Not. R. Astron. Soc.* **374**, 1377.

Tegmark, M., Taylor, A. N., and Heavens, A. F. (1997). *Astrophys. J.* **480**, 22.

Trotta, R. (2007a). *Mon. Not. R. Astron. Soc.* **378**, 72.

Trotta, R. (2007b). *Mon. Not. R. Astron. Soc.* **378**, 819.

van Waerbeke, L., and Mellier, Y. (2003). In *Proceedings of Aussois Winter School.* [astro-ph/0305089].

van Waerbeke, L., *et al.* (2000). *Astron. Astrophys.* **358**, 30.

van Waerbeke, L., *et al.* (2002). *Astron. Astrophys.* **393**, 369.

van Waerbeke, L., Mellier, Y., and Hoekstra, H. (2005). *Astron. Astrophys.* **429**, 75.

White, M. (2004). *Astropart. Phys.* **22**, 211.

Wittman, D., *et al.* (2000). *Nature (London)* **405**, 143.

Zhang, P., and Pen, U.-L. (2005). *Phys. Rev. Lett.* **95**, 241302.

Index

Printed in the United States
by Baker & Taylor Publisher Services

Printed in the United States
by Baker & Taylor Publisher Services